T0203883

Real and Complex Singularities

PURE AND APPLIED MATHEMATICS

A Program of Monographs, Textbooks, and Lecture Notes

LECTURE NOTES IN PURE AND APPLIED MATHEMATICS

1. *N. Jacobson*, Exceptional Lie Algebras
2. *L.-Å. Lindahl and F. Poulsen*, Thin Sets in Harmonic Analysis
3. *I. Satake*, Classification Theory of Semi-Simple Algebraic Groups
4. *F. Hirzebruch et al.*, Differentiable Manifolds and Quadratic Forms
5. *I. Chavel*, Riemannian Symmetric Spaces of Rank One
6. *R. B. Burckel*, Characterization of C(X) Among Its Subalgebras
7. *B. R. McDonald et al.*, Ring Theory
8. *Y.-T. Siu*, Techniques of Extension on Analytic Objects
9. *S. R. Caradus et al.*, Calkin Algebras and Algebras of Operators on Banach Spaces
10. *E. O. Roxin et al.*, Differential Games and Control Theory
11. *M. Orzech and C. Small*, The Brauer Group of Commutative Rings
12. *S. Thomier*, Topology and Its Applications
13. *J. M. Lopez and K. A. Ross*, Sidon Sets
14. *W. W. Comfort and S. Negrepontis*, Continuous Pseudometrics
15. *K. McKennon and J. M. Robertson*, Locally Convex Spaces
16. *M. Carmeli and S. Malin*, Representations of the Rotation and Lorentz Groups
17. *G. B. Seligman*, Rational Methods in Lie Algebras
18. *D. G. de Figueiredo*, Functional Analysis
19. *L. Cesari et al.*, Nonlinear Functional Analysis and Differential Equations
20. *J. J. Schäffer*, Geometry of Spheres in Normed Spaces
21. *K. Yano and M. Kon*, Anti-Invariant Submanifolds
22. *W. V. Vasconcelos*, The Rings of Dimension Two
23. *R. E. Chandler*, Hausdorff Compactifications
24. *S. P. Franklin and B. V. S. Thomas*, Topology
25. *S. K. Jain*, Ring Theory
26. *B. R. McDonald and R. A. Morris*, Ring Theory II
27. *R. B. Mura and A. Rhemtulla*, Orderable Groups
28. *J. R. Graef*, Stability of Dynamical Systems
29. *H.-C. Wang*, Homogeneous Branch Algebras
30. *E. O. Roxin et al.*, Differential Games and Control Theory II
31. *R. D. Porter*, Introduction to Fibre Bundles
32. *M. Altman*, Contractors and Contractor Directions Theory and Applications
33. *J. S. Golan*, Decomposition and Dimension in Module Categories
34. *G. Fairweather*, Finite Element Galerkin Methods for Differential Equations
35. *J. D. Sally*, Numbers of Generators of Ideals in Local Rings
36. *S. S. Miller*, Complex Analysis
37. *R. Gordon*, Representation Theory of Algebras
38. *M. Goto and F. D. Grosshans*, Semisimple Lie Algebras
39. *A. I. Arruda et al.*, Mathematical Logic
40. *F. Van Oystaeyen*, Ring Theory
41. *F. Van Oystaeyen and A. Verschoren*, Reflectors and Localization
42. *M. Satyanarayana*, Positively Ordered Semigroups
43. *D. L Russell*, Mathematics of Finite-Dimensional Control Systems
44. *P.-T. Liu and E. Roxin*, Differential Games and Control Theory III
45. *A. Geramita and J. Seberry*, Orthogonal Designs
46. *J. Cigler, V. Losert, and P. Michor*, Banach Modules and Functors on Categories of Banach Spaces
47. *P.-T. Liu and J. G. Sutinen*, Control Theory in Mathematical Economics
48. *C. Byrnes*, Partial Differential Equations and Geometry
49. *G. Klambauer*, Problems and Propositions in Analysis
50. *J. Knopfmacher*, Analytic Arithmetic of Algebraic Function Fields
51. *F. Van Oystaeyen*, Ring Theory
52. *B. Kadem*, Binary Time Series
53. *J. Barros-Neto and R. A. Artino*, Hypoelliptic Boundary-Value Problems
54. *R. L. Sternberg et al.*, Nonlinear Partial Differential Equations in Engineering and Applied Science
55. *B. R. McDonald*, Ring Theory and Algebra III
56. *J. S. Golan*, Structure Sheaves Over a Noncommutative Ring
57. *T. V. Narayana et al.*, Combinatorics, Representation Theory and Statistical Methods in Groups
58. *T. A. Burton*, Modeling and Differential Equations in Biology
59. *K. H. Kim and F. W. Roush*, Introduction to Mathematical Consensus Theory

60. *J. Banas and K. Goebel,* Measures of Noncompactness in Banach Spaces
61. *O. A. Nielson,* Direct Integral Theory
62. *J. E. Smith et al.,* Ordered Groups
63. *J. Cronin,* Mathematics of Cell Electrophysiology
64. *J. W. Brewer,* Power Series Over Commutative Rings
65. *P. K. Kamthan and M. Gupta,* Sequence Spaces and Series
66. *T. G. McLaughlin,* Regressive Sets and the Theory of Isols
67. *T. L. Herdman et al.,* Integral and Functional Differential Equations
68. *R. Draper,* Commutative Algebra
69. *W. G. McKay and J. Patera,* Tables of Dimensions, Indices, and Branching Rules for Representations of Simple Lie Algebras
70. *R. L. Devaney and Z. H. Nitecki,* Classical Mechanics and Dynamical Systems
71. *J. Van Geel,* Places and Valuations in Noncommutative Ring Theory
72. *C. Faith,* Injective Modules and Injective Quotient Rings
73. *A. Fiacco,* Mathematical Programming with Data Perturbations I
74. *P. Schultz et al.,* Algebraic Structures and Applications
75. *L Bican et al.,* Rings, Modules, and Preradicals
76. *D. C. Kay and M. Breen,* Convexity and Related Combinatorial Geometry
77. *P. Fletcher and W. F. Lindgren,* Quasi-Uniform Spaces
78. *C.-C. Yang,* Factorization Theory of Meromorphic Functions
79. *O. Taussky,* Ternary Quadratic Forms and Norms
80. *S. P. Singh and J. H. Burry,* Nonlinear Analysis and Applications
81. *K. B. Hannsgen et al.,* Volterra and Functional Differential Equations
82. *N. L. Johnson et al.,* Finite Geometries
83. *G. I. Zapata,* Functional Analysis, Holomorphy, and Approximation Theory
84. *S. Greco and G. Valla,* Commutative Algebra
85. *A. V. Fiacco,* Mathematical Programming with Data Perturbations II
86. *J.-B. Hiriart-Urruty et al.,* Optimization
87. *A. Figa Talamanca and M. A. Picardello,* Harmonic Analysis on Free Groups
88. *M. Harada,* Factor Categories with Applications to Direct Decomposition of Modules
89. *V. I. Istrătescu,* Strict Convexity and Complex Strict Convexity
90. *V. Lakshmikantham,* Trends in Theory and Practice of Nonlinear Differential Equations
91. *H. L. Manocha and J. B. Srivastava,* Algebra and Its Applications
92. *D. V. Chudnovsky and G. V. Chudnovsky,* Classical and Quantum Models and Arithmetic Problems
93. *J. W. Longley,* Least Squares Computations Using Orthogonalization Methods
94. *L. P. de Alcantara,* Mathematical Logic and Formal Systems
95. *C. E. Aull,* Rings of Continuous Functions
96. *R. Chuaqui,* Analysis, Geometry, and Probability
97. *L. Fuchs and L. Salce,* Modules Over Valuation Domains
98. *P. Fischer and W. R. Smith,* Chaos, Fractals, and Dynamics
99. *W. B. Powell and C. Tsinakis,* Ordered Algebraic Structures
100. *G. M. Rassias and T. M. Rassias,* Differential Geometry, Calculus of Variations, and Their Applications
101. *R.-E. Hoffmann and K. H. Hofmann,* Continuous Lattices and Their Applications
102. *J. H. Lightbourne III and S. M. Rankin III,* Physical Mathematics and Nonlinear Partial Differential Equations
103. *C. A. Baker and L. M. Batten,* Finite Geometrics
104. *J. W. Brewer et al.,* Linear Systems Over Commutative Rings
105. *C. McCrory and T. Shifrin,* Geometry and Topology
106. *D. W. Kueke et al.,* Mathematical Logic and Theoretical Computer Science
107. *B.-L. Lin and S. Simons,* Nonlinear and Convex Analysis
108. *S. J. Lee,* Operator Methods for Optimal Control Problems
109. *V. Lakshmikantham,* Nonlinear Analysis and Applications
110. *S. F. McCormick,* Multigrid Methods
111. *M. C. Tangora,* Computers in Algebra
112. *D. V. Chudnovsky and G. V. Chudnovsky,* Search Theory
113. *D. V. Chudnovsky and R. D. Jenks,* Computer Algebra
114. *M. C. Tangora,* Computers in Geometry and Topology
115. *P. Nelson et al.,* Transport Theory, Invariant Imbedding, and Integral Equations
116. *P. Clément et al.,* Semigroup Theory and Applications
117. *J. Vinuesa,* Orthogonal Polynomials and Their Applications
118. *C. M. Dafermos et al.,* Differential Equations
119. *E. O. Roxin,* Modern Optimal Control
120. *J. C. Díaz,* Mathematics for Large Scale Computing

121. *P. S. Milojević,* Nonlinear Functional Analysis
122. *C. Sadosky,* Analysis and Partial Differential Equations
123. *R. M. Shortt,* General Topology and Applications
124. *R. Wong,* Asymptotic and Computational Analysis
125. *D. V. Chudnovsky and R. D. Jenks,* Computers in Mathematics
126. *W. D. Wallis et al.,* Combinatorial Designs and Applications
127. *S. Elaydi,* Differential Equations
128. *G. Chen et al.,* Distributed Parameter Control Systems
129. *W. N. Everitt,* Inequalities
130. *H. G. Kaper and M. Garbey,* Asymptotic Analysis and the Numerical Solution of Partial Differential Equations
131. *O. Arino et al.,* Mathematical Population Dynamics
132. *S. Coen,* Geometry and Complex Variables
133. *J. A. Goldstein et al.,* Differential Equations with Applications in Biology, Physics, and Engineering
134. *S. J. Andima et al.,* General Topology and Applications
135. *P Clément et al.,* Semigroup Theory and Evolution Equations
136. *K. Jarosz,* Function Spaces
137. *J. M. Bayod et al.,* p-adic Functional Analysis
138. *G. A. Anastassiou,* Approximation Theory
139. *R. S. Rees,* Graphs, Matrices, and Designs
140. *G. Abrams et al.,* Methods in Module Theory
141. *G. L. Mullen and P. J.-S. Shiue,* Finite Fields, Coding Theory, and Advances in Communications and Computing
142. *M. C. Joshi and A. V. Balakrishnan,* Mathematical Theory of Control
143. *G. Komatsu and Y. Sakane,* Complex Geometry
144. *I. J. Bakelman,* Geometric Analysis and Nonlinear Partial Differential Equations
145. *T. Mabuchi and S. Mukai,* Einstein Metrics and Yang–Mills Connections
146. *L. Fuchs and R. Göbel,* Abelian Groups
147. *A. D. Pollington and W. Moran,* Number Theory with an Emphasis on the Markoff Spectrum
148. *G. Dore et al.,* Differential Equations in Banach Spaces
149. *T. West,* Continuum Theory and Dynamical Systems
150. *K. D. Bierstedt et al.,* Functional Analysis
151. *K. G. Fischer et al.,* Computational Algebra
152. *K. D. Elworthy et al.,* Differential Equations, Dynamical Systems, and Control Science
153. *P.-J. Cahen, et al.,* Commutative Ring Theory
154. *S. C. Cooper and W. J. Thron,* Continued Fractions and Orthogonal Functions
155. *P. Clément and G. Lumer,* Evolution Equations, Control Theory, and Biomathematics
156. *M. Gyllenberg and L. Persson,* Analysis, Algebra, and Computers in Mathematical Research
157. *W. O. Bray et al.,* Fourier Analysis
158. *J. Bergen and S. Montgomery,* Advances in Hopf Algebras
159. *A. R. Magid,* Rings, Extensions, and Cohomology
160. *N. H. Pavel,* Optimal Control of Differential Equations
161. *M. Ikawa,* Spectral and Scattering Theory
162. *X. Liu and D. Siegel,* Comparison Methods and Stability Theory
163. *J.-P. Zolésio,* Boundary Control and Variation
164. *M. Křížek et al.,* Finite Element Methods
165. *G. Da Prato and L. Tubaro,* Control of Partial Differential Equations
166. *E. Ballico,* Projective Geometry with Applications
167. *M. Costabel et al.,* Boundary Value Problems and Integral Equations in Nonsmooth Domains
168. *G. Ferreyra, G. R. Goldstein, and F. Neubrander,* Evolution Equations
169. *S. Huggett,* Twistor Theory
170. *H. Cook et al.,* Continua
171. *D. F. Anderson and D. E. Dobbs,* Zero-Dimensional Commutative Rings
172. *K. Jarosz,* Function Spaces
173. *V. Ancona et al.,* Complex Analysis and Geometry
174. *E. Casas,* Control of Partial Differential Equations and Applications
175. *N. Kalton et al.,* Interaction Between Functional Analysis, Harmonic Analysis, and Probability
176. *Z. Deng et al.,* Differential Equations and Control Theory
177. *P. Marcellini et al.* Partial Differential Equations and Applications
178. *A. Kartsatos,* Theory and Applications of Nonlinear Operators of Accretive and Monotone Type
179. *M. Maruyama,* Moduli of Vector Bundles
180. *A. Ursini and P. Aglianò,* Logic and Algebra
181. *X. H. Cao et al.,* Rings, Groups, and Algebras
182. *D. Arnold and R. M. Rangaswamy,* Abelian Groups and Modules
183. *S. R. Chakravarthy and A. S. Alfa,* Matrix-Analytic Methods in Stochastic Models

184. *J. E. Andersen et al.*, Geometry and Physics
185. *P.-J. Cahen et al.*, Commutative Ring Theory
186. *J. A. Goldstein et al.*, Stochastic Processes and Functional Analysis
187. *A. Sorbi*, Complexity, Logic, and Recursion Theory
188. *G. Da Prato and J.-P. Zolésio*, Partial Differential Equation Methods in Control and Shape Analysis
189. *D. D. Anderson*, Factorization in Integral Domains
190. *N. L. Johnson*, Mostly Finite Geometries
191. *D. Hinton and P. W. Schaefer*, Spectral Theory and Computational Methods of Sturm–Liouville Problems
192. *W. H. Schikhof et al.*, p-adic Functional Analysis
193. *S. Sertöz*, Algebraic Geometry
194. *G. Caristi and E. Mitidieri*, Reaction Diffusion Systems
195. *A. V. Fiacco*, Mathematical Programming with Data Perturbations
196. *M. Křížek et al.*, Finite Element Methods: Superconvergence, Post-Processing, and A Posteriori Estimates
197. *S. Caenepeel and A. Verschoren*, Rings, Hopf Algebras, and Brauer Groups
198. *V. Drensky et al.*, Methods in Ring Theory
199. *W. B. Jones and A. Sri Ranga*, Orthogonal Functions, Moment Theory, and Continued Fractions
200. *P. E. Newstead*, Algebraic Geometry
201. *D. Dikranjan and L. Salce*, Abelian Groups, Module Theory, and Topology
202. *Z. Chen et al.*, Advances in Computational Mathematics
203. *X. Caicedo and C. H. Montenegro*, Models, Algebras, and Proofs
204. *C. Y. Yıldırım and S. A. Stepanov*, Number Theory and Its Applications
205. *D. E. Dobbs et al.*, Advances in Commutative Ring Theory
206. *F. Van Oystaeyen*, Commutative Algebra and Algebraic Geometry
207. *J. Kakol et al.*, p-adic Functional Analysis
208. *M. Boulagouaz and J.-P. Tignol*, Algebra and Number Theory
209. *S. Caenepeel and F. Van Oystaeyen*, Hopf Algebras and Quantum Groups
210. *F. Van Oystaeyen and M. Saorin*, Interactions Between Ring Theory and Representations of Algebras
211. *R. Costa et al.*, Nonassociative Algebra and Its Applications
212. *T.-X. He*, Wavelet Analysis and Multiresolution Methods
213. *H. Hudzik and L. Skrzypczak*, Function Spaces: The Fifth Conference
214. *J. Kajiwara et al.*, Finite or Infinite Dimensional Complex Analysis
215. *G. Lumer and L. Weis*, Evolution Equations and Their Applications in Physical and Life Sciences
216. *J. Cagnol et al.*, Shape Optimization and Optimal Design
217. *J. Herzog and G. Restuccia*, Geometric and Combinatorial Aspects of Commutative Algebra
218. *G. Chen et al.*, Control of Nonlinear Distributed Parameter Systems
219. *F. Ali Mehmeti et al.*, Partial Differential Equations on Multistructures
220. *D. D. Anderson and I. J. Papick*, Ideal Theoretic Methods in Commutative Algebra
221. *Á. Granja et al.*, Ring Theory and Algebraic Geometry
222. *A. K. Katsaras et al.*, p-adic Functional Analysis
223. *R. Salvi*, The Navier-Stokes Equations
224. *F. U. Coelho and H. A. Merklen*, Representations of Algebras
225. *S. Aizicovici and N. H. Pavel*, Differential Equations and Control Theory
226. *G. Lyubeznik*, Local Cohomology and Its Applications
227. *G. Da Prato and L. Tubaro*, Stochastic Partial Differential Equations and Applications
228. *W. A. Carnielli et al.*, Paraconsistency
229. *A. Benkirane and A. Touzani*, Partial Differential Equations
230. *A. Illanes et al.*, Continuum Theory
231. *M. Fontana et al.*, Commutative Ring Theory and Applications
232. *D. Mond and M. J. Saia*, Real and Complex Singularities
233. *V. Ancona and J. Vaillant*, Hyperbolic Differential Operators

Additional Volumes in Preparation

121. *P. S. Milojević,* Nonlinear Functional Analysis
122. *C. Sadosky,* Analysis and Partial Differential Equations
123. *R. M. Shortt,* General Topology and Applications
124. *R. Wong,* Asymptotic and Computational Analysis
125. *D. V. Chudnovsky and R. D. Jenks,* Computers in Mathematics
126. *W. D. Wallis et al.,* Combinatorial Designs and Applications
127. *S. Elaydi,* Differential Equations
128. *G. Chen et al.,* Distributed Parameter Control Systems
129. *W. N. Everitt,* Inequalities
130. *H. G. Kaper and M. Garbey,* Asymptotic Analysis and the Numerical Solution of Partial Differential Equations
131. *O. Arino et al.,* Mathematical Population Dynamics
132. *S. Coen,* Geometry and Complex Variables
133. *J. A. Goldstein et al.,* Differential Equations with Applications in Biology, Physics, and Engineering
134. *S. J. Andima et al.,* General Topology and Applications
135. *P Clément et al.,* Semigroup Theory and Evolution Equations
136. *K. Jarosz,* Function Spaces
137. *J. M. Bayod et al.,* p-adic Functional Analysis
138. *G. A. Anastassiou,* Approximation Theory
139. *R. S. Rees,* Graphs, Matrices, and Designs
140. *G. Abrams et al.,* Methods in Module Theory
141. *G. L. Mullen and P. J.-S. Shiue,* Finite Fields, Coding Theory, and Advances in Communications and Computing
142. *M. C. Joshi and A. V. Balakrishnan,* Mathematical Theory of Control
143. *G. Komatsu and Y. Sakane,* Complex Geometry
144. *I. J. Bakelman,* Geometric Analysis and Nonlinear Partial Differential Equations
145. *T. Mabuchi and S. Mukai,* Einstein Metrics and Yang–Mills Connections
146. *L. Fuchs and R. Göbel,* Abelian Groups
147. *A. D. Pollington and W. Moran,* Number Theory with an Emphasis on the Markoff Spectrum
148. *G. Dore et al.,* Differential Equations in Banach Spaces
149. *T. West,* Continuum Theory and Dynamical Systems
150. *K. D. Bierstedt et al.,* Functional Analysis
151. *K. G. Fischer et al.,* Computational Algebra
152. *K. D. Elworthy et al.,* Differential Equations, Dynamical Systems, and Control Science
153. *P.-J. Cahen, et al.,* Commutative Ring Theory
154. *S. C. Cooper and W. J. Thron,* Continued Fractions and Orthogonal Functions
155. *P. Clément and G. Lumer,* Evolution Equations, Control Theory, and Biomathematics
156. *M. Gyllenberg and L. Persson,* Analysis, Algebra, and Computers in Mathematical Research
157. *W. O. Bray et al.,* Fourier Analysis
158. *J. Bergen and S. Montgomery,* Advances in Hopf Algebras
159. *A. R. Magid,* Rings, Extensions, and Cohomology
160. *N. H. Pavel,* Optimal Control of Differential Equations
161. *M. Ikawa,* Spectral and Scattering Theory
162. *X. Liu and D. Siegel,* Comparison Methods and Stability Theory
163. *J.-P. Zolésio,* Boundary Control and Variation
164. *M. Křížek et al.,* Finite Element Methods
165. *G. Da Prato and L. Tubaro,* Control of Partial Differential Equations
166. *E. Ballico,* Projective Geometry with Applications
167. *M. Costabel et al.,* Boundary Value Problems and Integral Equations in Nonsmooth Domains
168. *G. Ferreyra, G. R. Goldstein, and F. Neubrander,* Evolution Equations
169. *S. Huggett,* Twistor Theory
170. *H. Cook et al.,* Continua
171. *D. F. Anderson and D. E. Dobbs,* Zero-Dimensional Commutative Rings
172. *K. Jarosz,* Function Spaces
173. *V. Ancona et al.,* Complex Analysis and Geometry
174. *E. Casas,* Control of Partial Differential Equations and Applications
175. *N. Kalton et al.,* Interaction Between Functional Analysis, Harmonic Analysis, and Probability
176. *Z. Deng et al.,* Differential Equations and Control Theory
177. *P. Marcellini et al.* Partial Differential Equations and Applications
178. *A. Kartsatos,* Theory and Applications of Nonlinear Operators of Accretive and Monotone Type
179. *M. Maruyama,* Moduli of Vector Bundles
180. *A. Ursini and P. Aglianò,* Logic and Algebra
181. *X. H. Cao et al.,* Rings, Groups, and Algebras
182. *D. Arnold and R. M. Rangaswamy,* Abelian Groups and Modules
183. *S. R. Chakravarthy and A. S. Alfa,* Matrix-Analytic Methods in Stochastic Models

184. *J. E. Andersen et al.*, Geometry and Physics
185. *P.-J. Cahen et al.*, Commutative Ring Theory
186. *J. A. Goldstein et al.*, Stochastic Processes and Functional Analysis
187. *A. Sorbi*, Complexity, Logic, and Recursion Theory
188. *G. Da Prato and J.-P. Zolésio*, Partial Differential Equation Methods in Control and Shape Analysis
189. *D. D. Anderson*, Factorization in Integral Domains
190. *N. L. Johnson*, Mostly Finite Geometries
191. *D. Hinton and P. W. Schaefer*, Spectral Theory and Computational Methods of Sturm–Liouville Problems
192. *W. H. Schikhof et al.*, p-adic Functional Analysis
193. *S. Sertöz*, Algebraic Geometry
194. *G. Caristi and E. Mitidieri*, Reaction Diffusion Systems
195. *A. V. Fiacco*, Mathematical Programming with Data Perturbations
196. *M. Křížek et al.*, Finite Element Methods: Superconvergence, Post-Processing, and A Posteriori Estimates
197. *S. Caenepeel and A. Verschoren*, Rings, Hopf Algebras, and Brauer Groups
198. *V. Drensky et al.*, Methods in Ring Theory
199. *W. B. Jones and A. Sri Ranga*, Orthogonal Functions, Moment Theory, and Continued Fractions
200. *P. E. Newstead*, Algebraic Geometry
201. *D. Dikranjan and L. Salce*, Abelian Groups, Module Theory, and Topology
202. *Z. Chen et al.*, Advances in Computational Mathematics
203. *X. Caicedo and C. H. Montenegro*, Models, Algebras, and Proofs
204. *C. Y. Yıldırım and S. A. Stepanov*, Number Theory and Its Applications
205. *D. E. Dobbs et al.*, Advances in Commutative Ring Theory
206. *F. Van Oystaeyen*, Commutative Algebra and Algebraic Geometry
207. *J. Kakol et al.*, p-adic Functional Analysis
208. *M. Boulagouaz and J.-P. Tignol*, Algebra and Number Theory
209. *S. Caenepeel and F. Van Oystaeyen*, Hopf Algebras and Quantum Groups
210. *F. Van Oystaeyen and M. Saorin*, Interactions Between Ring Theory and Representations of Algebras
211. *R. Costa et al.*, Nonassociative Algebra and Its Applications
212. *T.-X. He*, Wavelet Analysis and Multiresolution Methods
213. *H. Hudzik and L. Skrzypczak*, Function Spaces: The Fifth Conference
214. *J. Kajiwara et al.*, Finite or Infinite Dimensional Complex Analysis
215. *G. Lumer and L. Weis*, Evolution Equations and Their Applications in Physical and Life Sciences
216. *J. Cagnol et al.*, Shape Optimization and Optimal Design
217. *J. Herzog and G. Restuccia*, Geometric and Combinatorial Aspects of Commutative Algebra
218. *G. Chen et al.*, Control of Nonlinear Distributed Parameter Systems
219. *F. Ali Mehmeti et al.*, Partial Differential Equations on Multistructures
220. *D. D. Anderson and I. J. Papick*, Ideal Theoretic Methods in Commutative Algebra
221. *Á. Granja et al.*, Ring Theory and Algebraic Geometry
222. *A. K. Katsaras et al.*, p-adic Functional Analysis
223. *R. Salvi*, The Navier-Stokes Equations
224. *F. U. Coelho and H. A. Merklen*, Representations of Algebras
225. *S. Aizicovici and N. H. Pavel*, Differential Equations and Control Theory
226. *G. Lyubeznik*, Local Cohomology and Its Applications
227. *G. Da Prato and L. Tubaro*, Stochastic Partial Differential Equations and Applications
228. *W. A. Carnielli et al.*, Paraconsistency
229. *A. Benkirane and A. Touzani*, Partial Differential Equations
230. *A. Illanes et al.*, Continuum Theory
231. *M. Fontana et al.*, Commutative Ring Theory and Applications
232. *D. Mond and M. J. Saia*, Real and Complex Singularities
233. *V. Ancona and J. Vaillant*, Hyperbolic Differential Operators

Additional Volumes in Preparation

Real and Complex Singularities

the sixth workshop at São Carlos

edited by

David Mond
University of Warwick
Coventry, England

Marcelo José Saia
University of São Paulo
São Carlos, São Paulo, Brazil

CRC Press
Taylor & Francis Group
Boca Raton London New York

CRC Press is an imprint of the
Taylor & Francis Group, an **informa** business

CRC Press
Taylor & Francis Group
6000 Broken Sound Parkway NW, Suite 300
Boca Raton, FL 33487-2742

First issued in paperback 2020

© 2003 by Taylor & Francis Group, LLC
CRC Press is an imprint of Taylor & Francis Group, an Informa business

No claim to original U.S. Government works

ISBN 13: 978-0-367-45456-2 (pbk)
ISBN 13: 978-0-8247-4091-7 (hbk)

Visit the Taylor & Francis Web site at
http://www.taylorandfrancis.com

and the CRC Press Web site at
http://www.crcpress.com

Library of Congress Cataloging-in-Publication Data
A catalog record for this book is available from the Library of Congress.

Preface

The *Workshop on Real and Complex Singularities* is a series, initiated in 1990 by Maria A. S. Ruas, of biennial workshops organized by the Singularity Theory group at the Institute of Mathematical Sciences and Computation, of the University of São Paulo, São Carlos (ICMC-USP, São Carlos), Brazil. Its main purpose is to bring together specialists in singularity theory and related fields.

This book contains papers presented at the 6*th Workshop on Real and Complex Singularities*. It focuses on the rôle of singularity theory in algebraic geometry, mathematical physics, differential geometry, and dynamical systems.

The meeting consisted of 13 plenary sections and 30 sections divided into three categories. The first deals with algebro-geometric aspects of singularity theory, the second is dedicated to singularity theory itself and the third is concerned with applications of singularity theory to mathematical physics, dynamical systems, and geometry. Three mini-courses were given also, one by Abramo Hefez, another by Terry Gaffney and Steven Kleiman, and one by Claus Hertling.

The papers presented here are a selection of those submitted to the editors. They are grouped into three categories: the first consists of the notes of the mini-course *Irreducible Plane Curve Singularities* by Abramo Hefez. This is in fact a small book on this subject, covering the following topics: rings of power series and Hensel's Lemma; the Preparation Theorem; the Hilbert-Rückert Basis Theorem; algebroid plane curves; Newton-Puiseux Theorem; plane analytic curves; intersection of curves; the semigroup of values; Apéry sequences; resolution of singularities; Max Noether's formula; Milnor's number and the conductor; contact among two plane branches; intersection indices of two curves.

The second group, dedicated to singularity theory, starts with the paper of C.T.C. Wall on *Openness and multitransversality*, followed by the paper of D. Barlet and A. Jeddi, *The distribution $\int_A f^s \square$ and the real asymptotic spectrum*, and the papers of I. Scherbak, *Deformations of boundary singularities and non-crystallographic Coxeter groups*, S. Izumiya and K. Maruyama, *Transversal Whitney topology and singularities of Haefliger foliations*, V. S. Kulikov, *On a conjecture of Chisini for coverings of the plane with A-D-E-singularities*, R. G. W. Atique and D. Mond, *Not all codimension 1 germs have good real pictures*, J. Seade, *On the topology of hypersurface singularities*, V. H. J. Pérez, *Polar multiplicities and equisingularity of map germs from $\mathbb{C}^3$ to $\mathbb{C}^4$*, D. Hacon, C. Mendez de Jesus, and M. C. Romero Fuster, *Topological invariants of stable maps from a surface to the plane from a global viewpoint* and R. Bulajich, L. Kushner and S. López de Medrano, *Cubics in $\mathbb{R}$ and $\mathbb{C}$*.

The third category includes applications of singularity theory to mathematical physics, dynamical systems, and differential geometry. We have the articles *Indices of Newton non-degenerate vector fields and a conjecture of Loewner for surfaces in $\mathbb{R}^4$*, by C. Gutierrez and M. A. S. Ruas, *Generic singularities of H-directions* by L. F. Mello, *Vertices of curves on constant curvature manifolds* by S. R. Costa and C. C. Pansonato, and *Projections of hypersurfaces in $\mathbb{R}^4$ to planes* by A. C. Nabarro.

Last, but not least, we have the notes of Claus Hertling's mini-course *Frobenius manifolds and hypersurface singularities*. This introduces the general theory of Frobenius manifolds, and then shows how to endow the base-space of a miniversal deformation of a hypersurface singularity with the structure of a Frobenius manifold. As an application, the author constructs global moduli spaces for isolated hypersurface singularities.

Thanks are due to many people and institutions. We start by thanking the members of the organizing committee, Ângela Sitta, Miriam Manoel, and Ton Marar and the members of the scientific committee: Takuo Fukuda, Terry Gaffney, Carlos Gutierrez, Steven Kleiman, and Alexandre Varchenko for their support. We also thank Maria Ruas and all the staff of the ICMC. Without their help we could not have organized the workshop.

The workshop was funded by FAPESP, CNPq, CAPES, USP, and SBM, whose support we gratefully acknowledge.

It is a pleasure to thank the speakers and the participants whose presence was the real success of the 6*th* Workshop.

We thank the staff members of Marcel Dekker, Inc., involved with the preparation of this book, and all those who have contributed in whatever way to these proceedings. All the papers here have been refereed.

David Mond

Marcelo José Saia

Contents

Preface   *iii*
Contributors   *vii*

1. Irreducible Plane Curve Singularities   1
    Abramo Hefez

2. Openness and Multitransversality   121
    C. T. C. Wall

3. The Distribution $\int_A f^s \square$ and the Real Asymptotic Spectrum   137
    Daniel Barlet and Ahmed Jeddi

4. Deformations of Boundary Singularities and Non-Crystallographic
    Coxeter Groups   151
    Ina Scherbak

5. Transversal Whitney Topology and Singularities of Haefliger Foliations   165
    Shyuchi Izumiya and Kunihide Maruyama

6. . On a Conjecture of Chisini for Coverings of the Plane with
    A-D-E-Singularities   175
    Valentine S. Kulikov

7. Not All Codimension 1 Germs Have Good Real Pictures   189
    David Mond and Roberta G. Wik Atique

8. On the Topology of Hypersurface Singularities   201
    José Seade

9. Polar Multiplicities and Equisingularity of Map Germs from $\mathbf{C}^3$ to $\mathbf{C}^4$ 207
    Victor Hugo Jorge Pérez

10. Topological Invariants of Stable Maps from a Surface to the Plane
    from a Global Viewpoint   227
    D. Hacon, C. Mendez de Jesus, and M. C. Romero Fuster

11. Cubics in $\mathbf{R}$ and $\mathbf{C}$ 237
    Radmila Bulajich, Leon Kushner, and Santiago López de Medrano

12. Indices of Newton Non-Degenerate Vector Fields and a Conjecture of
    Loewner for Surfaces in $\mathbf{R}^4$ 245
    Carlo Gutierrez and Maria Aparecida Soares Ruas

13. Generic Singularities of H-Directions   255
    Luis Fernando O. Mello

14.   Vertices of Curves on Constant Curvature Manifolds   267
      Claudia C. Pansonato and Sueli I. R. Costa

15.   Projections of Hypersurfaces in $\mathbf{R}^4$ to Planes   283
      Ana Claudia Nabarro

16.   Frobenius Manifolds and Hypersurface Singularities   301
      Claus Hertling

Contributors

Daniel Barlet Université Henri Poincaré and Institut Elie Cartan, Vandoeuvre-les-Nancy, France

Radmila Bulajich Universidad Autónoma del Edo. de Morelos, Morelos, Mexico

Sueli I. R. Costa Instituto di Matématica – Unicamp, Campinas – SP, Brazil

Carlos Gutierrez Vídalon Universidade de São Paulo, São Carlos, SP, Brazil

D. Hacon PUC-RIO, Gávea – Rio de Janeiro, Brazil

Abramo Hefez Universidade Federal Fluminense, Niterói, RJ, Brazil

Claus Hertling Universitat Bonn, Bonn, Germany

Shyuichi Izumiya Hokkaido University, Sapporo, Japan

Ahmed Jeddi Université Henri Poincaré and Institut Elie Cartan, Vandoeuvre-les-Nancy, France

Valentine S. Kulikov Moscow State University of Printing, Moscow, Russia

Leon Kushner Universidad Nacional Autónoma de México, México DF, Mexico

Santiago López de Medrano Universidad Nacional Autónoma de México, México DF, Mexico

Kunihide Maruyama* Hokkaido University, Sapporo, Japan

Luis Fernando O. Mello Escola Federal de Engenharia de Itajubá, Itajubá, MG, Brazil

C. Mendes de Jesus PUC-RIO, Gávea – Rio de Janeiro, Brazil

David Mond University of Warwick, Coventry, United Kingdom

Ana Claudia Nabarro Universidade de São Paulo, São Carlos, SP, Brazil

Claudia C. Pansonato CCNE – UFSM, Santa Maria-RS, Brazil

Victor Hugo Jorge Pérez Universidade Estadual de Maringá, Maringá (PR), Brazil

---

Current affiliation: NEC Nogawaryou, Kawasaki, Japan.

M. C. Romero Fuster   Universitat de València, Burjasot (València), Spain

Ina Scherbak   Tel Aviv University, Ramat Aviv, Israel

José Seade   Universidad Nacional Autónoma de México DF, Morelos, Mexico

Maria Aparecida Soares Ruas   Universidade de São Paulo, São Carlos, SP, Brazil

C. T. C. Wall   The University of Liverpool, Liverpool, United Kingdom

Roberta G. Wik Atique   Universidade de São Paulo, São Carlos, SP, Brazil

Real and Complex Singularities

Irreducible Plane Curve Singularities

ABRAMO HEFEZ   Instituto de Matemática, Universidade Federal Fluminense, R. Mario Santos Braga s/n, 24020-140, Niterói, R.J. Brasil. E-mail: hefez@mat.uff.br

To my parents from whom I learned the essence of life.

INTRODUCTION

The main objective of these notes is to introduce the reader to the local study of singularities of plane curves from an algebraic point of view. This small book is a kind of *formulaire* where the working singularist can find the basic facts and formulas, with their complete proofs, that exist in this context.

The subject, singularities of curves, has motivated for more than a century innumerable research work and is still a fertile field of investigation.

To motivate the framework in which we will place ourselves, suppose that a non-constant polynomial $f(X,Y) \in \mathbb{C}[X,Y]$ is given and consider the algebraic complex plane curve

$$C = C_f = \left\{ (x,y) \in \mathbb{C}^2; \; f(x,y) = 0 \right\}.$$

The local study of the curve C in the neighborhood of a point $P = (a,b) \in C$ depends whether the curve is regular or singular at P.

In case that C is regular at P, that is, when the partial derivatives f_X and f_Y are not simultaneously zero at P, say $f_Y(P) \neq 0$, then the Implicit Function Theorem tells us that in a neighborhood of P, we may explicit Y in the relation $f(X,Y) = 0$, as

1

a power series in X convergent in a neighborhood of a in $\mathbb{C}$. That is, in a neighborhood of P, the curve C is the graph of an analytic function of one variable.

When C is singular at P, that is, when $f(P) = f_X(P) = f_Y(P) = 0$, Newton, in his 1676 triangular correspondence with Leibniz via Oldenburg (see [5]), proposes a solution for the problem expanding Y as a power series with fractional exponents in X. The point of view of Newton was purely formal without any concern about convergence of series and without worrying about multivalued functions that appear in this context, things that only later would be clarified by Riemann and Puiseux. Newton's approach was constructive, giving an algorithm to determine such an expansion using what is called nowadays *Newton's polygon* of the curve at P.

In 1850, M.V. Puiseux published a long article [11], where he studies from the point of view of functions of a complex variable the solutions of the equation $f(X, Y) = 0$, where f is a polynomial function in two complex variables, in the neighborhood of an arbitrary point. Puiseux succeeds to justify from an analytic point of view Newton's manipulation on fractional exponents.

When the curve is regular, its behavior in a small neighborhood of P is entirely known: it is locally isomorphic to the curve given by $Y = 0$.

When the point P is singular, the situation is quite different. Consider the following examples, $f = Y^2 - X^3$ and $g = Y^2 - X^2(X+1)$, where $P = (0,0)$:

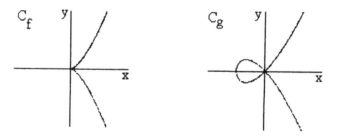

Figure 1

How to study a curve in a small neighborhood of a singular point? In particular, how one could recognize algebraically the local reducibility of a singular curve, as for example the above curve C_g at the point $P = (0,0)$? Certainly the ring $\mathbb{C}[X, Y]$ is not suitable for this purpose, since the polynomial g is irreducible in it. The idea is to enlarge the ring $\mathbb{C}[X, Y]$ in such a way that the local reducibility of C_g at P appears. As a first attempt, we may consider the local ring

$$\mathbb{C}[X, Y]_{\mathcal{M}} = \left\{ \frac{\varphi}{\psi}; \ \varphi, \psi \in \mathbb{C}[X, Y], \ \psi \notin \mathcal{M} \right\},$$

where $\mathcal{M} = \langle X, Y \rangle$ is the maximal ideal of $\mathbb{C}[X, Y]$, corresponding to the point $P = (0,0)$.

This is effectively a local ring, that is, a ring with a unique maximal ideal (in this case, $\langle X, Y \rangle \mathbb{C}[X, Y]_{\mathcal{M}}$), which in a sense contains localized information at $P = (0,0)$ about all curves that contain P. Unfortunately, this is not sufficient since g is still irreducible in this ring. Another attempt is to complete the local ring $\mathbb{C}[X,Y]_{\mathcal{M}}$ with respect to the topology associated to its maximal ideal, leading to the ring of formal power series in two variables $\mathbb{C}[[X,Y]]$, where finally g splits into two irreducible factors as follows:

$$g = (Y + X + \text{terms of degree} \geq 2)(Y - X + \text{terms of degree} \geq 2).$$

It is in this ring that we get all infinitesimal algebraic information about an algebraic curve at the point $P = (0,0)$. In this way we retake Newton's point of view, going back to the origins.

This book is a compilation of results of many mathematicians, starting in the seventeenth century with Isaac Newton, passing through the nineteenth century with the contribution of K. Weierstrass, M.V. Puiseux and M. Noether and ending in the twentieth century with the contribution of O. Zariski, S.S. Abhyankar, R. Apéry and A. Azevedo, among many others. The reader interested in more modern material is invited to consult [8].

The book is divided into eight chapters, which we describe below.

In Chapter 1, we introduce the rings of formal power series studying their units and automorphisms. We end the chapter with the fundamental Hensel's Lemma.

In Chapter 2, we study the Weierstrass Preparation Theorem that reduces the study of formal power series to that of certain polynomials. Most algebraic properties of the rings of formal power series are deduced from this Theorem. We also include, for the convenience of the reader, a section on elimination whose results will be used essentially in Chapter 4.

In Chapter 3, we introduce our main object of study, the algebroid irreducible plane curves or plane branches, and study their parametrization given by the Newton-Puiseux Theorem. This theorem is a fundamental tool for the study of plane branches defined over fields of characteristic zero. At the end of the chapter, we discuss some properties of germs of plane analytic curves.

In Chapter 4, we introduce the local ring of a plane branch, studying some of its properties. In the sequel we make a parenthetical presentation of some results in Linear Algebra that will be essential for what will follow. After this, we define the intersection index of two curves at a point and deduce its properties, showing several ways it may be computed.

Chapter 5 is dedicated to the resolution of singularities of irreducible algebroid plane curves by means of a sequence of quadratic transformations. In this way, we get a finite sequence of multiplicities associated to these successive transforms of the singularity, called the multiplicity sequence of the singularity and leading to the notion of equiresolubility. At the end of the chapter we establish a formula due to Max Noether that relates the intersection index of two curves with that of their quadratic transforms.

In Chapter 6, we introduce the semigroup of values associated to a plane branch. This is a semigroup of natural numbers with conductor and it is an important arithmetic invariant associated to the branch, leading to the notion of equisingularity due to Zariski. It is shown that equiresolubility and equisingularity for plane branches are equivalent. We also show that the characteristic integers associated to a Puiseux parametrization of the curve we introduced in Chapter 3 characterize as well the equisingularity classes.

In Chapter 7, we study some arithmetic associated to the sub-semigroups of the natural numbers. We explore the notion of strongly increasing semigroup and introduce the concept of symmetric semigroup, properties that are shared by the semigroups of values of plane branches. At the end of the chapter we apply these notions to branches, relating Milnor's number introduced in [10] to the conductor of the semigroup of the branch.

Finally, in Chapter 8 we study the contact among two plane branches and show how the intersection index of two such branches can be also computed by their Puiseux expansions. We also describe all possible intersection indices two curves, each belonging to a given equisingularity class. Finally, we discuss briefly the notion of maximal contact among branches.

Along these notes, to refer to a certain result, we use the convention that the numbers that appear after the words Theorem, Proposition, etc., indicate first the number of the chapter and then the relative position of the result in the chapter. To the problems are associated three numbers which represent in order, the chapter, the section and the number of the problem itself.

I would like to thank the Organizing Committee of the Workshop for the invitation to teach this mini-course and subsequently the Editors of this volume for inviting me to publish these notes. Finally, it is a pleasure to thank M.E. Hernandes and V. Bayer for sharing my interest on the subject and for reading carefully an early version of the manuscript.

1 Power Series

In this chapter we introduce the rings of formal power series and study some of their basic algebraic properties.

1.1 RINGS OF POWER SERIES

Let K be a field and $X_1, \ldots, X_r$ indeterminates over K. Denote by

$$\mathcal{R} = K[[X_1, \ldots, X_r]]$$

the set of all formal sums of the type

$$f = \sum_{i=0}^{\infty} P_i = P_0 + P_1 + P_2 + \cdots,$$

where each P_i is a homogeneous polynomial of degree i in the indeterminates $X_1, \ldots, X_r$, with coefficients in K, we will consider the zero polynomial as a homogeneous polynomial of any degree. The elements of $\mathcal{R}$ will be called *formal power series* in the indeterminates $X_1, \ldots, X_r$ with coefficients in K.

Let $f = P_0 + P_1 + \cdots$ and $g = Q_0 + Q_1 + \cdots$ be elements of $\mathcal{R}$. By definition we have

$$f = g \iff P_i = Q_i, \ \forall i \in \mathbb{N}.$$

In $\mathcal{R}$ we define the following operations:

$$f + g = \sum_{i=0}^{\infty}(P_i + Q_i) \quad \text{and} \quad f \cdot g = \sum_{i=0}^{\infty}\sum_{j=0}^{i} P_j Q_{i-j}.$$

It is easy to verify that, with these operations, $\mathcal{R}$ is a unitary commutative ring, called the *ring of formal power series* in r indeterminates with coefficients in K.

The ring $\mathcal{R}$ has as subrings the field K and the ring of polynomials $K[X_1, \ldots, X_r]$. The elements of $\mathcal{R}$ may be represented more explicitly in the form

$$f = \sum_{i=0}^{\infty} \sum_{i_1 + \cdots + i_r = i} a_{i_1, \ldots, i_r} X_1^{i_1} \ldots X_r^{i_r}; \qquad a_{i_1, \ldots, i_r} \in K.$$

If K is the field of real or complex numbers, we may consider the subring $A = K\{X_1, \ldots, X_r\}$ of $\mathcal{R}$ consisting of the absolutely convergent power series in the neighborhood of the origin $(0, \ldots, 0)$. In other words, the elements of A are the series $f = \sum_{i=0}^{\infty} \sum_{i_1 + \cdots + i_r = i} a_{i_1, \ldots, i_r} X_1^{i_1} \ldots X_r^{i_r}$ for which there exists a positive real number ρ (depending on f) such that the series

$$\sum_{i=0}^{\infty} \sum_{i_1 + \cdots + i_r = i} |a_{i_1, \ldots, i_r}| \rho^{i_1 + \cdots + i_r}$$

is convergent.

The following result will describe the invertible elements of $\mathcal{R}$.

PROPOSITION 1.1 *The element $f = \sum_{i=0}^{\infty} P_i$ in $\mathcal{R}$ is invertible if and only if $P_0 \neq 0$.*

Proof: Let $g = Q_0 + Q_1 + \cdots$, and consider the equation

$$1 = f \cdot g = P_0 Q_0 + (P_1 Q_0 + P_0 Q_1) + \cdots.$$

This equation is equivalent to the system of equations

$$
\begin{aligned}
&P_0 Q_0 = 1, \\
&P_1 Q_0 + P_0 Q_1 = 0, \\
&\cdots \\
&P_n Q_0 + P_{n-1} Q_1 + \cdots + P_0 Q_n = 0, \\
&\cdots
\end{aligned}
\tag{1.1}
$$

It follows that f is invertible if and only if the system (1.1) has a solution in the Q_i's. In this case, $f^{-1} = g$.

If f is invertible, then there exists Q_0 such that $P_0 Q_0 = 1$, and consequently, $P_0 \neq 0$.

Conversely, suppose that $P_0 \neq 0$. Then the system (1.1) has a solution given by the following recursive relations:

$$
\begin{aligned}
&Q_0 = P_0^{-1}, \qquad Q_1 = -P_0^{-1} P_1 Q_0, \\
&Q_n = -P_0^{-1}(P_n Q_0 + \cdots + P_1 Q_{n-1}).
\end{aligned}
$$

$\square$

Two elements f and g in $\mathcal{R}$ will be called *associated* if there exists a unit (i.e. an invertible element) u such that $f = u \cdot g$.

DEFINITION 1.2 *Let $f \in \mathcal{R} \setminus \{0\}$. Suppose that*

$$f = P_n + P_{n+1} + \cdots,$$

where every P_j is a homogeneous polynomial of degree j and $P_n \neq 0$. The homogeneous polynomial P_n is called the initial form *of f. The integer n is called the* multiplicity *of f and is denoted by* $\mathrm{mult}(f)$. *If $f = 0$, we put $\mathrm{mult}(f) = \infty$.*

According to the Proposition (1.1), we have that $f \in \mathcal{R}$ is invertible if and only if $\mathrm{mult}(f) = 0$.

The multiplicity of power series has the following properties:

PROPOSITION 1.3 *Let $f, g \in \mathcal{R}$. We have*

i) $\mathrm{mult}(f \cdot g) = \mathrm{mult}(f) + \mathrm{mult}(g)$

ii) $\mathrm{mult}(f \pm g) \geq \min\{\mathrm{mult}(f), \mathrm{mult}(g)\}$, *with equality sign holding whenever* $\mathrm{mult}(f) \neq \mathrm{mult}(g)$.

Proof: We will leave this easy verification as an exercise to the reader.

The notion of multiplicity of power series will play a role similar to that of the degree for polynomials.

PROPOSITION 1.4 *The ring $\mathcal{R}$ is a domain.*

Proof: If $f, g \in \mathcal{R} \setminus \{0\}$, then from the Proposition (1.3.i), we have
$\mathrm{mult}(f \cdot g) = \mathrm{mult}(f) + \mathrm{mult}(g) < \infty$. This implies that $f \cdot g \neq 0$. □

We will denote by $\mathcal{M}_\mathcal{R} = \langle X_1, \ldots, X_r \rangle$ the ideal of $\mathcal{R}$ generated by $X_1, \ldots, X_r$. We denote by $\mathcal{M}_\mathcal{R}^i$ the i-th power of the ideal $\mathcal{M}_\mathcal{R}$, and put $\mathcal{M}_\mathcal{R}^0 = \mathcal{R}$.

PROPOSITION 1.5 *The ideal $\mathcal{M}_\mathcal{R}$ is the unique maximal ideal of $\mathcal{R}$ and is such that*

$$\bigcap_{i \in \mathbb{N}} \mathcal{M}_\mathcal{R}^i = \{0\}.$$

Proof: For the first assertion, it suffices to show that $\mathcal{R} \setminus \mathcal{M}_\mathcal{R}$ is the set of all invertible elements of $\mathcal{R}$, which follows from the Proposition (1.1)

To finish the proof of the proposition, notice that if every P_i, $i \in \mathbb{N}$, is a homogeneous polynomial of degree i, then

$$\sum_{i \in \mathbb{N}} P_i \in \mathcal{M}_\mathcal{R}^j \quad \Longleftrightarrow \quad P_0 = P_1 = \cdots = P_{j-1} = 0.$$

This implies the result. □

PROBLEMS

1.1) a) Let $K[[X]]$ be the ring of power series in one indeterminate X, with coefficients in K. Show that in $K[[X]]$ we have the identity:

$$(1 - X) \sum_{i=0}^{\infty} X^i = 1.$$

b) Show that in $K[[X_1, \ldots, X_r]]$ we have the following equality:

$$(1 - X_1) \cdots (1 - X_r) \sum_{i=0}^{\infty} \sum_{i_1 + \cdots + i_r = i} X_1^{i_1} \ldots X_r^{i_r} = 1.$$

c) Find the first five homogeneous components of the inverse of
$f = 1 + X + Y + X^2 + Y^2 + X^3 + Y^3$ in $K[[X, Y]]$.

1.2) a) Prove the Proposition (1.3)

b) Show that if f and g are associated in $\mathcal{R}$, then $\mathrm{mult}(f) = \mathrm{mult}(g)$.

1.3) Let $g_1, \ldots, g_s \in \mathcal{M}_\mathcal{R}$ and let $P(Y_1, \ldots, Y_s)$ be a homogeneous polynomial of degree n.

a) Let $i_1, \ldots, i_s \in \mathbb{N}$. Show that

$$\mathrm{mult}(g_1^{i_1} \cdots g_s^{i_s}) = i_1 \mathrm{mult}(g_1) + \cdots + i_s \mathrm{mult}(g_s) \geq i_1 + \cdots + i_s.$$

b) Show that $\mathrm{mult}(P(g_1, \ldots, g_s)) \geq n$.

1.4) a) Show that if $f \in K[[X]] \setminus \{0\}$, then there exists an integer $m \geq 0$ and an invertible element $u \in K[[X]]$ such that $f = X^m u$.

b) Show that $K[[X]]$ is a principal ideal domain and that every non-zero ideal of it is generated by X^m, for some $m \in \mathbb{N}$.

c) Conclude that $K[[X]]$ is a unique factorization domain.

d) Let $F = Y^n + a_1(X) Y^{n-1} + \cdots + a_n(X) \in K[[X]][Y]$ such that X divides $a_i(X)$, for $i = 1, \ldots, n$, but X^2 doesn't divide $a_n(X)$. Show that F is irreducible in $K[[X]][Y]$ (Eisenstein's criterion).

1.5) Let $\mathcal{M}_\mathcal{R}$ be the maximal ideal of $\mathcal{R} = K[[X_1, \ldots, X_r]]$. Show that

a) $f \in \mathcal{M}_\mathcal{R}^j \iff \mathrm{mult}(f) \geq j$.

b) $\mathrm{mult}(f) = \max\{i; f \in \mathcal{M}_\mathcal{R}^i\}$.

1.2 METRIC ON POWER SERIES RINGS

To deal with some questions concerning infinite sets of power series, specially when we want to give a meaning to infinite sums of series, it will be convenient do define a metric on such sets. This is what we are going to do next.

As usual, denote by $\mathcal{R}$ the ring $K[[X_1, \ldots, X_r]]$ and by $\mathcal{M}_\mathcal{R}$ its maximal ideal. Let $\rho > 1$ be a real number and consider the following map

$$\begin{aligned} d: \quad &\mathcal{R} \times \mathcal{R} &\longrightarrow \quad &\mathbb{R}. \\ &(f, g) &\mapsto \quad &d(f, g) = \rho^{-\mathrm{mult}(f-g)} \end{aligned}$$

PROPOSITION 1.6 *The pair $(\mathcal{R}, d)$ is a complete metric space.*

Proof: To verify that d is a metric on $\mathcal{R}$, first remark that $d(f, g) \geq 0$. Next we have that $d(f, g) = 0$ if and only if $\mathrm{mult}(f - g) = \infty$, which is equivalent to $f = g$.

The map d is obviously symmetric. Consider now the following inequality:

$$\mathrm{mult}(f - g) = \mathrm{mult}((f - h) - (g - h)) \geq \min\{\mathrm{mult}(f - h), \mathrm{mult}(g - h)\} := m,$$

which implies the triangular inequality, because

$$d(f, g) = \rho^{-mult(f-g)} \leq \rho^{-m} \leq \rho^{-\mathrm{mult}(f-h)} + \rho^{-\mathrm{mult}(g-h)} = d(f, h) + d(h, g).$$

To show that $(\mathcal{R}, d)$ is a complete metric space, let $(f_n)_{n \in \mathbb{N}}$ be a Cauchy sequence in $\mathcal{R}$. Then, for every $n \in \mathbb{N}$, there exists an integer $\nu(n)$ such that

$$d(f_\ell, f_m) < \rho^{-n}, \quad \forall \, \ell, m \geq \nu(n),$$

and this implies that

$$f_\ell - f_m \in \mathcal{M}_\mathcal{R}^n, \quad \forall \, \ell, m \geq \nu(n).$$

We may obviously choose the $\nu(n)$ in such a way that they form an increasing sequence, so we have

$$f_{\nu(n)} - f_{\nu(n+1)} \in \mathcal{M}_\mathcal{R}^n, \quad \forall \, n \in \mathbb{N}.$$

Now, for every $i \in \mathbb{N}$, write

$$f_{\nu(i)} = P_{i,0} + P_{i,1} + \cdots,$$

where each $P_{i,j}$ is a homogeneous polynomial of degree j, and define

$$f = P_{1,0} + P_{2,1} + \cdots + P_{i+1,i} + \cdots$$

It is not difficult to verify that

$$f - f_{\nu(i)} \in \mathcal{M}_\mathcal{R}^i, \quad \forall \, i \in \mathbb{N},$$

which in turn implies easily that

$$\lim_{n \to \infty} f_{\nu(n)} = f,$$

This shows that the Cauchy sequence (f_n) has a subsequence $(f_{\nu(n)})$ that converges to the element $f \in \mathcal{R}$, hence the sequence itself converges to f. $\qquad \square$

Although, by definition, we only may add a finite number of power series, it is possible, in some circumstances, to sum infinite families of power series.

DEFINITION 1.7 *Let* $\mathcal{F} = \{f_\lambda; \, \lambda \in \Lambda\}$ *be a given family of power series in* $\mathcal{R}$. *We will say that the family* $\mathcal{F}$ *is* summable *if, for every integer* $n \in \mathbb{N}$, *we have*

$$\# \{\lambda \in \Lambda; \; \mathrm{mult}(f_\lambda) \leq n\} < \infty. \qquad (1.2)$$

PROPOSITION 1.8 *Let* $\{f_\lambda; \, \lambda \in \Lambda\}$, *and* $\{g_\lambda; \, \lambda \in \Lambda\}$ *be two summable families in* $\mathcal{R}$ *and let* $A, B \in \mathcal{R}$. *Then the following holds.*

i) *The sum* $\sum_{\lambda \in \Lambda} f_\lambda$ *is well defined in* $\mathcal{R}$.

ii) *The family* $\{A f_\lambda + B g_\lambda; \, \lambda \in \Lambda\}$ *is summable and*

$$\sum_{\lambda \in \Lambda} (A f_\lambda + B g_\lambda) = A \sum_{\lambda \in \Lambda} f_\lambda + B \sum_{\lambda \in \Lambda} g_\lambda.$$

iii) *The family* $\{f_\lambda g_\mu; \ (\lambda, \mu) \in \Lambda \times \Lambda\}$ *is summable and*

$$(\sum_{\lambda \in \Lambda} f_\lambda)(\sum_{\lambda \in \Lambda} g_\lambda) = \sum_{(\lambda, \mu) \in \Lambda \times \Lambda} f_\lambda g_\mu.$$

Proof: (i) The set

$$\Lambda_n = \{\lambda \in \Lambda; \ \mathrm{mult}(f_\lambda) \leq n\}$$

is finite, since (1.2) holds. Put

$$h_n = \sum_{\lambda \in \Lambda_n} f_\lambda.$$

Since for $m > n$,

$$\Lambda_m = \Lambda_n \cup \{\lambda \in \Lambda; \ n < \mathrm{mult}(f_\lambda) \leq m\},$$

it follows that, for all $m \geq n$, we have $\mathrm{mult}(h_m - h_n) > n$, which implies that $(h_n)_{n \in \mathbb{N}}$ is a Cauchy sequence, hence convergent to an element of $\mathcal{R}$. The limit will be denoted by $\sum_{\lambda \in \Lambda} f_\lambda$.
(ii) Given $n \in \mathbb{N}$, we have that the set

$$\Lambda_n'' = \{\lambda \in \Lambda; \ \mathrm{mult}(Af_\lambda + Bg_\lambda) \leq n\},$$

is finite because there are finitely many λ's satisfying the following inequality

$$\min \{\mathrm{mult}(A) + \mathrm{mult}(f_\lambda), \mathrm{mult}(B) + \mathrm{mult}(g_\lambda)\} \leq \mathrm{mult}(Af_\lambda + Bg_\lambda) \leq n.$$

Now, let Λ_n be as above and put

$$\Lambda_n' = \{\lambda \in \Lambda; \ \mathrm{mult}(g_\lambda) \leq n\}.$$

Defining $\phi_n = \sum_{\lambda \in \Lambda_n''}(Af_\lambda + Bg_\lambda)$ and $\psi_n = A\sum_{\lambda \in \Lambda_n} f_\lambda + B\sum_{\lambda \in \Lambda_n'} g_\lambda$, we have that $\psi_n - \phi_n \in \mathcal{M}_\mathcal{R}^{n+1}$, hence $\lim_{n \to \infty}(\psi_n - \phi_n) = 0$, and the result follows considering the following equalities:

$$\lim_{n \to \infty} \psi_n = A \sum_{\lambda \in \Lambda} f_\lambda + B \sum_{\lambda \in \Lambda} g_\lambda \quad \text{and} \quad \lim_{n \to \infty} \phi_n = \sum_{\lambda \in \Lambda}(Af_\lambda + Bg_\lambda).$$

(iii) Let Λ_n and Λ_n' be as above, and define

$$\Lambda_n'' = \{(\lambda, \mu) \in \Lambda \times \Lambda; \ \mathrm{mult}(f_\lambda g_\mu) \leq n\}.$$

This last set is obviously finite. Now, defining

$$\Phi_n = \sum_{(\lambda, \mu) \in \Lambda_n''} f_\lambda g_\mu \quad \text{and} \quad \Psi_n = (\sum_{\lambda \in \Lambda_n} f_\lambda)(\sum_{\mu \in \Lambda_n'} g_\mu),$$

we have as above,

$$\lim_{n \to \infty}(\Psi_n - \Phi_n) = 0,$$

which together with

$$\lim_{n \to \infty} \Psi_n = (\sum_{\lambda \in \Lambda} f_\lambda)(\sum_{\lambda \in \Lambda} g_\lambda), \quad \text{and} \quad \lim_{n \to \infty} \Phi_n = \sum_{(\lambda, \mu) \in \Lambda \times \Lambda} f_\lambda g_\mu,$$

imply the result. □

Let $\{P_i; \; i \in \mathbb{N}\}$ be a family of homogeneous polynomials, with $P_i \in K[Y_1, \dots Y_s]$ of degree i, and let $g_1, \dots, g_s \in \mathcal{M}_{\mathcal{R}}$. the family

$$\mathcal{F} = \{P_i(g_1, \dots, g_s); \; i \in \mathbb{N}\}$$

is summable, because of the inequalities (see Problem 1.3)

$$\text{mult}(P_i(g_1, \dots, g_s)) \geq i, \; \forall i \in \mathbb{N}.$$

If $f = \sum_{i \in \mathbb{N}} P_i(Y_1, \dots, Y_s) \in K[[Y_1, \dots, Y_s]]$, then the sum of the family $\mathcal{F}$ will be denoted by $f(g_1, \dots, g_s)$, and will be called the substitution of $Y_1, \dots, Y_s$ by $g_1, \dots, g_s$ in f. In particular, it is possible to substitute $Y_1, \dots, Y_r$ by $0, \dots, 0$ in f, getting P_0.

The following result follows immediately from the Proposition (1.8).

COROLLARY 1.9 *Given* $g_1, \dots, g_s \in \mathcal{M}_{\mathcal{R}}$, $f, h \in K[[Y_1, \dots, Y_s]]$ *and* $a \in K$, *then*

(i) $(f + ah)(g_1, \dots, g_s) = f(g_1, \dots, g_s) + ah(g_1, \dots, g_s)$.

(ii) $(f \cdot h)(g_1, \dots, g_s) = f(g_1, \dots, g_s) \cdot h(g_1, \dots, g_s)$.

(iii) *The substitution map*

$$S_{g_1, \dots, g_s} : \quad K[[Y_1, \dots, Y_s]] \quad \longrightarrow \quad K[[X_1, \dots, X_r]]$$
$$f \quad \mapsto \quad S_{g_1, \dots, g_s}(f) = f(g_1, \dots, g_s)$$

is a homomorphism of K-algebras .

In the next section we will see that any homomorphism of K-algebras is a substitution map.

Notice that the condition $g_1, \dots, g_s \in \mathcal{M}_{\mathcal{R}}$ is essential in order to make substitutions. For example, if $f(X) = 1 + X + X^2 + \cdots \in K[[X]]$, what would be $f(1)$? More generally, given $g_1, \dots, g_s \in \mathcal{R}$, such that for some i, $g_i(0) = c \in K \setminus \{0\}$, then taking $f(Y_1, \dots, Y_s) = 1 + c^{-1}Y_i + c^{-2}Y_i^2 + \cdots$, we have that $f(g_1, \dots, g_s)$ is not defined as an element of $\mathcal{R}$, since otherwise we would have

$$f(g_1, \dots, g_s)(0, \dots, 0) = 1 + 1 + 1 + \cdots \in \mathcal{R},$$

which is a contradiction.

Let $P_i \in K[X_1, \dots, X_r]$, $i \in \mathbb{N}$ be a family of homogeneous polynomials with $\deg(P_i) = i$. For all $(j_1, \dots, j_r) \in \mathbb{N}^r$ the family

$$\mathcal{G} = \left\{ \frac{\partial^{j_1 + \cdots + j_r} P_i}{\partial X_1^{j_1} \cdots \partial X_r^{j_r}}; \; i \in \mathbb{N} \right\}$$

is summable. The sum of the family $\mathcal{G}$ is what we call the partial derivative of $f = \sum_{i \in \mathbb{N}} P_i$ of order j_1 in X_1, etc., and of order j_r in X_r, and is denoted by

$$\frac{\partial^{j_1 + \cdots + j_r} f}{\partial X_1^{j_1} \cdots \partial X_r^{j_r}}.$$

PROBLEMS

2.1) Suppose that $f \in \mathcal{R}$ is such that $f(0) = a \neq 0$. Using the identity in Problem 1.1 (a), show that

$$f^{-1} = a^{-1} \sum_{i=0}^{\infty} (1 - a^{-1}f)^i.$$

2.2) Let $f_i \in \mathcal{R}$, $i \in \mathbb{N}$, be such that $\lim_i \operatorname{mult}(f_i) = \infty$.
a) Show that (f_i) is a Cauchy sequence in $\mathcal{R}$, whose limit is $0 \in \mathcal{R}$.
b) Show that the family $\mathcal{F} = \{f_i; \ i \in \mathbb{N}\}$ is summable.

2.3) Let $\mathcal{F} = \{f_i; \ i \in \mathbb{N}\}$ be a summable family of power series in $\mathcal{R}$.
a) Show that the infinite sum $\sum_{i=1}^{\infty} f_i$ is independent of the order in which we write the elements f_i.
b) Show that $\operatorname{mult}(\sum_{i=0}^{\infty} f_i) \geq \min\{\operatorname{mult}(f_i); i \in \mathbb{N}\}$.

2.4) Show that with the metric we defined in $\mathcal{R}$, we have:
a) The operations of addition and multiplication are continuous.
b) The substitution homomorphism $S_{g_1, \ldots, g_s}$ is continuous.

2.5) Show that $K[X_1, \ldots, X_r]$ is dense in $K[[X_1, \ldots, X_r]]$. Show also that $K[[X_1, \ldots, X_{r-1}]]$ is closed in $K[[X_1, \ldots, X_r]]$.

2.6) a) Show that if $f \in K[[Y_1, \ldots, Y_s]]$ and $g_1, \ldots, g_s \in K[[X_1, \ldots, X_r]]$, are such that $\operatorname{mult}(g_i) \geq 1$, for $i = 1, \ldots, s$, then

$$\operatorname{mult}(S_{g_1, \ldots, g_s}(f)) = \operatorname{mult}(f(g_1, \ldots, g_s)) \geq \operatorname{mult}(f) \cdot \min\{\operatorname{mult}(g_i); i \in \mathbb{N}\}.$$

b)   Show that $f \in K[[X]] \setminus \{0\}$ and $g \in K[[X_1, \ldots, X_r]] \setminus \{0\}$, with $\operatorname{mult}(g) \geq 1$, then

$$\operatorname{mult}(f(g)) = \operatorname{mult}(f) \cdot \operatorname{mult}(g).$$

2.7) Show that if the characteristic of K is zero, then the multiplicity of $f \in \mathcal{R} \setminus \{0\}$ may be determined as follows:

$$\operatorname{mult}(f) = \min\left\{i_1 + \cdots + i_r; \ \frac{\partial^{i_1 + \cdots + i_r} f}{\partial X_1^{i_1} \cdots \partial X_r^{i_r}}(0) \neq 0\right\}.$$

1.3   HOMOMORPHISMS

Let $\mathcal{R} = K[[X_1, \ldots, X_r]]$ and $\mathcal{S} = K[[Y_1, \ldots, Y_s]]$. Denoting respectively their maximal ideals by $\mathcal{M}_{\mathcal{R}}$ and $\mathcal{M}_{\mathcal{S}}$, we have the following result:

PROPOSITION 1.10 *Let* $T : \mathcal{S} \longrightarrow \mathcal{R}$ *be a homomorphism of K-algebras. Then*

i) $T(\mathcal{M}_{\mathcal{S}}) \subset \mathcal{M}_{\mathcal{R}}.$

ii)   *T is continuous.*

iii)   *There exist $g_1, \ldots, g_s \in \mathcal{M}_{\mathcal{R}}$ such that $T = S_{g_1, \ldots, g_s}.$*

Proof: (i) Let $f \in \mathcal{M}_\mathcal{S}$ and suppose that $T(f) \notin \mathcal{M}_\mathcal{R}$. Then we may write $T(f) = c + g$, where $c \in K \setminus \{0\}$ and $g \in \mathcal{M}_\mathcal{R}$. It then follows that

$$T(f - c) = g,$$

where $f - c$ is invertible in $\mathcal{S}$, while g is not invertible in $\mathcal{R}$, which is a contradiction.
(ii) It follows immediately from (i) that $T(\mathcal{M}_\mathcal{S}^n) \subset \mathcal{M}_\mathcal{R}^n$, which easily implies that T is continuous.
(iii) From (i) we have that $T(Y_j) = g_j \in \mathcal{M}_\mathcal{R}$, for $j = 1, \ldots, s$. Now let $f = \sum_{i \in \mathbb{N}} P_i$. Since

$$T(\sum_{i=0}^{n} P_i) = \sum_{i=0}^{n} P_i(g_1, \ldots, g_s),$$

$\lim_{n \to \infty} \sum_{i=0}^{n} P_i = f$, $\lim_{n \to \infty} \sum_{i=0}^{n} P_i(g_1, \ldots, g_s) = f(g_1, \ldots, g_s)$ and T is continuous, the result follows. $\square$

So, given a homomorphism $T : \mathcal{S} \longrightarrow \mathcal{R}$, there exist $g_1, \ldots, g_s \in \mathcal{M}_\mathcal{R}$ such that $T = S_{g_1, \ldots, g_s}$. Let us now see which additional conditions we must impose on $g_1, \ldots, g_s$ to guarantee that T is a K-isomorphism.

Initially, observe that every K-isomorphism from $\mathcal{S}$ to $\mathcal{R}$ must preserve multiplicities. Indeed, from Problem 2.6(a), it follows that for all $f \in \mathcal{S}$,

$$\mathrm{mult}(f) \leq \mathrm{mult}(T(f)) \leq \mathrm{mult}(T^{-1}(T(f))) = \mathrm{mult}(f),$$

which proves our assertion.

Since $T(Y_i) = g_i$, for $i = 1, \ldots, s$, it follows that $\mathrm{mult}(g_i) = 1$, for all $i = 1, \ldots, s$. Let $L_1, \ldots, L_s$ respectively the initial forms of $g_1, \ldots, g_s$, which are homogeneous polynomials of degree one in $X_1, \ldots, X_r$. Suppose for a moment that there exists a K-linear non-trivial dependency relation,

$$a_1 L_1 + \cdots + a_s L_s = 0.$$

Then if we take $f = a_1 Y_1 + \cdots + a_s Y_s$, it would follow that

$$\mathrm{mult}(T(f)) = \mathrm{mult}(S_{g_1, \ldots, g_s}(f)) > \mathrm{mult}(f),$$

which would imply that $S_{g_1, \ldots, g_s}$ is not a K-isomorphism.

Consequently, a necessary condition for $S_{g_1, \ldots, g_s}$ to be a K-isomorphism from $\mathcal{S}$ onto $\mathcal{R}$ is that the initial forms of the g_i's ought to be linear forms, linearly independent over K. This, in particular, shows that $s \leq r$, and by arguing in the same way with T^{-1} instead of T, we get that $r = s$.

In order to prove that the above conditions are also sufficient, we will need the following lemma.

LEMMA 1.11 *A subring A of $\mathcal{R}$ is dense in $\mathcal{R}$ if and only if given any homogeneous polynomial $P \in K[X_1, \ldots, X_r]$, there exists an element in A whose initial form is P.*

Proof: Suppose that A is dense in $\mathcal{R}$, and let P be a homogeneous polynomial of degree d in $K[X_1, \ldots, X_r]$. Let n be an integer greater than d. Because A is dense in $\mathcal{R}$, the polynomial P is the limit of a sequence of elements in A, therefore, there exists an $f \in \mathcal{R}$ such that $\mathrm{mult}(f - P) \geq n$. Since $n > d$, it follows that P is the initial form of f.

Conversely, suppose that A has the property of the statement of the lemma and let f be an element of $\mathcal{R}$. We are going to construct an infinite sequence (f_i) of elements in A such that $\lim f_i = f$.

For $i = 0$, take $f_0 = 0$. Suppose that we have constructed the elements $f_0, f_1, \ldots, f_{n-1}$ of A such that

$$\mathrm{mult}(f - f_i) \geq i, \quad i = 0, \ldots, n - 1.$$

If $\mathrm{mult}(f - f_{n-1}) \geq n$, put $f_n = f_{n-1}$.

Suppose that $\mathrm{mult}(f - f_{n-1}) = n - 1$. If G_{n-1} is the initial form of $f - f_{n-1}$, there exists an element $h_{n-1} \in \mathcal{R}$, whose initial form is G_{n-1}. Defining

$$f_n = f_{n-1} + h_{n-1},$$

it follows that $f_n \in A$ is such that

$$\mathrm{mult}(f - f_n) = \mathrm{mult}(f - f_{n-1} - h_{n-1}) \geq n.$$

In this way we have constructed a sequence in A such that $\lim f_i = f$, showing that A is dense in $\mathcal{R}$. $\qquad\square$

In particular, we have that $K[X_1, \ldots, X_r]$ is dense in $K[[X_1, \ldots, X_r]]$.

If $r = s$ and if $L_1, \ldots, L_r$ are linear forms in $\mathcal{R}$, linearly independent over K, then $T = S_{L_1, \ldots, L_r}$ is a K-isomorphism from S onto $\mathcal{R}$.

Indeed, if

$$L_i = a_{i,1}X_1 + \cdots + a_{i,r}X_r, \quad i = 1, \ldots, r,$$

and if M is the matrix $(a_{i,j})$, then $T^{-1} = S_{L'_1, \ldots, L'_r}$, where

$$L'_i = b_{i,1}Y_1 + \cdots + b_{i,r}Y_r, \quad i = 1, \ldots, r,$$

and $(b_{i,j})$ is the inverse matrix of $(a_{i,j})$.

PROPOSITION 1.12 *Suppose that $r = s$ and let $g_1, \ldots, g_r \in \mathcal{R}$ with initial forms, $L_1, \ldots, L_r$, K-linearly independent. Then $S_{g_1, \ldots, g_r}$ is a K-isomorphism from S onto $\mathcal{R}$.*

Proof: We are going initially to prove that $S_{g_1, \ldots, g_r}$ is injective.

Notice that if $0 \neq f = P_n + P_{n+1} + \cdots$, with each P_i a homogeneous polynomial of degree i and $P_n \neq 0$, then the initial term of $S_{g_1, \ldots, g_r}(f)$ is $S_{L_1, \ldots, L_r}(P_n)$. Indeed, it suffices to note that this last element is non-zero because

$$0 \neq P_n = S^{-1}_{L_1, \ldots, L_r}\left(S_{L_1, \ldots, L_r}(P_n)\right).$$

From this it follows immediately that

$$\mathrm{mult}\left(S_{g_1, \ldots, g_r}(f)\right) = \mathrm{mult}(f), \tag{1.3}$$

implying the injectivity of $S_{g_1, \ldots, g_r}$.

We are going now to prove that $S_{g_1, \ldots, g_r}$ is surjective. For this, it is enough to prove that its image $A = K[[g_1, \ldots, g_r]]$ is closed and dense in $\mathcal{R}$.

Let $P \in K[Y_1, \ldots, Y_s]$ be an arbitrary homogeneous polynomial. Consider

$$Q = S_{L_1,\ldots,L_r}^{-1}(P) \in \mathcal{R},$$

then $P = S_{L_1,\ldots,L_r}(Q)$ is the initial form of $S_{g_1,\ldots,g_r}(Q) \in A$, which in view of Lemma (1.11) implies that A is dense in $\mathcal{R}$.

We will prove now that A is closed in $\mathcal{R}$. Let $h \in \mathcal{R}$ be such that

$$h = \lim_i f_i(g_1, \ldots, g_r),$$

where $f_i \in \mathcal{R}$. We must prove that $h \in A$.

From (1.3) and from the fact that $S_{g_1,\ldots,g_r}$ is a homomorphism, we have that

$$\mathrm{mult}(f_i - f_j) = \mathrm{mult}\left(S_{g_1,\ldots,g_r}(f_i) - S_{g_1,\ldots,g_r}(f_j)\right),$$

hence (f_i) is a Cauchy sequence in $\mathcal{R}$ and therefore there exists $f \in \mathcal{R}$ such that

$$f = \lim_i f_i.$$

Since $S_{g_1,\ldots,g_r}$ is continuous (see the Proposition (1.10.ii), or Problem 2.4(b)), it follows that

$$h = \lim_i f_i(g_1, \ldots, g_r) = S_{g_1,\ldots,g_r}\left(\lim_i f_i\right) = S_{g_1,\ldots,g_r}(f) \in A.$$

$\square$

PROBLEMS

3.1) Let $T : \mathcal{S} \longrightarrow \mathcal{R}$ be a homomorphism of K-algebras. Show that T is completely determined by the elements $T(Y_1), \ldots, T(Y_s)$, and that for all $f(Y_1, \ldots, Y_s) \in \mathcal{S}$, we have

$$T(f(Y_1, \ldots, Y_s)) = f(T(Y_1), \ldots, T(Y_s)).$$

3.2) Show that a K-endomorphism of $\mathcal{R}$ is an automorphism if and only if it preserves multiplicities.

3.3) Let $\mathcal{M}$ be the maximal ideal of $K[[X, Y]]$. Show that if $h_1, h_2 \in \mathcal{M}^2$, and if $g_1 = aX + bY + h_1$ and $g_2 = cX + dY + h_2$, with $ad - bc \neq 0$, then S_{g_1,g_2} is a K-automorphism of $K[[X, Y]]$.

1.4 THE LAURENT FIELD

Let $K((X))$ be the field of fractions of the ring of formal power series in one variable $K[[X]]$. Given $h = f/g \in K((X)) \setminus \{0\}$, since we may write $f = X^n u$ and $g = X^m v$, with $n, m \in \mathbb{N}$ and u and v units in $K[[X]]$, we have that

$$h = X^{n-m}uv^{-1} = X^r w,$$

where $r \in \mathbb{Z}$ and w is a unit in $K[[X]]$.

This shows that any element h of $K((X))$ is of the form

$$a_{-m}X^{-m} + a_{-m+1}X^{-m+1} + \cdots + a_{-1}X^{-1} + a_0 + a_1X + a_2X^2 + \cdots,$$

where $m \in \mathbb{N}$ and the a_i's are elements of K. The elements of $K((X))$ are called *Laurent formal power series*.

Given $h \in K((X))$, written as above, we define the *polar part* of h as being

$$\wp(h) = a_{-m}X^{-m} + a_{-m+1}X^{-m+1} + \cdots + a_{-1}X^{-1}.$$

We therefore have that $h - \wp(h) \in K[[X]]$.

Writing an element $h \in K((X)) \setminus \{0\}$ as $X^r w$, where $r \in \mathbb{Z}$ and w is a unit in $K[[X]]$, then we define the multiplicity of h as

$$\mathrm{mult}(h) = r.$$

We also put $\mathrm{mult}(0) = \infty$.

As in the case of $K[[X]]$ we define the distance between two elements h and k in $K((X))$ as

$$d(h, k) = \rho^{-\mathrm{mult}(h-k)},$$

where ρ is a real number greater than one. Remark that

$$d(h, k) < 1 \Longleftrightarrow \wp(h) = \wp(k).$$

It can be proved, in the same way as in the Proposition (1.6), that $(K((X)), d)$ is a complete metric space.

Let $\mathcal{F} = \{h_\lambda;\ \lambda \in \Lambda\}$ be a family of elements in $K((X))$, we define

$$\mathcal{F}^+ = \{h_\lambda - \wp(h_\lambda);\ \lambda \in \Lambda\} \subset K[[X]].$$

We will say that the family $\mathcal{F}$ is summable in $K((X))$ if $\mathcal{F}^+$ is summable in $K[[X]]$ and $\wp(h_\lambda) = 0$, except possibly for finitely many λ's. We then have a result analogous to that of the Proposition (1.8) concerning sums and products of summable families.

PROPOSITION 1.13 *Given a K-endomorphism T of $K((X))$, we have that $T(X) \in K[X]$, with $T(0) = 0$.*

Proof: Let $T(X) = X^r w(X)$. We must have $r > 0$, because otherwise $T(X + X^2 + X^3 + \cdots)$ wouldn't be defined. On the other hand, if $w(X) = \sum_{i \in \mathbb{N}} a_i X^i \in K[[X]] \setminus K[X]$, then $T(X^{-1}) = \sum_{i \in \mathbb{N}} a_i X^{-i}$ wouldn't be defined. $\qquad\square$

COROLLARY 1.14 *If T is an automorphism of $K((X))$, then $T(X) = aX$ for some $a \in K \setminus \{0\}$.*

Proof: If T is an automorphism, then $T(X) = P(X) \in K[X]$, with $P(0) = 0$. On the other hand, we have that $T^{-1}(X) = Q(X) \in K[X]$, so that

$$X = P(Q(X)) = Q(P(X)).$$

It then follows that $P(X)$ is an automorphism of K and consequently it is a linear polynomial. Since $P(0) = 0$, we must have $P(X) = aX$ with $a \in K \setminus \{0\}$. $\qquad\square$

PROBLEMS

4.1) Show that $(K((X)), d)$ is a complete metric space. Show the analog of the Proposition (1.8) for $K((X))$.

4.2) Let T be an endomorphism of $K((X))$.
a) Show that $T(K[[X]]) \subset K[[X]]$.
b) Show that $T(\mathcal{M}^i_{K[[X]]}) \subset \mathcal{M}^i_{K[[X]]}$, for all $i \in \mathbb{N}$.
c) Show that T is continuous.
d) Show that $T(X)$ determines T.

1.5 HENSEL'S LEMMA

In this section we will establish an important reducibility criterion in $K[[X]][Y]$, where K is a field and X and Y are indeterminates. For convenience we will define the degree of the zero polynomial as $-\infty$.

LEMMA 1.15 *Let p and q be two non-constant relatively prime polynomials in $K[Y]$ of degrees respectively r and s. Given a polynomial F in $K[Y]$, of degree less than $r + s$, there exist two uniquely determined polynomials $g, h \in K[Y]$, with $\deg h < r$ and $\deg g < s$, such that*

$$F = gp + hq.$$

Proof: Since p and q are relatively prime in $K[Y]$, their GCD is 1 and consequently there exist polynomials $\varphi, \psi \in K[Y]$ such that $1 = \varphi p + \psi q$. Therefore, we have

$$F = F\varphi p + F\psi q. \tag{1.4}$$

From the division algorithm, we have

$$F\psi = p\rho + h, \tag{1.5}$$

where $\rho, h \in K[Y]$ and $\deg h < \deg p = r$.

If we substitute (1.5) in (1.4), putting $g = F\varphi + \rho q$, we get that $F = gp + hq$. Since $\deg p = r$, we have

$$\deg g + \deg p = \deg gp = \deg(F - hq) < r + s,$$

it follows that $\deg g < s$.
Uniqueness: Suppose that $gp + hq = g'p + h'q$, with h and h' such that $\deg h, \deg h' < r$, and g and g' such that $\deg g, \deg g' < s$.

Therefore, $(g - g')p = (h' - h)q$. Since p and q are relatively prime, it follows that p divides $h' - h$. For degree reasons, this implies that $h' - h = 0$, which in turn implies that $g - g' = 0$. $\square$

THEOREM 1.16 (Hensel's Lemma) *Let $f \in K[[X]][Y]$ be monic and such that $f(0, Y) = p(Y)q(Y)$, where $p(Y), q(Y) \in K[Y]$ are relatively prime and non-constant, of degrees respectively r and s. Then there exist two uniquely determined polynomials $g, h \in K[[X]][Y]$, of degrees respectively r and s, such that $f = gh$, with $g(0, Y) = p(Y)$, and $h(0, Y) = q(Y)$.*

Proof: If $n = \deg_Y f$, since f is monic, then $n = \deg f(0, Y) = r + s$.

Let us write

$$f = f_0(Y) + X f_1(Y) + X^2 f_2(Y) + \cdots,$$

where $f_0(Y) = f(0, Y)$ has degree n and each $f_i(Y)$, $i \geq 1$, if not zero, is a polynomial in $K[Y]$ of degree less than n.

We wish to determine

$$g(X, Y) = p(Y) + X g_1(Y) + X^2 g_2(Y) + \cdots,$$

and

$$h(X, Y) = q(Y) + X h_1(Y) + X^2 h_2(Y) + \cdots,$$

where the $g_i(Y)'s$, $i \geq 1$, are zero or have degrees less than r and the $h_i(Y)'s$, $i \geq 1$, are zero or have degrees less than s, and are such that

$$f = gh. \tag{1.6}$$

From (1.6) it follows, for $i \geq 1$, that

$$f_i(Y) = p(Y) h_i(Y) + g_1(Y) h_{i-1}(Y) + \cdots + g_i(Y) q(Y). \tag{1.7}$$

We may now solve the equation (1.7) in the $g_i(Y)$ and $h_i(Y)$, recursively. Indeed, suppose that we have determined $g_j(Y)$ and $h_j(Y)$ of degrees respectively less than s and r, for all $j \leq i - 1$, then from (1.7), we get

$$p(Y) h_i(Y) + q(Y) g_i(Y) = f_i(Y) - g_1(Y) h_{i-1}(Y) - \cdots - g_{i-1}(Y) h_1(Y),$$

which in virtue of Lemma (1.15) may be solved in a unique way in $g_i(Y)$ and $h_i(Y)$ of degrees respectively less than s and r, if non-zero.

$\square$

The proof we gave for Hensel's Lemma is constructive since it is possible to determine the series g and h step by step.

COROLLARY 1.17 *Let K be an algebraically closed field and $u = u(X)$ a unit in $K[[X]]$. If n is a positive integer which is not a multiple of* $\mathrm{char}(K)$, *then there exists an invertible element v in $K[[X]]$ such that $u = v^n$.*

Proof: Since u is invertible we have that $u(0) \neq 0$. Defining

$$f(X, Y) = Y^n - u(X) \in K[[X]][Y],$$

we have that

$$f(0, Y) = Y^n - u(0) = \prod_{i=1}^{n} (Y - a_i),$$

where the a_i, $i = 1, \ldots, n$, are the n-th roots of $u(0)$ (all distinct, from the hypothesis on $\mathrm{char}(K)$). From Hensel's Lemma, there exist $g_1, \ldots, g_n \in K[[X]][Y]$ such that $g_i(Y, 0) = Y - a_i$, and

$$f = g_1 \cdots g_n.$$

Since the degree of f as a polynomial in Y is n, it follows that the degree of each g_i in Y is 1, and consequently, $g_i = Y - a_i(X)$, where $a_i(0) = a_i \neq 0$. Since $g_1(X, a_1(X)) = 0$, it follows that $f(X, a_1(X)) = 0$, and therefore,

$$u = (a_1(X))^n.$$

The result follows taking $v = a_1(X)$. $\square$

PROBLEMS

5.1) Find the three initial terms of an element $v \in \mathbb{C}[[X]]$ such that $v^3 = 1 + X$.

5.2) Decide whether each one of the following polynomials is reducible or irreducible in $\mathbb{C}[[X]][Y]$.
i) $f = X - 2Y + X^3 + Y^5 + X^2Y^3$,
ii) $g = Y - XY - Y^2 + X^2Y^3$,
iii) $h = Y^5 + X^2Y^4 - 3XY^3 + 2X^7Y^2 - X^5Y + 8X$.
Any of these series is reducible in $\mathbb{C}[[X, Y]]$?

5.3) Let K be an algebraically closed field and T an indeterminate over K. Show that if $\psi(T) \in K[[T]]$ with $r := \mathrm{mult}(\psi(T))$ not a multiple of $\mathrm{char}(K)$, then there exists an automorphism ρ of $K[[T]]$ such that $\rho(\psi(T)) = T^r$.

2   The preparation theorem

In this chapter we will continue to focus on the algebraic properties of the rings of formal power series. Our main goal will be to present a proof of the Weierstrass Preparation Theorem and some of its many consequences.

2.1   THE DIVISION THEOREM

In what follows we continue to denote by $\mathcal{R}$ the ring $K[[X_1, \ldots, X_r]]$ and by $\mathcal{M}_\mathcal{R}$ its maximal ideal.

DEFINITION 2.1 *We will say that $f \in \mathcal{R}$ is* regular of order m, *with respect to the indeterminate X_j, if $f(0, \ldots, X_j, \ldots, 0)$ is divisible exactly by X_j^m.*

We say shortly that f is *regular* in X_j when f is regular with respect to X_j of order $n = \mathrm{mult}(f)$. In this case,

$$\mathrm{mult}(f) = \mathrm{mult}\left(f(0, \ldots, X_j, \ldots, 0)\right).$$

The following result is a direct consequence of the definitions.

LEMMA 2.2 *Given $f, g \in \mathcal{R}$, then $f \cdot g$ is regular with respect to X_j of a certain order, if and only if, f and g are regular with respect to X_j, of some orders.*

The following theorem, proved by Stickelberger in 1887, will play a fundamental role in the theory of singularities. This theorem, inspired in the Weierstrass Preparation Theorem of 1860, has as a corollary the Preparation Theorem itself.

We will denote in what follows by $\mathcal{R}'$ the ring $K[[X_1, \ldots, X_{r-1}]]$ and by $\mathcal{M}_{\mathcal{R}'}$ its maximal ideal.

THEOREM 2.3 (The Division Theorem) *Let $F \in \mathcal{M}_\mathcal{R} \subset \mathcal{R}$, regular of order m with respect to X_r. Given any $G \in \mathcal{R}$, there exist $Q \in \mathcal{R}$ and $R \in \mathcal{R}'[X_r]$ with $R = 0$ or $\deg_{X_r}(R) < m$, uniquely determined by F and G, such that*

$$G = FQ + R.$$

Proof: Write $G = \sum_{i=0}^{\infty} A_i X_r^i$ as an element of $\mathcal{R}'[[X_r]]$, and let

$$R_{-1} = \sum_{i=0}^{m-1} A_i X_r^i.$$

To prove the theorem, we are going to construct recursively polynomials q_i, R_i, $i \geq 0$, in $K[X_1, \ldots, X_r]$, not necessarily homogeneous, with $R_i = 0$ or $\deg_{X_r} R_i < m$, for all i, such that $\mathrm{mult}(q_i) \geq i$ and $\mathrm{mult}(R_i) \geq 1 + i$, in such a way that if we put $Q = q_0 + q_1 + \cdots$, and $R = R_{-1} + R_0 + R_1 + \cdots$, we will have $G = FQ + R$. The uniqueness of Q and R will follow from our process of construction of the polynomials q_i, R_i, $i \geq 0$.

Write now F as a sum of homogeneous polynomials:

$$F = F_n + \cdots + F_m + F_{m+1} + \cdots.$$

Since F is regular in X_r of order m, we have that

$$P = F_n + \cdots + F_m = cX_r^m + (\text{terms in } X_r \text{ of degree} < m), \qquad (2.1)$$

where $c \in K \setminus \{0\}$ and, necessarily,

$$1 \leq n = \mathrm{mult}(P) \leq m.$$

Write $H = G - R_{-1}$ as a sum of homogeneous polynomials:

$$G - R_{-1} = H = H_m + H_{m+1} + \cdots$$

Since P is the form (2.1), with $c \in K \setminus \{0\}$, there exists a unique $q_0 \in K$ such that

$$\deg_{X_r}(H_m - q_0 P) < m.$$

Note for future reference that

$$q_0 \begin{cases} = 0, & \text{if } H_m(0, \ldots, 0, X_r) = 0, \\ \neq 0 & \text{if } H_m(0, \ldots, 0, X_r) \neq 0 \end{cases} \qquad (2.2)$$

It is clear that $\mathrm{mult}(q_0) \geq 0$ and that $\mathrm{mult}(R_0) \geq 1$, where

$$R_0 = H_m - q_0 P.$$

The leading coefficient c of P in X_r being invertible, and looking at P as a polynomial in $K[X_1, \ldots, X_{r-1}][X_r]$, by the division algorithm, there exists a unique polynomial $q_1 \in K[X_1, \ldots, X_{r-1}][X_r]$ such that

$$\deg_{X_r}(H_{m+1} - q_0 F_{m+1} - q_1 P) < m.$$

It is clear that $\mathrm{mult}(q_1) \geq 1$, because otherwise the above inequality would be false. It then follows that $\mathrm{mult}(R_1) \geq 2$, where

$$R_1 = H_{m+1} - q_0 F_{m+1} - q_1 P.$$

By similar arguments as used above, there exists a unique polynomial $q_2 \in K[X_1, \ldots, X_{r-1}][X_r]$ such that

$$\deg_{X_r}(H_{m+2} - q_0 F_{m+2} - q_1 F_{m+1} - q_2 P) < m.$$

It is clear that $\text{mult}(q_2) \geq 2$, because otherwise the above inequality would be false. It then follows that $\text{mult}(R_2) \geq 3$, where

$$R_2 = H_{m+2} - q_0 F_{m+2} - q_1 F_{m+1} - q_2 P.$$

In this way, we have constructed the sequences $q_0, q_1, \ldots$ and $R_0, R_1, \ldots$, such that

$$H - F \cdot (q_0 + q_1 + \cdots) = R_0 + R_1 + \cdots,$$

and therefore,

$$G = F \cdot (q_0 + q_1 + \cdots) + R_{-1} + R_0 + R_1 + \cdots,$$

and the result follows defining $Q = q_0 + q_1 + \cdots$ and $R = R_{-1} + R_0 + R_1 + \cdots$, which are well defined as elements of $\mathcal{R}$. $\qquad\square$

Notice that the proof we gave for the Division Theorem is constructive, allowing to construct the series Q and R until any desired order.

THEOREM 2.4 (Weierstrass Preparation Theorem) *Let $F \in \mathcal{R}$ be a given series, regular with respect to X_r of order m. Then there exist $U \in \mathcal{R}$ invertible and $A_1, \ldots, A_m \in \mathcal{M}_{\mathcal{R}'}$, uniquely determined by F, such that*

$$F \cdot U = X_r^m + A_1 X_r^{m-1} + A_2 X_r^{m-2} + \cdots + A_m.$$

Moreover, if F is regular in X_r, that is, $m = \text{mult}(F)$, then $\text{mult}(A_i) \geq i$, for $i = 1, \ldots, m$.

Proof: The proof of the existence follows from the Division Theorem taking $G = X_r^m, U = Q$ and $A_1 X_r^{m-1} + A_2 X_r^{m-2} + \cdots + A_m = -R$.

The fact that U is invertible follows from (2.2) because in this case $q_0 \neq 0$. Since X_r^m divides F, we must have $A(0) = \cdots = A_m(0) = 0$.

Finally, if F is regular in X_r, then

$$\text{mult}(X_r^m + A_1 X_r^{m-1} + A_2 X_r^{m-2} + \cdots + A_m) = \text{mult}(F \cdot U) = \text{mult}(F) = m,$$

and it follows that $\text{mult}(A_i) \geq i$, for all $i = 1, \ldots, m$.

The uniqueness follows immediately from the uniqueness in the Division Theorem. $\qquad\square$

COROLLARY 2.5 *(The Implicit Function Theorem) Let F be an element of $\mathcal{R}$ such that $F(0, \ldots, 0) = 0$ and $\dfrac{\partial F}{\partial X_r}(0, \ldots, 0) \neq 0$. Then there exists*

$$\varphi(X_1, \ldots, X_{r-1}) \in \mathcal{M}_{\mathcal{R}'},$$

such that

$$F(X_1, \ldots, X_{r-1}, \varphi(X_1, \ldots, X_{r-1})) = 0$$

as an element of $\mathcal{R}'$.

Proof: The condition $\dfrac{\partial F}{\partial X_r}(0,\ldots,0) \neq 0$ is equivalent to say that F is regular in X_r and of multiplicity 1, then from the Preparation Theorem there exists a unit $U \in \mathcal{R}$ such that $F \cdot U = X_r + A_1$, with $A_1 \in \mathcal{M}_{\mathcal{R}'}$. The result now follows taking $\varphi(X_1,\ldots,X_{r-1}) = -A_1$, and noting that $U(0,\ldots,0) \neq 0$. $\qquad\qquad\square$

The condition on F to be regular is not so restrictive as it may appear at a first glance. Assuming K infinite, we will show that after composing F with a linear automorphism of $\mathcal{R}$, we may transform it into a regular series in one of the indeterminates, chosen arbitrarily. In the case of K finite, it is not possible to guarantee the existence of linear automorphisms, but it is always possible to find a K-automorphism of $\mathcal{R}$ that transforms F into a regular series in one of the indeterminates (cf. [16, Cor. Lemma 3, Ch. VII, Vol 2]).

LEMMA 2.6 *Let K be an infinite field. Given a finite family $\mathcal{F}$ of non-zero homogeneous polynomials in $K[Y_1,\ldots,Y_r]$, there exists a linear transformation $T : K[X_1,\ldots,X_r] \longrightarrow K[Y_1,\ldots,Y_r]$, such that for every $F \in \mathcal{F}$, of degree n, there exists $c_F \in K \setminus \{0\}$ such that*

$$F(T(X_1,\ldots,X_r)) = c_F X_r^n + \text{ (terms of lower degree in } X_r) \,.$$

Proof: Since $\mathcal{F}$ is finite and K is infinite, there exists $(\alpha_1,\ldots,\alpha_r) \in K^r$ such that (reader, verify this!)
$$F(\alpha_1,\ldots,\alpha_r) \neq 0, \quad \forall F \in \mathcal{F}.$$

Using the transformation T defined by

$$\begin{pmatrix} Y_1 \\ Y_2 \\ \vdots \\ Y_r \end{pmatrix} = \begin{pmatrix} 1 & 0 & \cdots & 0 & \alpha_1 \\ 0 & 1 & \cdots & 0 & \alpha_2 \\ \vdots & \vdots & \vdots & \vdots & \vdots \\ 0 & 0 & \cdots & 0 & \alpha_r \end{pmatrix} \begin{pmatrix} X_1 \\ X_2 \\ \vdots \\ X_r \end{pmatrix},$$

with the $\alpha_i, i = 1,\ldots,r$, as above, we have that

$$Y_1^{m_1} \cdots Y_r^{m_r} = (X_1 + \alpha_1 X_r)^{m_1} \cdots (X_{r-1} + \alpha_{r-1} X_r)^{m_{r-1}} (\alpha_r X_r)^{m_r}$$

$$= \alpha_1^{m_1} \cdots \alpha_r^{m_r} X_r^{m_1 + \cdots + m_r} + \text{(terms of lower degree in } X_r).$$

Hence, for every $F \in \mathcal{F}$, we have that

$$F(T(X_1,\ldots,X_r)) = F(\alpha_1,\ldots,\alpha_r) X_r^n + \text{(terms of lower degree in } X_r),$$

and the result follows taking $c_F = F(\alpha_1,\ldots,\alpha_r)$. $\qquad\qquad\square$

We get immediately from this the following Corollary:

COROLLARY 2.7 *Let K be an infinite field. Given a finite family $\mathcal{F}$ of non-zero elements in $\mathcal{R} = K[[X_1,\ldots,X_r]]$, there exists a linear automorphism T of $\mathcal{R}$ such that all the elements of $\mathcal{F} \circ T$, are regular in the last indeterminate.*

From the above Corollary we obtain the following Corollary of the Weierstrass Preparation Theorem.

COROLLARY 2.8 *Let $F \in \mathcal{R} \setminus \{0\}$ be of multiplicity n. There exist a K-automorphism T of $\mathcal{R}$, a unit $U \in \mathcal{R}$ and $A_1, \ldots, A_n \in \mathcal{R}'$ such that $\mathrm{mult}(A_i) \geq i$, for $i = 1, \ldots, n$, and*

$$T(F) \cdot U = X_r^n + A_1 X_r^{n-1} + \cdots + A_n.$$

The "polynomial" $X_r^n + A_1 X_r^{n-1} + A_2 X_r^{n-2} + \cdots + A_n$ associated to F, after possibly a linear change of coordinates, will be considered a preparation of F for its study. The study of a polynomial is much simpler than the study of a power series. More generally, given a finite number of non-invertible series in $\mathcal{R}$, we have seen that there exists a change of coordinates that allows to prepare simultaneously all these series.

PROBLEMS

1.1) Given $f, g \in \mathcal{R}$, show that $f \cdot g$ is regular with respect to X_j, if and only if, f and g are regular with respect to X_j.

2.2 FACTORIZATION OF POWER SERIES

In the present section we will study the factorization in the ring $\mathcal{R}$.

DEFINITION 2.9 *A* pseudo-polynomial *(resp. a* Weierstrass polynomial *) in X_r is a power series in $\mathcal{R}$ of the form*

$$P(X_1, \ldots, X_r) = X_r^n + A_1 X_r^{n-1} + \cdots + A_n \in \mathcal{R}'[X_r],$$

such that $n \geq 1$ and $\mathrm{mult}(A_i) \geq 1$ (resp. $\mathrm{mult}(A_i) \geq i$), for $i = 1, \ldots, n$.

LEMMA 2.10 *Let $F_1, \ldots, F_s$ be monic polynomials in $\mathcal{R}'[X_r]$. Then $F_1 \cdots F_s$ is a pseudo-polynomial (resp. a Weierstrass polynomial) if and only if each $F_i, i = 1, \ldots, s$, is a pseudo-polynomial (resp. a Weierstrass polynomial).*

Proof: It suffices to prove the result for $s = 2$.
Let $F_1 = X_r^m + A_1 X_r^{m-1} + \cdots + A_m$ and $F_2 = X_r^n + B_1 X_r^{n-1} + \cdots + B_n$, then

$$F_1 \cdot F_2 = X_r^{m+n} + (A_1 + B_1) X_r^{m+n-1} + \cdots +$$

$$(A_i + A_{i-1} B_1 + \cdots + A_1 B_{i-1} + B_i) X_r^{m+n-i} + \cdots + A_m B_n. \qquad (2.3)$$

If F_1 and F_2 are pseudo-polynomials (resp. Weierstrass polynomials), then

$$\mathrm{mult}(A_i + A_{i-1} B_1 + \cdots + A_1 B_{i-1} + B_i) \geq 1 \quad (\text{resp. } \geq i),$$

which proves that $F_1 \cdot F_2$ is a pseudo-polynomial (resp. Weierstrass polynomial).
Conversely, suppose that $F_1 \cdot F_2$ is a Weierstrass polynomial, we have from (2.3) that

$$\mathrm{mult}(F_1) + \mathrm{mult}(F_2) = \mathrm{mult}(F_1 \cdot F_2) = \mathrm{mult}((F_1 \cdot F_2)(0, \ldots, 0, X_r)) = m + n.$$

Since $\mathrm{mult}(F_1) \leq m$, $\mathrm{mult}(F_2) \leq n$, it follows from the above equality that $\mathrm{mult}(F_1) = m$ and $\mathrm{mult}(F_2) = n$. This, in particular, implies that F_1 and F_2 are Weierstrass polynomials because otherwise, $\mathrm{mult}(F_1) < m$ or $\mathrm{mult}(F_2) < n$. $\square$

LEMMA 2.11 *Let $F \in \mathcal{R}'[X_r]$ be a pseudo-polynomial. Then F is reducible in $\mathcal{R}$ if and only if F is reducible in $\mathcal{R}'[X_r]$.*

Proof: Suppose that F is reducible in $\mathcal{R}$, then there exist F_1, F_2 in $\mathcal{R}$, non unitary, such that $F = F_1 \cdot F_2$. Since F is a pseudo-polynomial, it is regular of a certain order with respect to X_r, then from the Lemma (2.2), we have that F_1 and F_2 are regular of certain orders greater or equal to 1. From the Preparation Theorem, there exist units $U_1, U_2 \in \mathcal{R}$ such that $H_1 = F_1 \cdot U_1$ and $H_2 = U_2 \cdot F_2$ are pseudo-polynomials of degrees greater or equal to 1. Putting $U = U_1 \cdot U_2$, we have that U is invertible and

$$F \cdot U = (F_1 U_1)(F_2 U_2) = H_1 \cdot H_2.$$

Since H_1 and H_2 are pseudo-polynomials, we have from the Lemma (2.10) that $F \cdot U$ is a pseudo-polynomial. Since we also have $F \cdot 1 = F$, with F a pseudo-polynomial, from the uniqueness in the Preparation Theorem it follows that $F = H_1 \cdot H_2$, that is, F is reducible in $\mathcal{R}'[X_r]$.

Conversely, suppose that $F \in \mathcal{R}'[X_r]$ is a reducible pseudo-polynomial of degree d. Then there exist H_1 and H_2, monics of degrees respectively m and n, in $\mathcal{R}'[X_r]$ such that

$$F = H_1 H_2,$$

with $m, n \geq 1$ and $m + n = d$. Since from the Lemma (2.10), H_1 and H_2 are pseudo-polynomials of degrees greater or equal to 1, it follows that they are not invertible in $\mathcal{R}$. Then F is reducible in $\mathcal{R}$. $\square$

THEOREM 2.12 *The ring $\mathcal{R}$ is a unique factorization domain.*

Proof: We will prove, by induction on r, that every irreducible element of $\mathcal{R}$ is prime.

The ring $K[[X]]$ is a unique factorization domain (see Problem 1.1.4).

Suppose now, by induction hypothesis, that $\mathcal{R}'$ is a unique factorization domain. Let $F, G, H \in \mathcal{R}$ with F irreducible and $F \mid G \cdot H$, we want to show that $F \mid G$ or $F \mid H$. If G or H is invertible in $\mathcal{R}$, the result follows immediately. Suppose that G and H are not invertible in $\mathcal{R}$, then from the Preparation Theorem, changing coordinates if necessary, we have that F, G and H are associated in $\mathcal{R}$ to Weierstrass polynomials. We then may assume that F, G and H are Weierstrass polynomials and therefore from the Lemma (2.11), F is irreducible in $\mathcal{R}'[X_r]$, hence prime because this last ring is a unique factorization domain (Gauss' Lemma). Therefore from $F \mid G \cdot H$, it follows that $F \mid G$ or $F \mid H$. $\square$

COROLLARY 2.13 *Suppose that $F \in \mathcal{R}'[X_r]$ is a pseudo-polynomial (resp. a Weierstrass polynomial) with respect to the indeterminate X_r. If $F = F_1 \cdots F_s$ is the decomposition of F into irreducible factors in $\mathcal{R}$, then we may choose a decomposition where each F_i is a pseudo-polynomial (resp. a Weierstrass polynomial).*

Proof: From the Theorem (2.12) and Gauss' Lemma we have that $\mathcal{R}'[X_r]$ is a unique factorization domain. Let $F = F_1 \cdots F_s$ be a decomposition of F into irreducible factors in $\mathcal{R}'[X_r]$, which by the Lemma (2.11) is also a decomposition into irreducible factors in $\mathcal{R}$. Since F is monic, we may suppose that the F_i's are monic. The result now follows from the Lemma (2.10). $\square$

2.3 HILBERT-RÜCKERT BASIS THEOREM

This section may be omitted in a first reading.

The ring A will be called noetherian if every ideal in A is finitely generated. The following theorem is fundamental.

THEOREM 2.14 (Hilbert's Basis Theorem) *If A is a noetherian ring, then $A[X]$ is noetherian.*

Proof: Let $f = a_n X^n + a_{n-1} X^{n-1} + \cdots + a_0 \in A[X]$. We will call the term $a_n X^n$ the *leading term* of f and the element a_n the *leading coefficient* of f.

Let I be an ideal in $A[X]$. We want to show that I is finitely generated.

Choose an element f_1 of I of minimal degree. If $I = \langle f_1 \rangle$, the result is proved. If not, choose an element $f_2 \in I \setminus \langle f_1 \rangle$ of minimal degree. If $I = \langle f_1, f_2 \rangle$, the result follows. If not, choose an element $f_3 \in I \setminus \langle f_1, f_2 \rangle$ of minimal degree. If $I = \langle f_1, f_2, f_3 \rangle$, the result follows. If not, we continue with this procedure. We will show that this procedure must stop after finitely many steps.

Let J be the ideal in A generated by the leading coefficients a_i of the f_i's we have chosen above. Since A is noetherian, we may suppose that $J = \langle a_1, \ldots, a_m \rangle$, for some m. We are going to prove that $I = \langle f_1, \ldots, f_m \rangle$.

Suppose by absurd that $I \neq \langle f_1, \ldots, f_m \rangle$. Consider the element $f_{m+1} \in I \setminus \langle f_1, \ldots, f_m \rangle$, which we know to be of minimal degree. We then have that $a_{m+1} = \sum_{i=1}^{m} c_i a_i$, for $c_i \in A$. By construction

$$\deg(f_{m+1}) \geq \deg(f_i), \quad i = 1, \ldots, m.$$

Then we have that

$$g = \sum_{i=1}^{m} c_i f_i X^{\deg(f_{m+1}) - \deg(f_i)} \in \langle f_1, \ldots, f_m \rangle.$$

Since the leading term of g is equal to that of f_{m+1}, it follows that the degree of $f_{m+1} - g$ is smaller than that of f_{m+1} and at the same time,

$$f_{m+1} - g \in I \setminus \langle f_1, \ldots, f_m \rangle,$$

which is a contradiction. $\square$

The Hilbert Basis Theorem, together with the Division Theorem, allow to prove the following result.

THEOREM 2.15 (Rückert's Basis Theorem) *The ring $\mathcal{R}$ is noetherian.*

Proof: The proof will be done by induction on the number of indeterminates r. If $r = 1$, the result follows from Problem 1.1.4.

Suppose that $r \geq 2$ and that $\mathcal{R}'$ is noetherian. Let $I \subset \mathcal{R}$ be a non-zero ideal and let $G \in I \setminus \{0\}$ be any element. After coordinates change, if necessary, we may suppose that G is regular with respect to X_r, since the property to be proved is invariant by K-automorphisms of $\mathcal{R}$.

Since $\mathcal{R}'$ is noetherian, by Hilbert's Basis Theorem, $\mathcal{R}'[X_r]$ is also noetherian. Therefore, there exist $G_1, \ldots, G_m \in \mathcal{R}$ such that

$$I \cap \mathcal{R}'[X_r] = \langle G_1, \ldots, G_m \rangle.$$

Let $F \in I$. From the Division Theorem, we may write

$$F = gG + R; \quad g \in \mathcal{R}, \quad R \in \mathcal{R}'[X_r]. \tag{2.4}$$

Since $F, G \in I$, it follows that $R \in I \cap \mathcal{R}'[X_r]$. Therefore, there exist $g_1, \ldots, g_m \in \mathcal{R}'[X_r] \subset \mathcal{R}$ such that

$$R = g_1 G_1 + \cdots + g_m G_m,$$

and therefore from (2.4), it follows that

$$F = gG + g_1 G_1 + \cdots + g_m G_m,$$

which shows that

$$I = \langle G, G_1, \ldots, G_m \rangle.$$

$\square$

2.4   ELIMINATION

Let A be a unique factorization domain, and let

$$f = a_0 Y^n + a_1 Y^{n-1} + \cdots + a_n,$$

and

$$g = b_0 Y^m + b_1 Y^{m-1} + \cdots + b_m,$$

be two polynomials in $A[Y]$. We want to determine a criterion to decide when f and g have a non-constant common factor in $A[Y]$. An object that will play an important role to answer this question is the *resultant* of f and g which is the element of A defined as follows:

$$R_Y(f,g) = \det \begin{pmatrix} a_0 & a_1 & . & . & . & . & . & a_n & 0 & . & 0 \\ 0 & a_0 & . & . & . & . & . & . & a_n & . & 0 \\ . & . & . & . & . & . & . & . & . & . & . \\ 0 & . & . & a_0 & . & . & . & . & . & . & a_n \\ b_0 & b_1 & . & . & b_m & 0 & . & . & . & . & 0 \\ 0 & b_0 & . & . & . & b_m & . & . & . & . & 0 \\ . & . & . & . & . & . & . & . & . & . & . \\ 0 & . & . & . & . & . & b_0 & . & . & . & b_m \end{pmatrix}$$

If we consider the a_i's and the b_j's as indeterminates, we have that $R_Y(f,g)$ is a bi-homogeneous polynomial of degree m in the a_i's and of degree n in the b_j's. Observe also that in the development of the determinant we have the monomial $a_0^m b_m^n$ corresponding to the principal diagonal.

LEMMA 2.16 *The polynomials f and g in $A[Y]$ admit a non-constant common factor if and only if there exist non-zero polynomials $p, q \in A[Y]$, of degrees respectively less than the degrees of f and g, such that*

$$qf = pg. \tag{2.5}$$

Proof: Indeed, if f and g have a non-constant common factor h, then $f = ph$ and $g = qh$. Consequently (2.5) follows immediately.

Conversely, suppose that (2.5) is true with $\deg(p) < \deg(f)$ and $\deg(q) < \deg(g)$. If f and g don't have any non-constant common factor, then since $A[Y]$ is factorial, we have that f divides p, and therefore, $\deg(f) \leq \deg(p)$, a contradiction because $\deg(p) < \deg(f)$. $\square$

PROPOSITION 2.17 *Consider the polynomials* $f = a_0 Y^n + a_1 Y^{n-1} + \cdots + a_n$ *and* $g = b_0 Y^m + b_1 Y^{m-1} + \cdots + b_m$ *in* $A[Y] \setminus A$. *Then* $R_Y(f, g) = 0$, *if and only if,* $a_0 = b_0 = 0$ *or* f *and* g *have a non-constant common factor in* $A[Y]$.

Proof: If $a_0 = b_0 = 0$, it is clear that $R_Y(f, g) = 0$. Suppose now that $a_0 \neq 0$ or $b_0 \neq 0$, say $a_0 \neq 0$. It will be sufficient to prove that $R_Y(f, g) = 0$ if and only if f and g have a non-constant common factor.

From the Lemma (2.16), f and g admit a non-constant common factor if and only if there exist $p, q \in A[Y]$ with $\deg(p) < \deg(f)$ and $\deg(q) < \deg(g)$ such that

$$qf = pg.$$

On the other hand, the above condition is equivalent to say that the polynomials

$$Y^{m-1}f, Y^{m-2}f, \ldots, Yf, f, Y^{n-1}g, \ldots, Yg, g,$$

are linearly dependent over the field of fractions F of A. Writing the above polynomials as vectors in F^{n+m}, the determinant of the matrix whose rows are these vectors is precisely $R_Y(f, g)$. Therefore the result follows immediately. $\square$

COROLLARY 2.18 *Let* $P, Q \in \mathcal{R}[Y]$ *be pseudo-polynomials with respect to the indeterminate* Y. *The series* P *and* Q *admit a non-unitary common factor in* $\mathcal{R}[[Y]]$ *if and only if* $R_Y(P, Q) = 0$.

Proof: From the Corollary (2.13), P and Q admit a non-unitary common factor in $\mathcal{R}[[Y]]$, if and only if they admit a non-unitary common factor in $\mathcal{R}[Y]$. The result then follows from the proposition. $\square$

COROLLARY 2.19 *Let* $\mathbb{C}$ *be the field of complex numbers and let* $A = \mathbb{C}\{\mathbf{X}\}$, *where* $\mathbf{X} = (X_1, \ldots, X_{n-1})$. *Let* $f = a_0(\mathbf{X})Y^n + \cdots + a_n(\mathbf{X})$ *and* $g = b_0(\mathbf{X})Y^m + \cdots + b_m(\mathbf{X})$ *be elements in* $\mathbb{C}\{\mathbf{X}\}[Y]$ *and let* U *be a neighborhood of* 0 *in* $\mathbb{C}^{n-1}$, *where the* $a_i(\mathbf{X})$'s *and the* $b_j(\mathbf{X})$'s *converge absolutely in a neighborhood of* $0 \in \mathbb{C}^{n-1}$. *Denote by* $R(\mathbf{X})$ *the resultant* $R_Y(f, g)$. *We have that* $R(\alpha) = 0$, *where* $\alpha \in U$, *if and only if, either* $a_0(\alpha) = b_0(\alpha) = 0$, *or* $f(\alpha, Y)$ *and* $g(\alpha, Y)$ *admit a common root in* $\mathbb{C}$.

The above Corollary may be used to determine the points of intersection of two algebraic plane curves, as one may see for example in Problem 4.1.

PROPOSITION 2.20 *Let* $f, g \in A[Y] \setminus A$. *Then there exist* $p, q \in A[Y]$, *with* $\deg(p) < \deg(f)$ *and* $\deg(q) < \deg(g)$, *such that*

$$qf + pg = R_Y(f, g).$$

Proof: If f and g have a non-constant common factor, the result follows from the Proposition (2.17) and from the Lemma (2.16).

Suppose that f and g have no non-constant common factor. Then from the Proposition (2.17), we have that $R_Y(f, g) \neq 0$. We may write

$$\begin{pmatrix} a_0 & a_1 & . & . & . & . & . & a_n & 0 & . & 0 \\ 0 & a_0 & . & . & . & . & . & . & a_n & . & 0 \\ . & . & . & . & . & . & . & . & . & . & . \\ 0 & . & . & a_0 & . & . & . & . & . & . & a_n \\ b_0 & b_1 & . & . & b_m & 0 & . & . & . & . & 0 \\ 0 & b_0 & . & . & . & b_m & . & . & . & . & 0 \\ . & . & . & . & . & . & . & . & . & . & . \\ 0 & . & . & . & . & b_0 & . & . & . & b_m \end{pmatrix} \begin{pmatrix} Y^{n+m-1} \\ Y^{n+m-2} \\ \vdots \\ Y^n \\ Y^{n-1} \\ Y^{n-2} \\ \vdots \\ 1 \end{pmatrix} = \begin{pmatrix} Y^{m-1}f \\ Y^{m-2}f \\ \vdots \\ f \\ Y^{n-1}g \\ Y^{n-2}g \\ \vdots \\ g \end{pmatrix}$$

Applying Cramer's rule in the field of fractions of A, we have that

$$\frac{\det M}{R_Y(f, g)} = 1, \tag{2.6}$$

where

$$M = \begin{vmatrix} a_0 & a_1 & . & . & . & . & a_n & 0 & . & y^{m-1}f \\ 0 & a_0 & . & . & . & . & . & a_n & . & y^{m-2}f \\ . & . & . & . & . & . & . & . & . & . \\ 0 & . & . & a_0 & . & . & . & . & . & f \\ b_0 & b_1 & . & . & b_m & 0 & . & . & . & y^{n-1}g \\ 0 & b_0 & . & . & . & b_m & . & . & . & y^{n-2}g \\ . & . & . & . & . & . & . & . & . & . \\ 0 & . & . & . & . & b_0 & . & . & . & g \end{vmatrix}.$$

Computing $\det M$ by expanding it with respect to the last column and calling A_i (respectively, B_j) the algebraic complement of $Y^{m-i}f$ (respectively, of $Y^{n-j}g$), we have

$$\det M = A_1 Y^{m-1}f + A_2 Y^{m-2}f + \cdots + A_m f + B_1 Y^{n-1}g + \cdots + B_n g =$$

$$(A_1 Y^{m-1} + \cdots + A_m)f + (B_1 Y^{n-1} + \cdots + B_n)g.$$

The result now follows in view of (2.6), putting $q = A_1 Y^{m-1} + \cdots + A_m$, and $p = B_1 Y^{n-1} + \cdots + B_n$. $\qquad\square$

Let $\mathbf{X} = (X_1, \ldots, X_n)$ and $\mathbf{Y} = (Y_1, \ldots, Y_m)$ be two sets of indeterminates and consider the ring

$$\mathcal{R}'' = A[\mathbf{X}, \mathbf{Y}] = A[X_1, \ldots, X_n, Y_1, \ldots, Y_m].$$

Consider the polynomials

$$F = a_0(Y - X_1) \cdots (Y - X_n) \in \mathcal{R}''[Y],$$

and

$$G = b_0(Y - Y_1) \cdots (Y - Y_m) \in \mathcal{R}''[Y].$$

The resultant $R_Y(F, G)$ is therefore a polynomial in $\mathcal{R}''$.

LEMMA 2.21 *The polynomial $R_Y(F,G)$ is homogeneous of degree nm in $\mathcal{R}''$.*

Proof: We may write

$$R_Y(F,G)(\mathbf{X},\mathbf{Y}) = a_0^m b_0^n \det \begin{pmatrix} 1 & S_1 & . & . & . & . & . & S_n & 0 & . & 0 \\ 0 & 1 & . & . & . & . & . & . & S_n & . & 0 \\ . & . & . & . & . & . & . & . & . & . & . \\ 0 & . & . & 1 & . & . & . & . & . & . & S_n \\ 1 & S_1' & . & . & S_m' & 0 & . & . & . & . & 0 \\ 0 & 1 & . & . & . & S_m' & . & . & . & . & 0 \\ . & . & . & . & . & . & . & . & . & . & . \\ 0 & . & . & . & . & . & 1 & . & . & . & S_m' \end{pmatrix}$$

where the S_i's and the S_j''s are the elementary symmetric function in the indeterminates X_i's and Y_j's, respectively.

Substituting, in the above expression, X_i by TX_i and Y_j by TY_j, where T is a new indeterminate, we get

$$R_Y(F,G)(T\mathbf{X},T\mathbf{Y}) =$$

$$a_0^m b_0^n \det \begin{pmatrix} 1 & TS_1 & . & . & . & . & . & T^n S_n & 0 & . & 0 \\ 0 & 1 & . & . & . & . & . & . & T^n S_n & . & 0 \\ . & . & . & . & . & . & . & . & . & . & . \\ 0 & . & . & 1 & . & . & . & . & . & . & T^n S_n \\ 1 & TS_1' & . & . & T^m S_m' & 0 & . & . & . & . & 0 \\ 0 & 1 & . & . & . & T^m S_m' & . & . & . & . & 0 \\ . & . & . & . & . & . & . & . & . & . & . \\ 0 & . & . & . & . & . & 1 & . & . & . & T^m S_m' \end{pmatrix}$$

Multiplying in the above determinant, the second row by T, the third by T^2, ..., the m-th by T^{m-1}, the $(m+2)$-th by T, ..., the $(m+n)$-th by T^{m+n-1}, we get

$$T^M R_Y(F,G)(T\mathbf{X},T\mathbf{Y}) = T^N R_Y(F,G)(\mathbf{X},\mathbf{Y}),$$

where $M = (1 + \cdots + m - 1) + (1 + \cdots + n - 1)$, and $N = 1 + \cdots + (m+n-1)$, from which the result follows trivially. $\square$

We will use the above Lemma to prove the following result:

PROPOSITION 2.22 *Let $f = a_0 Y^n + \cdots + a_n, g = b_0 Y^m + \cdots + b_m \in K[Y] \setminus K$, where K is a field. Let E be a field extension of K that contains the roots $\alpha_1, \ldots, \alpha_n$ and $\beta_1, \ldots, \beta_m$, respectively of f and of g. We have that*

$$R_Y(f,g) = a_0^m b_0^n \prod_{i=1}^{n} \prod_{j=1}^{m} (\alpha_i - \beta_j) = a_0^m \prod_{i=1}^{n} g(\alpha_i) = (-1)^{nm} b_0^n \prod_{j=1}^{m} f(\beta_j).$$

Proof: Consider the polynomials F, G and $R_Y(F,G)$ as defined in the Lemma (2.21). Put $X_i = Y_j$. From the Proposition (2.17), the polynomial $R_Y(F,G)$ vanishes, since F and G will have a non-constant common factor. This implies that $R_Y(F,G)$, as a polynomial in $\mathcal{R}''$, is divisible by $X_i - Y_j$, for all i and j. Since $X_i - Y_j$ and $X_r - Y_s$, for $(i,j) \neq (r,s)$ are relatively prime, we have that $R(F,G)$ is divisible by

$$P = a_0^m b_0^n \prod_{i=1}^{n} \prod_{j=1}^{m} (X_i - Y_j),$$

which is a homogeneous polynomial in $\mathcal{R}''$ of degree mn, and therefore of the same degree as $R_Y(F, G)$.

Note that the term of lower degree in $R_Y(F, G)$ which only contains the indeterminates $\mathbf{Y}$, is $a_0^m S_1'^n$, that corresponds to the term given by the principal diagonal of the determinant.

From the expressions of F and G, it follows that

$$P = a_0^m \prod_{i=1}^{n} g(X_i) = (-1)^{nm} b_0^n \prod_{j=1}^{m} f(Y_j).$$

Note that the term of lower degree in $P = a_0^m \prod_{i=1}^{n} g(X_i)$, which only contains the indeterminates $\mathbf{Y}$, is also $a_0^m S_1'^n$.

It then follows that

$$R_Y(F, G) = a_0^m b_0^n \prod_{i=1}^{n} \prod_{j=1}^{m} (X_i - Y_j) = a_0^m \prod_{i=1}^{n} g(X_i) = (-1)^{nm} b_0^n \prod_{j=1}^{m} f(Y_j).$$

Substituting the indeterminates X_i by α_i and Y_j by β_j, the result follows. $\quad\square$

DEFINITION 2.23 *Let $P \in A[Y]$, and let P_Y be the derivative of P with respect to Y. We define the* discriminant *of P as*

$$D_Y(P) = R_Y(P, P_Y).$$

From the Corollary (2.18), we have that, if $P \in \mathcal{R}[Y]$ is a Weierstrass polynomial whose decomposition in irreducible factors in $\mathcal{R}[[Y]]]$ is

$$P = P_1^{n_1} \cdots P_r^{n_s},$$

where the P_i's are irreducible and not associated in pairs, then

$$D_Y(P) \neq 0 \quad \text{if and only if} \quad n_1 = n_2 = \cdots = n_s = 1. \qquad (2.7)$$

In the case in which $n_1 = \cdots = n_s = 1$, we say that P is *reduced*. If $P = P_1^{n_1} \cdots P_r^{n_s}$, as above, we define the *reduction* of P as

$$\mathrm{red}(P) = P_1 \cdots P_r.$$

PROPOSITION 2.24 *Let $f = a_0 Y^n + \cdots + a_n \in K[Y] \setminus K$, where K is a field that contains the roots $\alpha_1, \ldots, \alpha_n$ of f. We have that*

$$D_Y(f) = a_0^{2n-1} \prod_{i \neq j} (\alpha_i - \alpha_j)$$

Proof: Since $f_Y(\alpha_i) = a_0 \prod_{j \neq i} (\alpha_i - \alpha_j)$, the result follows from the Proposition (2.22), because

$$D_Y(f) = R_Y(f, f_Y) = (-1)^{n(n-1)} a_0^{n-1} \prod_{i=1}^{n} f_Y(\alpha_i) = a_0^{2n-1} \prod_{i \neq j} (\alpha_i - \alpha_j).$$

PROBLEMS

4.1) Using resultants, solve the following system:

$$\begin{cases} Y^3 - Y^2 - 3X^2Y - X^2 = 0, \\ Y^3 - Y^2 + 3X^2 = 0. \end{cases}$$

4.2) Let A be a domain and let $f, g, h \in A[Y]$. Show that

$$R_Y(f, gh) = R_Y(f, g)R_Y(f, h).$$

Suggestion: Make the computations in the algebraic closure of the field of fractions of A and use the Proposition (2.22).

4.3) Determine the discriminant with respect to X of
a) $f = aX^2 + bX + c$ b) $f = X^3 + bX + c$.

3   Plane curves

In this chapter we will introduce our main objects of study: *algebroid irreducible plane curves* or plane branches. Although several results we prove along these notes are valid over arbitrary algebraically closed base fields, we will only focus them in characteristic zero, since we will systematically use Newton-Puiseux parametrization that is only obtainable in this context.

3.1   ALGEBROID PLANE CURVES

As we mentioned in the introduction of these notes, the study of singularities of an algebraic plane curve or an analytic curve in $\mathbb{C}^2$, given locally by an equation

$$f(X, Y) = 0,$$

where f is a polynomial or an analytic function in the neighborhood of the origin, may be realized by studying the algebraic properties of $f(X, Y)$ as an element of $K[[X, Y]]$. This motivates the following definition:

DEFINITION 3.1 *An* algebroid plane curve (f) *is the equivalence class of a non-invertible element f of $K[[X, Y]] \setminus \{0\}$, modulo the relation of associates.*

This means that

$$(f) = \{u \cdot f; \ u \text{ is a unit in } K[[X, Y]]\}.$$

Therefore, by definition, we have

$$(f) = (g) \iff \exists \text{ a unit } u \in K[[X, Y]] \text{ such that } g = u \cdot f$$

We will return, in the last section of this chapter, on the geometrical motivations of these notions.

Since the multiplicity of a formal power series is left invariant when we multiply it by a unit, we may define the multiplicity of an algebroid plane curve (f) as being the multiplicity of f. An algebroid curve of multiplicity one will be called *regular*. When this multiplicity is greater than one, we will say that the curve is *singular*.

Let (f) be an algebroid plane curve. We say that the curve (f) is *irreducible* if the formal power series f is irreducible in $K[[X,Y]]$. Note that this notion is independent of the representative f of (f). An irreducible algebroid plane curve will also be called a *branch*.

Let (f) be an algebroid plane curve and consider the decomposition of f into irreducible factors in $K[[X,Y]]$,

$$f = f_1 f_2 \cdots f_r.$$

The algebroid plane curves (f_j), for $j = 1, \ldots, r$, above defined, are called the *branches* of the curve (f). The curve (f) will be called *reduced*, if $(f_i) \neq (f_j)$ for $i \neq j$, that is, when f_i and f_j are not associated if $i \neq j$.

Several properties of an algebroid plane curve are preserved after we change coordinates in $K[[X,Y]]$ through a K-automorphism. This motivates the next fundamental definition.

DEFINITION 3.2 *Two algebroid plane curves (f) and (g) will be said* equivalent, *writing in such case $(f) \sim (g)$, if there exists a K-automorphism Φ of $K[[X,Y]]$ such that*

$$(\Phi(f)) = (g).$$

In other words, (f) and (g) are equivalent, if there exist a K-automorphism Φ and a unit u of $K[[X,Y]]$ such that

$$\Phi(f) = u \cdot g.$$

In what follows we will be concerned with the properties of irreducible algebroid plane curves which are invariant modulo the equivalence relation $\sim$. For instance, the reducible or irreducible character of a curve, its multiplicity, among several other properties, are preserved by equivalence of curves.

Since any curve is equivalent to a Weierstrass polynomial (see Corollary (2.7), under this point of view, we may suppose that the curve is given by a Weierstrass polynomial.

Concerning regular curves, we have the following result:

PROPOSITION 3.3 *If (f) and (g) are two regular curves, then $(f) \sim (g)$.*

Proof: Since $\operatorname{mult}(f) = 1$, it follows that $f = aX + bY + \cdots$, with $a \neq 0$ or $b \neq 0$. If $a \neq 0$, we may consider the K-automorphism Φ defined as follows:

$$
\begin{array}{rccc}
\Phi: & K[[X,Y]] & \longrightarrow & K[[X,Y]], \\
& X & \mapsto & f \\
& Y & \mapsto & Y
\end{array}
$$

that tells us that $(f) \sim (X)$. If instead, $b \neq 0$ we get analogously that $(f) \sim (Y)$. The same being true for (g), and since $(X) \sim (Y)$, we get that $(f) \sim (g)$. $\qquad\square$

It is a central problem in this theory, and still an open one, to perform the classification of algebroid plane curves modulo the equivalence relation $\sim$ and at the

time being there is no known efficient algorithm that allows to decide if two curves
are or aren't equivalent.

Just to have a feeling on how this question can be difficult to answer, consider the
following algebroid curves:

$$f = Y^2 - X^3,$$

and

$$g = (-X^3 - 6X^2 - 12X - 8)Y^3 + (-3X^3 - 12X^2 - 12X + 9)Y^2 +$$
$$(-3X^3 - 6X^2 - 12X)Y - X^3 + 4X^2.$$

Are these curves equivalent or not? It is not easy to answer to this questions a
priori since the curves have two different equations. The only thing we can easily
observe is that the two curves have the same multiplicity. Well, the series g was
chosen in order that

$$\Phi(f) = g,$$

where

$$\Phi(X, Y) = (X + 2Y + XY, 2X - 3Y),$$

and therefore, $f \sim g$.

Suppose now that (f) is an algebroid plane curve of multiplicity n, then

$$f = F_n + F_{n+1} + \cdots,$$

where each F_i is a homogeneous polynomial in $K[X, Y]$ of degree i and $F_n \neq 0$. We
call the curve (F_n) the *tangent cone* of the curve (f).

Since any homogeneous polynomial in two indeterminates with coefficients in an
algebraically closed field decomposes into linear factors (see problem 1.1 at the end
of the section), we may write

$$F_n = \prod_{i=1}^{s} (a_i X + b_i Y)^{r_i},$$

where $\sum_{i=1}^{s} r_i = n$, $a_i, b_j \in K$, for $i, j = 1, \ldots, s$, and $a_i b_j - a_j b_i \neq 0$, if $i \neq j$. So
the tangent cone of (f) consists of the linear forms (lines) $(a_i X + b_i Y)$, $i = 1, \ldots, s$,
each taken with multiplicity r_i, called the *tangent lines* of (f).

If the curve (f) is of multiplicity 1, that is if (f) is regular, then its tangent cone
(F_1) consists of one tangent line taken with multiplicity one.

EXAMPLE 3.4 The tangent cone to the curve $(Y^2 - X^3)$ is the line (Y), counted
with multiplicity 2, while the tangent cone to the curve $(Y^2 - X^2(X + 1))$ consists
of the two lines $(Y + X)$ and $(Y - X)$, each counted with multiplicity 1. The curve
$(Y - X^2)$ is regular and its tangent line is (Y).

PROBLEMS

1.1) Show that if K is an algebraically closed field, then any homogeneous polynomial
in two indeterminates with coefficients in K decomposes into a product of linear
factors.

1.2) Let $a, b, c, d \in K$. Show that $(aX + bY) = (cX + dY)$ if and only if $ad - bc = 0$.

1.3) Let $f, g \in K[[X, Y]]$ with initial forms respectively F_n and G_m. Show that if $(f) \sim (g)$, then $n = m$ and $(F_n) \sim (G_n)$.

1.4) Let (f) be a regular curve.
a) Show that (f) is irreducible.
b) Show that the tangent line to (f) is given by $(X f_X(0,0) + Y f_Y(0,0))$.

3.2   NEWTON-PUISEUX THEOREM

The technique of fractional power series expansions, without any concern about convergence questions, was used by Newton to study singular points of curves. In 1850, Puiseux in [11], vindicated Newton's method putting it into the context of functions of one complex variable, proving in this way the convergence of the series that appear. We will adopt here essentially Newton's point of view, placing ourselves in the framework of formal geometry and not bother about convergence of series.

Since any curve is equivalent to a curve defined by a Weierstrass polynomial in $K[[X]][Y]$, it is of great utility to determine the roots of this polynomial in the algebraic closure of $K((X))$. This is the strategy we will use to study algebroid irreducible plane curves. For this purpose we will characterize, only when $\text{char}(K) = 0$, the algebraic closure $\overline{K((X))}$ of $K((X))$.

From now on, in this chapter, we will assume that $\text{char}(K) = 0$.

We have seen in Chapter 1 that $K((X))$ is the set of formal Laurent series with coefficients in K, that is, the formal series of the form

$$b_{-m} X^{-m} + b_{-m+1} X^{-m+1} + \cdots + b_0 + b_1 X + b_2 X^2 + \cdots.$$

Clearly $\overline{K((X))}$ must contain the roots of the equations $Y^n - X = 0$, for all positive integer n, hence it must contain the elements of the form $X^{\frac{1}{n}}$, subject to relations of the type:
i) $X^{\frac{1}{1}} = X$,
ii) $(X^{\frac{m}{rn}})^r = X^{\frac{m}{n}}$, $\forall n, m, r \in \mathbb{Z}$ and $n, r > 0$.

In this way, we obtain extensions $K((X^{\frac{1}{n}}))$ of $K((X))$ which are finite and galoisian, as we will show next.

Recall the following definition from field theory.

DEFINITION 3.5 *Let F/k be a field extension. If the group*

$$G(F/k) = \{\sigma : F \to F \mid \sigma \text{ is a } k\text{-automorphism }\},$$

is finite and its fixed field is k, that is,

$$\{a \in F \mid \sigma(a) = a, \text{ for all } \sigma \in G(F/k)\} = k,$$

then the extension is called galoisian *and $G(F/k)$ is its* Galois group *.*

We denote by U_n the multiplicative group of the nth roots of unity in K. This is a cyclic group because it is a subgroup of the multiplicative group of a field and has order n since the polynomial $X^n - 1$ is separable over K (recall that K is algebraically closed of characteristic zero).

Since $K((X^{\frac{1}{n}}))$ is K-isomorphic to $K((X))$, we have from the Corollary (1.14) that any K-automorphism σ of $K((X^{\frac{1}{n}}))$ is such that $\sigma(X^{\frac{1}{n}}) = bX^{\frac{1}{n}}$, for some $b \in K \setminus \{0\}$.

LEMMA 3.6 *The field extension $K((X^{\frac{1}{n}}))/K((X))$ is finite and galoisian with Galois group isomorphic to the group U_n.*

Proof: Put $G = G(K((X^{\frac{1}{n}}))/K((X)))$, and let $\sigma \in G$. Since σ is a $K((X))$-automorphism, for some $b_\sigma \in K \setminus \{0\}$, we must have

$$\sigma(X^{\frac{1}{n}}) = b_\sigma X^{\frac{1}{n}},$$

and that

$$b_\sigma^n X = (\sigma(X^{\frac{1}{n}}))^n = \sigma((X^{\frac{1}{n}})^n) = \sigma(X) = X.$$

Hence $b_\sigma^n = 1$ and therefore $b_\sigma \in U_n$.

The mapping $h : G \to U_n$ defined by $h(\sigma) = b_\sigma$ is a group isomorphism. Indeed, if $\rho \in G$, then

$$b_{\rho \circ \sigma} X^{\frac{1}{n}} = \rho \circ \sigma(X^{\frac{1}{n}}) = \rho(b_\sigma X^{\frac{1}{n}}) = b_\sigma \rho(X^{\frac{1}{n}}) = b_\rho b_\sigma X^{\frac{1}{n}},$$

then $b_{\rho \circ \sigma} = b_\rho b_\sigma$ and therefore h is a group homomorphism.

Suppose now that $b_\sigma = b_\rho$. Since

$$\sigma(\sum b_i X^{\frac{i}{n}}) = \sum b_i b_\sigma^i X^{\frac{i}{n}} = \sum b_i b_\rho^i X^{\frac{i}{n}} = \rho(\sum b_i X^{\frac{i}{n}}),$$

it follows that $\sigma = \rho$ and therefore h is injective. It is clear from the definition of σ that h is surjective. Hence G is isomorphic to U_n.

We will now prove that the fixed field of G is precisely $K((X))$. Suppose that $\sum_{i \geq i_0} b_i X^{\frac{i}{n}} \in K((X^{\frac{1}{n}}))$ is invariant by the action of the elements of G, that is, for all $\xi \in U_n$ we have that

$$\sum_{i \geq i_0} b_i X^{\frac{i}{n}} = \sum_{i \geq i_0} b_i \xi^i X^{\frac{i}{n}}.$$

Hence $b_i = b_i \xi^i$ for all $i \geq i_0$. If ξ is a primitive root then $b_i = 0$ for all i not divisible by n, hence $\sum_{i \geq i_0} b_i X^{\frac{i}{n}} \in K((X))$, which concludes the proof. $\square$

The above Lemma also shows that the fields $K((X^{\frac{1}{n}}))$ are all contained in $\overline{K((X))}$. Hence we may define

$$K((X))^* = \bigcup_{n=1}^{\infty} K((X^{\frac{1}{n}})) \subset \overline{K((X))}.$$

It is clear that the elements of $K((X))^*$ may be written in the form

$$\alpha = b_1 X^{\frac{p_1}{q_1}} + b_2 X^{\frac{p_2}{q_2}} + \cdots, \tag{3.1}$$

with $b_1, b_2, \ldots \in K$, $p_i, q_i \in \mathbb{Z}$, $q_i > 0$, for all i, and $\frac{p_1}{q_1} < \frac{p_2}{q_2} < \cdots$, where the set $\{\frac{p_i}{q_i}; \ i \in \mathbb{N} \setminus \{0\}\}$ admits a common denominator.

If $b_1 \neq 0$, then the rational number $\frac{p_1}{q_1}$ is called the *multiplicity* of α and will be denoted by $\mathrm{mult}(\alpha)$. It is clear that given $\alpha, \beta \in K((X))^*$, we have that

$$\mathrm{mult}(\alpha \cdot \beta) = \mathrm{mult}(\alpha) + \mathrm{mult}(\beta),$$

and that

$$\mathrm{mult}(\alpha \pm \beta) \geq \min\{\mathrm{mult}(\alpha), \mathrm{mult}(\beta)\},$$

with equality holding whenever $\mathrm{mult}(\alpha) \neq \mathrm{mult}(\beta)$.

For commodity, we define

$$\mathrm{mult}(0) = \infty.$$

We also define

$$K[[X]]^* = \bigcup_{n=1}^{\infty} K[[X^{\frac{1}{n}}]].$$

Therefore, any element α of $K[[X]]^*$ is of the form (3.1) with $\mathrm{mult}(\alpha) \geq 0$.

LEMMA 3.7 $K((X))^*$ *is a subfield of* $\overline{K((X))}$.

Proof: If $f, g \in K((X))^*$, then there exist $r, s \in \mathbb{N}$ such that $f \in K((X^{\frac{1}{r}}))$ and $g \in K((X^{\frac{1}{s}}))$. Since $K((X^{\frac{1}{r}})) \subset K((X^{\frac{1}{rs}}))$ and $K((X^{\frac{1}{s}})) \subset K((X^{\frac{1}{rs}}))$, we have that $f + g$, $f \cdot g$ and $\frac{f}{g}$ (if $g \neq 0$) are in $K((X^{\frac{1}{rs}})) \subset K((X))^*$. $\square$

The next Theorem will play a fundamental role in the theory of plane curves in characteristic zero.

THEOREM 3.8 (Newton-Puiseux) *We have that* $\overline{K((X))} = K((X))^*$.

Proof: Since every element of $K((X))^*$ is algebraic over $K((X))$, it is enough to prove that $K((X))^*$ is algebraically closed. For this, it is sufficient to show that every polynomial in $K((X))^*[Y]$ of degree greater or equal to 2 is reducible.

Let

$$p(X, Y) = a_0(X)Y^n + a_1(X)Y^{n-1} + \cdots + a_n(X) \in K((X))^*[Y],$$

with $n \geq 2$ and $a_0(X) \neq 0$. We may, without loss of generality, suppose that $a_0(X) = 1$.

We will use a well known change of variables to eliminate in $p(X, Y)$ the term of degree $n - 1$. This is done considering the $K((X))^*$-isomorphism

$$\Phi: \quad \begin{aligned} K((X))^*[Y] &\longrightarrow K((X))^*[Z], \\ Y &\longmapsto Z - n^{-1}a_1(X) \end{aligned}$$

and taking

$$q(X, Z) = \Phi(p(X, Y)) = p(X, Z - n^{-1}a_1(X)) = Z^n + b_2(X)Z^{n-2} + \cdots + b_n(X),$$

where $b_i(X) \in K((X))^*$, for $i = 2, \ldots, n$.

If $b_i(X) = 0$, for all $i = 2, \ldots, n$, it follows that q is reducible in $K((X))^*[Z]$, and therefore p is reducible in $K((X))^*[Y]$.

Suppose now that there exists an index i such that $b_i(X) \neq 0$. The next step will be to perform another transformation on $q(X, Z)$ in order to make it an element of $K[[W]]^*[Z]$ for some W.

We denote by u_i the multiplicity of $b_i(X)$ and put

$$u = \min \left\{ \frac{u_i}{i} \mid 2 \leq i \leq n \right\}.$$

Let r be such that $u = \frac{u_r}{r}$ and consider the map

$$\Psi: \quad \begin{aligned} K((X))^*[Z] &\longrightarrow K((W))^*[Z]. \\ f(X, Z) &\longmapsto f(W^r, ZW^{u_r}) \end{aligned}$$

It is easy to verify that Ψ is an isomorphism of K-algebras and that it preserves the degrees as polynomials in Z.

Let

$$h(W, Z) = W^{-nu_r} \Psi(q(X, Z)) = W^{-nu_r} q(W^r, ZW^{u_r}) = Z^n + \sum_{i=2}^{n} c_i(W) Z^{n-i}, \quad (3.2)$$

where $c_i(W) = b_i(W^r) W^{-iu_r}$.

We have that

$$\mathrm{mult}(c_i) = ru_i - iu_r \geq 0,$$

with equality for $i = r$. So, $c_r(0) \neq 0$ and $c_i(W) \in K[[W]]^*$, $2 \leq i \leq n$.

Then there exists a positive integer k such that

$$h(W^k, Z) = Z^n + \sum_{i=2}^{n} c_i(W^k) Z^{n-i} \in K[[W]][Z].$$

Since $c_r(0) \neq 0$ and the characteristic of K is zero, then $h(0, Z)$ has at least two distinct roots. So, by Hensel's Lemma (Theorem (1.16)) there exist $h_1(W, Z), h_2(W, Z) \in K[[W]][Z]$ of degrees greater or equal to one in Z, such that

$$h(W^k, Z) = h_1(W, Z) h_2(W, Z).$$

From this and from (3.2) it follows that

$$\Psi(q(X, Z)) = W^{nu_r} h(W, Z) = W^{nu_r} h_1(W^{\frac{1}{k}}, Z).h_2(W^{\frac{1}{k}}, Z),$$

and therefore,

$$q(X, Z) = \Psi^{-1} \left(W^{nu_r} h_1(W^{\frac{1}{k}}, Z).h_2(W^{\frac{1}{k}}, Z) \right) =$$

$$X^{\frac{nu_r}{r}} \Psi^{-1} \left(h_1(W^{\frac{1}{k}}, Z) \right) .\Psi^{-1} \left(h_2(W^{\frac{1}{k}}, Z) \right).$$

Consequently, $q(X, Z)$ is reducible in $K((X))^*[Z]$. $\square$

The above Theorem is not true anymore if $\mathrm{char}(K) > 0$. In such case it is known that $K((X))^*$ is a proper subfield of $\overline{K((X))}$.

PROBLEMS

2.1) Is any of the series below an element of $K[[X]]^*$?

   a) $\sum_{i \in \mathbb{N}} X^{\frac{1}{i}}$ b) $\sum_{i \in \mathbb{N}} X^{\frac{1}{i!}}$.

2.2) Show that the map Ψ in the proof of the Theorem of Newton-Puiseux is a K-isomorphism and preserves degrees as polynomials in Z.

2.3) Let K be an algebraically closed field of characteristic zero and let n be an integer greater or equal to 2. Let

$$p(Y) = Y^n + a_2 Y^{n-2} + \cdots + a_n \in K[Y],$$

with some of the a_i's, $i = 2, \ldots, n$, non-zero. Then $p(Y)$ admits at least two distinct roots in K.

3.3   EXTENSIONS OF THE FIELD OF LAURENT SERIES

Since the Galois group of the extension $K((X^{\frac{1}{n}}))/K((X))$ is isomorphic to U_n, we have that an element ρ of U_n acts on an element

$$\alpha = \sum_{i \geq i_0} b_i \left(X^{\frac{1}{n}} \right)^i = \sum_{i \geq i_0} b_i X^{\frac{i}{n}}$$

in $K((X^{\frac{1}{n}}))$ in the following way:

$$\rho * \alpha = \sum_{i \geq i_0} b_i \left(\rho X^{\frac{1}{n}} \right)^i = \sum_{i \geq i_0} b_i \rho^i X^{\frac{i}{n}}.$$

LEMMA 3.9 *Let $\alpha \in K((X))^* \setminus K((X))$ and let $n = \min\{q \mid \alpha \in K((X^{\frac{1}{q}}))\}$. Considering α as an element of $K((X^{\frac{1}{n}}))$ we have that $\xi * \alpha \neq \rho * \alpha$, for all $\xi, \rho \in U_n$, with $\xi \neq \rho$.*

Proof: Since $\alpha \notin K((X))$, we have that $n \geq 2$.
   Write
$$\alpha = \varphi(X^{\frac{1}{n}}) = \sum_{i \geq i_0} b_i X^{\frac{i}{n}},$$

and suppose by absurdity that $\varphi(\rho X^{\frac{1}{n}}) = \varphi(\xi X^{\frac{1}{n}})$. Then $\rho^i b_i = \xi^i b_i$ for all i. Hence $\xi^i = \rho^i$ for all i such that $b_i \neq 0$.
   Let d be the GCD of n and of the i's for which $b_i \neq 0$. Then $d = 1$ because, otherwise, we would have $\alpha \in K((X^{\frac{1}{n'}}))$, where $n' = \frac{n}{d} < n$, which is a contradiction.
   It follows that there exist $b_{i_1} \neq 0, \ldots, b_{i_k} \neq 0$ and $v, v_1, \ldots, v_k \in \mathbb{Z}$ such that $vn + v_1 i_1 + \cdots + v_k i_k = 1$, and therefore

$$\xi = (\xi^n)^v (\xi^{i_1})^{v_1} \cdots (\xi^{i_k})^{v_k} = (\rho^n)^v (\rho^{i_1})^{v_1} \cdots (\rho^{i_k})^{v_k} = \rho,$$

contradicting the assumption that $\xi \neq \rho$.   □

The following result will describe the principal algebraic extensions of $K((X))$, that is, the fields $K((X))(\alpha)$, obtained by the adjunction to $K((X))$ of an algebraic element α. In this situation, from general field theory we know that

$$K((X))(\alpha) = K((X))[\alpha] = \{P(\alpha); \quad P \in K((X))[Y]\}.$$

THEOREM 3.10 *Let $\alpha \in K((X))^* \setminus K((X))$ and let us write $\alpha = \varphi(X^{\frac{1}{n}})$, where $n = \min\{q \in \mathbb{N} \mid \alpha \in K((X^{\frac{1}{q}}))\}$. Then*

i) $K((X))[\alpha] = K((X^{\frac{1}{n}}))$.

ii) *The minimal polynomial of α over $K((X))$ is given by*

$$g(X, Y) = \prod_{i=1}^{n} (Y - \alpha_i),$$

   where $\alpha_i = \varphi(\zeta^i X^{\frac{1}{n}})$, for some fixed generator ζ of the group U_n.

iii) *We have $g(X, Y) = Y^n + a_1(X)Y^{n-1} + \cdots + a_n(X) \in K((X))[Y]$, where*

$$\mathrm{mult}(a_i(X)) \geq i \cdot \mathrm{mult}(\alpha) = i\frac{\mathrm{mult}(a_n(X))}{n},$$

   with equality when $i = n$. In particular, if $\mathrm{mult}(\alpha) \geq 1$ (respectively, $\mathrm{mult}(\alpha) > 0$), then $g(X, Y) \in K[[X]][Y]$ and it is a Weierstrass polynomial (respectively, a pseudo-polynomial).

Proof:
(i) Let $G = G\left(K((X^{\frac{1}{n}}))/K((X))\right)$ and $G' = G\left(K((X^{\frac{1}{n}}))/K((X))[\alpha]\right)$.

By the Lemma (3.15), we have that

$$G' = \{\rho \in G \mid \rho * \alpha = \alpha\} = (1),$$

hence $K((X^{\frac{1}{n}})) = K((X))[\alpha]$.
(ii) Since the α_i's are the transforms of α by the elements of G, it follows that $g(X, Y) \in K((X))[Y]$; and since

$$\deg(g(X, Y)) = |G\left(K((X))[\alpha]/K((X))\right)|,$$

it follows that $g(X, Y)$ is irreducible in $K((X))[Y]$.
(iii) If S_i, $i = 1, \ldots, n$, are the elementary symmetric polynomials, then the coefficients of $g(X, Y)$ are given by

$$a_i(X) = (-1)^i S_i(\alpha_1, \ldots, \alpha_n) \in K((X)).$$

Since from (ii) we have that $\mathrm{mult}(\alpha_j) = \mathrm{mult}(\alpha)$ for all $j = 1, \ldots, n$, it follows that $\mathrm{mult}(a_n(X)) = n \cdot \mathrm{mult}(\alpha)$; and for all $i = 1, \ldots, n$,

$$\mathrm{mult}(a_i(X)) = \mathrm{mult}((-1)^i S_i(\alpha_1, \ldots, \alpha_n)) \geq i \, \mathrm{mult}(\alpha).$$

The other assertions now follow immediately. $\qquad \square$

COROLLARY 3.11 *Every finite extension of $K((X))$ is of the form $K((X^{\frac{1}{n}}))$, for some $n \in \mathbb{N} \setminus \{0\}$.*

Proof: Indeed, since $\mathrm{char}K((X)) = 0$, it follows that any finite extension of $K((X))$ has a primitive element and hence is of the form $K((X))[\alpha]$. The result follows now from the Theorem. $\qquad \square$

COROLLARY 3.12 *Let $f \in K((X))[Y]$ be an irreducible monic polynomial of degree $n \geq 1$, and let $\alpha \in K((X))^*$ be any root of f. Then*

i) $\min\{q \in \mathbb{N} \mid \alpha \in K((X^{\frac{1}{q}}))\} = n$.

ii) The α_i's being as in the Theorem (3.10), we have that

$$f(X,Y) = \prod_{i=1}^{n}(Y - \alpha_i).$$

iii) If $f \in K[[X]][Y]$ is a Weierstrass polynomial (respectively, a pseudo-polynomial), then $\mathrm{mult}(\alpha) \geq 1$ (respectively, $\mathrm{mult}(\alpha) > 0$). In particular, $\alpha \in K[[X]]^*$.

Proof: The items (i) and (ii) follow immediately from the Theorem (3.10). To prove (iii), suppose that $f = Y^n + a_1(X)Y^{n-1} + \cdots + a_n(X)$ is a Weierstrass polynomial. Then

$$-\alpha^n = a_n(X) + a_{n-1}(X)\alpha + \cdots + a_1(X)\alpha^{n-1},$$

hence for some $i_0 = 1, \ldots, n$,

$$n \cdot \mathrm{mult}(\alpha) \geq \min_i \left\{ \mathrm{mult}(a_i(X)) + (n-i)\mathrm{mult}(\alpha) \right\} =$$

$$= \mathrm{mult}(a_{i_0}(X)) + (n - i_0)\mathrm{mult}(\alpha) \geq i_0 + (n - i_0)\mathrm{mult}(\alpha).$$

It then follows that $i_0 \cdot \mathrm{mult}(\alpha) \geq i_0$ and hence $\mathrm{mult}(\alpha) \geq 1$. The proof in case f is a pseudo-polynomial is similar. $\qquad \square$

COROLLARY 3.13 *(Newton's Implicit Function Theorem) Let $f(X,Y) \in K[[X,Y]]$ be irreducible and of multiplicity n and suppose that $\frac{\partial^n f}{\partial Y^n}(0,0) \neq 0$. Then there exists*

$$\varphi(X^{\frac{1}{n}}) = \sum_{i \geq 1} b_i X^{\frac{i}{n}} \in K[[X^{\frac{1}{n}}]],$$

such that

$$f(X, \varphi(X^{\frac{1}{n}})) = 0.$$

Moreover, any $\alpha \in K[[X^{\frac{1}{n}}]]$ satisfying $f(X, \alpha) = 0$ is such that $\alpha = \varphi(\zeta X^{\frac{1}{n}})$, for some $\zeta \in U_n$.

Proof: Since the multiplicity of f is n and its nth partial derivative with respect to Y in $(0,0)$ doesn't vanish we have that f is regular with respect to Y. From the Weierstrass Preparation Theorem (Theorem (2.4)), it follows that f is associated to a pseudo-polynomial in $K[[X]][Y]$ of degree n. The result now follows immediately from the Corollary (3.12) above. $\qquad \square$

When $n = 1$, the above result is precisely the Implicit Function Theorem.

REMARK 3.14 Since the ring $K[[X]]$ is a unique factorization domain with field of fractions is $K((X))$, every irreducible polynomial in $K[[X]][Y]$ is irreducible in $K((X))[Y]$.

The following is an important necessary condition for the irreducibility a power series, which is of fundamental geometric importance.

LEMMA 3.15 (Unitangent Lemma) *Let $f \in K[[X,Y]]$ with $f(0,0) = 0$ be irreducible of multiplicity n. Then the initial form of f is of the type*

$$F_n = (aX + bY)^n,$$

with $a, b \in K$ and not simultaneously zero.

Proof: If necessary, we may perform a linear change of coordinates, which doesn't affect the type of the tangent cone (a linear form to a certain power), in such a way that f is regular in Y. By the Weierstrass Preparation Theorem there exist a Weierstrass polynomial $p = p(X,Y)$ in $K[[X]][Y]$ of degree n, and a unitary element u in $K[[X,Y]]$ such that $up = f$. Since f is irreducible, it follows that p is irreducible in $K[[X,Y]]$. Hence by the Lemma (2.6), we have that p is irreducible in $K[[X]][Y]$ and therefore from the Remark (3.14), p is irreducible in $K((X))[Y]$. From the Corollary (2.5), we have that

$$p(X,Y) = \prod_{k=1}^{n} (Y - \varphi(\zeta^k X^{\frac{1}{n}})),$$

where ζ is a primitive nth root of unity and

$$\varphi(X^{\frac{1}{n}}) = b_r X^{\frac{r}{n}} + b_{r+1} X^{\frac{r+1}{n}} + \cdots \in K((X^{\frac{1}{n}})),$$

with $b_r \neq 0$. Since p is a Weierstrass polynomial, from the Corollary (2.5.iii) we have $\mathrm{mult}(\varphi(X^{\frac{1}{n}})) \geq 1$, and therefore $r \geq n$. Since the initial form of p is the initial form of the polynomial

$$\begin{aligned}
q(X,Y) &= \prod_{k=1}^{n}(Y - \zeta^{kr} b_r X^{\frac{r}{n}}) = Y^n - (b_r X^{\frac{r}{n}} \sum_{k=1}^{n} \zeta^{kr}) Y^{n-1} + \cdots \\
&+ \left((-1)^i b_r^i X^{\frac{ir}{n}} \sum_{k=1}^{n} \zeta^{ikr}\right) Y^{n-i} + \cdots + (-1)^n b_r^n X^r,
\end{aligned}$$

we have that:
a)   If $r = n$, the initial form of p is $(Y - b_n X)^n$.
b)   If $r > n$, then $n - i + i\frac{r}{n} > n$, for all $i = 1, \ldots, n$. Hence the initial form of p is Y^n.

Since the initial form of f is the product of $u(0,0)(\neq 0)$ with the initial form of p, it follows that F_n is of the form $(aX + bY)^n$.   □

The Unitangent Lemma says that if f is irreducible, then the tangent cone to (f) consists of the only line $(aX + bY)$, counted with multiplicity.

Although we proved the Unitangent Lemma in characteristic zero since we used Newton-Puiseux' Theorem, the result is also true for any algebraically closed field K (see [6] for a systematic study of this subject in arbitrary characteristic).

EXAMPLE 3.16 The following polynomials are reducible in $K[[X,Y]]$.
- The *nodal cubic*: $Y^2 - X^2(X-1)$,
- *Descartes' Folium*: $3XY - (X^3 + Y^3)$,
- *Maclaurin's Trissectrix*: $X(X^2 + Y^2) - (Y^2 - 3X^2)$,
- *Nicomedes' Conchoid*: $(Y-1)^2(X^2 + Y^2) - 2Y^2$,
- *Pascal's Snail*: $(X^2 + Y^2)^2 - 2X(X^2 + Y^2) + 3X^2 - Y^2$.

PROBLEMS

3.1) Decide whether each of the following series is reducible or irreducible in $K[[X,Y]]$.
a) $X - Y + X^2 + Y^2 - XY^2 + X^2Y - X^3 + Y^3$.
b) $X^2 + 2XY + 2Y^2$.
c) $X - Y + XY + X^3 + Y^5$.
d) $X^6 - X^2Y^3 - Y^5$.
e) $X + Y + (X+Y)^2$.

3.4 PARAMETRIZATION AND CHARACTERISTIC EXPONENTS

We are going to introduce the notion of parametrization of plane branches. This will be a powerful tool to study the properties of curves.

Let $f = F_n + F_{n+1} + \cdots \in K[[X,Y]]$ be an irreducible power series of multiplicity n. By the Unitangent Lemma we have that $F_n = (aX + bY)^n$ for some $a, b \in K$. So, either f is regular in Y (when $b \neq 0$) or f is regular in X (when $a \neq 0$).

If f is regular in Y, then we may write it in the form:

$$f = a_0(X)Y^n + a_1(X)Y^{n-1} + \cdots + a_n(X) + Y^{n+1}h(X,Y), \qquad (3.3)$$

with $a_i(X) \in K[[X]]$, $\text{mult}(a_i(X)) \geq i$ for $i = 1, \ldots, n$, $a_0(0) \neq 0$ and $h(X,Y) \in K[[X,Y]]$.

LEMMA 3.17 *Let $f \in K[[X,Y]]$ be an irreducible power series of multiplicity n and regular in Y. Write f as in (3.3). Then*

$$\text{mult}(a_i(X)) \geq i\frac{\text{mult}(a_n(X))}{n}, \quad i = 0, \ldots, n.$$

Proof: Since f is regular in Y of order n, from the Weierstrass Preparation Theorem, we know that there exist a unit u in $K[[X,Y]]$ and $A_1, \ldots, A_n \in K[[X]]$, with $A_i(0) = 0$, such that

$$u \cdot f = P(X,Y) = Y^n + A_1Y^{n-1} + \cdots + A_n. \qquad (3.4)$$

Since the above right hand side is irreducible in $K[[X,Y]]$, we have from the Theorem (2.4.iii) that, for $i = 0, \ldots, n$,

$$\text{mult}(A_i(X)) \geq i\frac{\text{mult}(A_n(X))}{n}.$$

Let us write $u = u_0 + u_1Y + \cdots$, with $u_i \in K[[X]]$. So, $u_0(0) \neq 0$. From (3.4) we get that

$$u_0a_i + u_1a_{i+1} + \cdots + u_{n-i}a_n = A_i, \quad i = 1, \ldots, n.$$

In particular, we have

$$\text{mult}(a_n) = \text{mult}(u_0 a_n) = \text{mult}(A_n).$$

The proof will be done by descent on i. For $i = n$ the result is trivially true. Suppose that we have shown that $\text{mult}(a_j) \geq j\dfrac{\text{mult}(a_n(X))}{n}$, for $j > i$, hence

$$\text{mult}(a_i) = \text{mult}(A_i - (u_1 a_{i+1} + \cdots + u_{n-i} a_n)) \geq i\frac{\text{mult}(a_n(X))}{n},$$

proving the result. □

REMARK 3.18 The following conditions are clearly equivalent:

(i) The tangent cone of (f) is (Y^n),

(ii) For all $i \geq 1$, $\text{mult}(a_i(X)) > i$,

(iii) For some $i \geq 1$, $\text{mult}(a_i(X)) > i$.

Suppose now that f is irreducible of multiplicity n and regular in Y as in (3.3). Let $P(X,Y) \in K[[X]][Y]$ be the pseudo-polynomial of degree n associated to f, as in (3.4) above, and let $\alpha = \varphi(X^{\frac{1}{n}}) \in K[[X^{\frac{1}{n}}]]$, where

$$n = \min\{q \in \mathbb{N};\ \alpha \in K((X^{\frac{1}{q}}))\},$$

be such that $P(X,\alpha) = P(X,\varphi(X^{\frac{1}{n}})) = 0$ (see Corollary (2.5)). If we put $T = X^{\frac{1}{n}}$, then we have $\varphi(T) \in K[[T]]$ and

$$f(T^n, \varphi(T)) = 0.$$

In this situation we say that

$$\begin{cases} X = T^n \\ Y = \varphi(T) = \sum_{i \geq m} b_i T^i, & b_m \in K \setminus \{0\}, \end{cases} \tag{3.5}$$

is a *Puiseux parametrization* of the branch (f).

Any other root of P will give another Puiseux parametrization $(T^n, \psi(T))$ of (f), where $\psi(T) = \varphi(\zeta T)$, and ζ is an nth root of unity. These are the only parametrizations of (f) of the form $(T^n, \varphi(T))$.

Notice that the condition

$$n = \min\{q \in \mathbb{N}; \alpha \in K((X^{\frac{1}{q}}))\ \text{and}\ f(X,\alpha) = 0\},$$

implies that in any Puiseux parametrization as in (3.5), n and the indices i for which $b_i \neq 0$, are relatively prime.

Notice also that from the Theorem (3.10.iii) we have

$$\text{mult}_T(\varphi(T)) = n \cdot \text{mult}_X(\alpha) = \text{mult}_X(A_n(X)) = \text{mult}_X(a_n(X)) \geq n.$$

In particular, if the tangent cone of (f) is (Y^n), then from the Remark (3.18) we have that

$$\text{mult}_T(\varphi(T)) = \text{mult}(a_n(X)) > n. \qquad (3.6)$$

There are many other possible parametrizations of (f) by means of other series in $K[[T]]$. Let $(\psi_1(T), \psi_2(T))$ be a pair of non-zero and non-unit elements in $K[[T]]$. We say that $(\psi_1(T), \psi_2(T))$ is a *parametrization* of (f) if

$$f(\psi_1(T), \psi_2(T)) = 0,$$

as an element of $K[[T]]$. A parametrization $(\psi_1(T), \psi_2(T))$ of (f) will be called *primitive* if there is an automorphism ρ of $K[[T]]$ such that

$$(\rho(\psi_1(T)), \rho(\psi_2(T))) = (T^n, \varphi(T)),$$

where $(T^n, \varphi(T))$ is a Puiseux parametrization of (f).

If f is regular in X, then we have the same results as above interchanging the roles of X and Y.

There exists an algorithm in characteristic zero, due to Newton, to determine a Puiseux parametrization of an algebroid plane curve (f) (see [13] or [5]). This algorithm is implemented, for example, in the package MAPLE.

The relationship between the equation of a curve and its Puiseux parametrization is of high complexity as one may see in the following examples.

EXAMPLE 3.19 The curve given by

$$
\begin{aligned}
f = \ & Y^8 - 4X^3Y^6 - 8X^5Y^5 + (6X^6 - 26X^7)Y^4 + (-24X^9 + 16X^8)Y^3 + \\
& + (36X^{10} - 4X^9 - 20X^{11})Y^2 + (-8X^{11} + 16X^{12} - 8X^{13})Y + 21X^{14} + \\
& + X^{12} + 6X^{13} - X^{15},
\end{aligned}
$$

may be parametrized by

$$
\begin{cases}
X = T^8 \\
Y = T^{12} + T^{14} - T^{15}.
\end{cases}
$$

EXAMPLE 3.20 The curve

$$(f) = (Y^3 - 9X^3Y - X^4),$$

has a parametrization whose terms up to the power 19 in T are:

$$
\begin{cases}
X = \ & T^3 \\
Y = \ & T^4 + 3T^5 - 9T^7 + 27T^8 - 324T^{10} + 1215T^{11} - 18711T^{13} + \\
& + 75816T^{14} - 1301265T^{16} + 5484996T^{17} - 100048689T^{19} + \cdots
\end{cases}
$$

Let C be a plane branch defined by an irreducible power series f of multiplicity n and regular in Y with a Puiseux parametrization

$$
\begin{cases}
X = T^n \\
Y = \varphi(T) = \sum_{i \geq m} b_i T^i, \quad b_m \neq 0 \ \ m \geq n.
\end{cases}
$$

We define two sequences (ε_i) and (β_i) of natural numbers associated to (f) as follows:

$$\varepsilon_0 = \beta_0 = n$$
$$\beta_j = \min\{i \mid i \not\equiv 0 \bmod \varepsilon_{j-1} \text{ and } b_i \neq 0\}, \quad \text{if } \varepsilon_{j-1} \neq 1$$
$$\varepsilon_j = \mathrm{GCD}(\varepsilon_{j-1}, \beta_j) = \mathrm{GCD}(\beta_0, \dots, \beta_j).$$

Observe that if $\varepsilon_{j-1} \neq 1$, then the set $\{i; \ i \not\equiv 0 \bmod \varepsilon_{j-1} \text{ and } b_i \neq 0\}$ is not empty since the parametrization is primitive. Therefore, the β_j are well defined, with β_1 equal to the first exponent of T in $\varphi(T)$ which is not divisible by n, and with non-vanishing coefficient. We also have that ε_j divides ε_{j-1}, for all $j \geq 1$ and

$$n = \varepsilon_0 > \varepsilon_1 > \varepsilon_2 > \cdots$$

Consequently, for some $\gamma \in \mathbb{N}$, we must have $\varepsilon_\gamma = 1$, and therefore, the sequence of the β_j, $j \geq 1$ is increasing and stops in β_γ.

DEFINITION 3.21 *The* characteristic exponents *of C are the $(\gamma+1)$ natural numbers* $(\beta_0, \beta_1, \dots, \beta_\gamma)$.

EXAMPLE 3.22 *Let the parametrization $(T^4, T^6 + T^8 + T^{10} + T^{11})$ be given, then we have $\beta_0 = \varepsilon_0 = 4$, $\beta_1 = 6$, $\varepsilon_1 = 2$, $\beta_2 = 11$ and $\varepsilon_2 = 1$. So, in this case, $\gamma = 2$.*

Notice that the characteristic exponents of a plane branch C determine the integers ε_j, since $\varepsilon_j = \mathrm{GCD}(\beta_0, \beta_1, \dots, \beta_j)$.

With the above notation, we may write a parametrization of the branch C as follows:

$$\begin{cases} x = T^n \\ y = P(T^n) + \sum_{i=\beta_1}^{\beta_2 - 1} b_i T^i + \cdots + \sum_{i=\beta_{\gamma-1}}^{\beta_\gamma - 1} b_i T^i + \sum_{i \geq \beta_\gamma} b_i T^i, \end{cases}$$

where $P(T) \in K[T]$, and $b_{\beta_1} \cdots b_{\beta_\gamma} \neq 0$.

From the definition of the β_j, we may easily deduce that the coefficients of the above parametrization have the following property: if i and j are integers such that $\beta_{j-1} \leq i < \beta_j$, and if e_{j-1} doesn't divide i, then $b_i = 0$.

Conversely, given any increasing sequence of natural relatively prime integers $\beta_0, \dots, \beta_\gamma$, such that the integers defined by

$$\varepsilon_j = \mathrm{GCD}(\beta_0, \dots, \beta_j)$$

are strictly decreasing, then this sequence corresponds to the characteristic exponents of some plane branch C.

One also defines the *Puiseux pairs* (η_j, μ_j), $j = 1, \dots, \gamma$, of C as follows:

$$\eta_j = \frac{\varepsilon_{j-1}}{\varepsilon_j}, \quad \text{and} \quad \mu_j = \frac{\beta_j}{\varepsilon_j}.$$

Now, since $\varepsilon_j = \mathrm{GCD}(e_{j-1}, \beta_j)$, we have that

$$\mathrm{GCD}(\eta_j, \mu_j) = 1.$$

PROBLEMS

4.1) Show, for all $i, j = 1, \ldots, \gamma$, with $i < j$, that
a) $\eta_j > 1$.
b) $n = \eta_1 \eta_2 \cdots \eta_j \varepsilon_j = \eta_1 \eta_2 \cdots \eta_\gamma$.
c) $\varepsilon_j = \eta_{j+1} \cdots \eta_\gamma$.
d) $\eta_i \eta_{i+1} \ldots \eta_j = \frac{e_{i-1}}{e_j}$.
e) $\gamma \leq \log_2 n$.

4.2) a) Show that a set of pairs of natural numbers (η_j, μ_j), $j = 1, \ldots, \gamma$ is the set of Puiseux pairs of a branch C if and only if

$$\eta_1 < \mu_1, \ \eta_j > 1, \ \text{for} \ j \geq 1,$$
$$\mu_{j-1}\eta_j < \mu_j, \ \text{for} \ j \geq 2,$$
$$\text{GCD}(\eta_j, \mu_j) = 1, \ \text{for} \ j \geq 1.$$

b) Show that the characteristic exponents and the Puiseux pairs determine each other.

4.3) Find the characteristic exponents and the Puiseux pairs of the branches defined by the following parametrizations:
a) $x = T^6, \ \ y = T^8 + T^{10} + T^{12} + T^{13}$.
b) $x = T^6, \ \ y = T^9 + T^{10} + T^{11}$.
c) $x = T^6, \ \ y = T^{20} + T^{31}$.
d) $x = T^6, \ \ y = T^{20} + T^{28} + T^{31}$.
e) $x = T^6, \ \ y = T^{22} + T^{31}$.
f) $x = T^8, \ \ y = T^{12} + T^{13}$.
g) $x = T^{12}, \ \ y = T^{24} + T^{30} + T^{33} + T^{40}$.
h) $x = T^{18}, \ \ y = T^{24} + T^{30} + T^{33} + T^{40}$.

3.5   PLANE ANALYTIC CURVES

The algebraic properties of the ring of convergent power series with complex coefficients $\mathbb{C}\{X, Y\}$ are very similar to those of the ring $\mathbb{C}[[X, Y]]$. For instance, $\mathbb{C}\{X, Y\}$ is a domain whose invertible elements are the elements f such that $f(0,0) \neq 0$ (this may be proved by showing that the series f^{-1} constructed in the proof of Proposition (1.1)) is absolutely convergent in some neighborhood of the origin). From this it follows that $\mathcal{M} = \langle X, Y \rangle$ is the unique maximal ideal of the ring $\mathbb{C}\{X, Y\}$ and it has the property

$$\bigcap_{i \in \mathbb{N}} \mathcal{M}^i = \{0\}.$$

Hence, it is also possible to substitute non-unit convergent power series $g_1, g_2 \in \mathbb{C}\{X, Y\}$, into a convergent series $f \in \mathbb{C}\{X, Y\}$, getting a convergent series. In this context it is also possible to show that every automorphism of $\mathbb{C}\{X, Y\}$ is the result of the substitution of X and Y by g_1 and g_2, each of multiplicity one and with linearly independent initial forms over $\mathbb{C}$. On the other hand, the Division and Preparation Theorems as well as Hensel's Lemma are valid, implying that $\mathbb{C}\{X, Y\}$ is a noetherian unique factorization domain.

When we deal with a convergent series f we gain an extra geometric interpretation, more precisely, we have an analytic curve

$$C_f = \{(z, w) \in U; f(z, w) = 0\},$$

where U is a sufficiently small neighborhood of the origin of $\mathbb{C}^2$.

Let $f, g \in \mathbb{C}\{X, Y\}$ and let Φ be an automorphism of $\mathbb{C}\{X, Y\}$. To Φ there is associated a bi-analytic isomorphism $\tilde{\Phi} : V \longrightarrow U$ where U and V are open neighborhoods of the origin in $\mathbb{C}^2$ and suppose that f and g are convergent in U. It is easy to verity that

$$C_{f \cdot g} = C_f \cup C_g,$$

and that

$$\tilde{\Phi}(C_f) = C_{\Phi(f)}.$$

A point $P \in C_f$ is *singular* if

$$f(P) = f_X(P) = f_Y(P) = 0,$$

where f_X and f_Y are the partial derivatives of f with respect to X and to Y.

A point $P = (a, b) \in C_f$ is *regular* when it is not singular and in this case, we define the tangent line to C_f at P as

$$T_P C_f : f_X(P)(X - a) + f_Y(P)(Y - b) = 0.$$

PROPOSITION 3.23 *Suppose that $f \in \mathbb{C}\{X, Y\}$ is reduced, then there exists an open set U in $\mathbb{C}^2$, containing the origin, such that C_f is regular in $C_f \cap (U \setminus \{(0, 0)\})$.*

Proof: Indeed, after a linear change of variables, which doesn't affect the singular or regular character of a point, we may assume that f is a pseudo-polynomial in a neighborhood U' of $(0, 0)$. Let $g(X) = D_Y(f)$ be the discriminant of f; that is, the resultant $R_Y(f, f_Y)$. We know (see Section 2.4) that $g(X)$ is not identically zero, because f is reduced. Since the zeros of an analytic function in one variable $g(X)$ are isolated, there exists an open neighborhood V of the origin in $\mathbb{C}$ such that

$$R_Y(f(a, Y), f_Y(a, Y)) = g(a) \neq 0, \quad \forall a \in V \setminus \{0\},$$

and therefore by the Corollary (2.19), we have that f and f_Y do not vanish simultaneously in $V \times \mathbb{C} \setminus \{(0, 0)\}$. Taking $U = V \times \mathbb{C}$, the result is now proved. $\qquad \square$

We define the equivalence relation $\sim$ in $\mathbb{C}\{X, Y\}$ in a similar way as it was defined in $\mathbb{C}[[X, Y]]$: given $f, g \in \mathbb{C}\{X, Y\}$, we say that $(f) \sim (g)$ if and only if there exist a $\mathbb{C}$-automorphism Φ and a unit u of $\mathbb{C}\{X, Y\}$, such that

$$\Phi(f) = ug.$$

This implies that there exist open neighborhoods U and V of the origin of $\mathbb{C}^2$, such that C_f is defined in U and C_g is defined in V, and a bi-analytic map $\tilde{\Phi} : U \longrightarrow V$ such that

$$\tilde{\Phi}(C_f) = C_g.$$

Conversely, we will show below that if f and g are reduced elements in $\mathbb{C}\{X, Y\}$ and if there exists a bi-analytic map $\tilde{\Phi} : U \longrightarrow V$, where U and V are open neighborhoods of the origin of $\mathbb{C}^2$, such that $\tilde{\Phi}(C_f) = C_g$, then $(f) \sim (g)$.

REMARK 3.24 Newton-Puiseux' Theorem, in this context, says that if $f \in \mathbb{C}\{X, Y\}$ is an irreducible pseudo-polynomial of multiplicity n, then there exists a power series $\varphi(t) \in \mathbb{C}\{t\}$, convergent in a disc D centered at the origin of $\mathbb{C}$, such that

$$f(t^n, \varphi(t)) = 0, \quad \forall\, t \in D.$$

This gives us a local parametrization of C_f around the origin of $\mathbb{C}^2$, by analytic functions. On the other hand, there are n such parametrizations for C_f, namely, $(t^n, \varphi_i(t))$, $i = 0, \ldots, n-1$, with $\varphi_i(t) = \varphi(\zeta^i t)$, where ζ is a primitive nth root of the identity, in such a way that for any $z \in \mathbb{C}$ close to the origin, if w is an nth root of z, then $\varphi_i(w)$, $i = 0, \ldots, n-1$, are the roots of the polynomial $f(z, Y) = 0$. Now, since $D_Y(f) \neq 0$, we also have that the roots in $\mathbb{C}$ of $f(z, Y)$ for $z \neq 0$, close to the origin, are all distinct. This shows that for a non-zero z, close to zero, if $\varphi(z) = \varphi_i(z)$, then $i = 0$.

LEMMA 3.25 *Let $f \in \mathbb{C}\{X\}[Y]$ be an irreducible pseudo-polynomial of multiplicity n. Then there exist an open neighborhood U of $(0,0)$ in $\mathbb{C}^2$, a disc D centered at the origin of $\mathbb{C}$ and an analytic map*

$$\begin{array}{rccc} \Psi : & D & \longrightarrow & U, \\ & t & \mapsto & \Psi(t) = (t^n, \varphi(t)) \end{array}$$

such that f is convergent in U, $\Psi(D) = C_f \cap U$ and $\Psi : D \longrightarrow \Psi(D)$ is a homeomorphism with $\Psi(0) = (0,0)$.

Proof: Let D_1 be a small disc centered at the origin of $\mathbb{C}$ such that the coefficients of f in $\mathbb{C}\{X\}$ are convergent, and $D_Y(f)(z) \neq 0$ for all $z \in D_1 \setminus \{0\}$. Let φ and D be as in the Remark (3.24) with D small enough in order that $t^n \in D_1$ for $t \in D$. Let $U = D_1 \times \mathbb{C}$. It is clear that $\Psi(D) \subset C_f \cap U$. Let $(z, w) \in C_f \cap U$, then there exists $t \in D$ such that $z = t^n$ and $w = \varphi(\zeta^i t)$, for some i, hence Ψ is onto $C_f \cap U$. The map Ψ is injective because, if $\Psi(t) = \Psi(t')$, then $t' = \zeta^i t$, for some i, hence

$$\varphi(t) = \varphi(t') = \varphi(\zeta^i t) = \varphi_i(t),$$

which implies, after the Remark (3.24), that $i = 0$, and therefore $t = t'$.

The map Ψ is obviously analytic and moreover, $\Psi|_{D \setminus \{0\}}$ has a non-zero differential at all points, so it is an analytic embedding, giving a homeomorphism between $D \setminus \{0\}$ and $C_f \cap U \setminus \{(0,0)\}$, which by continuity extends to a homeomorphism between D and $C_f \cap U$. $\qquad\square$

COROLLARY 3.26 *If $f \in \mathbb{C}\{X, Y\}$ is irreducible, then C_f is homeomorphic to a disc, in a neighborhood of the origin.*

LEMMA 3.27 *If f and g are irreducible and not associated in $\mathbb{C}\{X, Y\}$, then there exists an open neighborhood U of the origin in $\mathbb{C}^2$ such that*

$$C_f \cap C_g \cap U = \{(0,0)\}.$$

Proof: We may reduce easily the problem to the case in which f and g are pseudo-polynomials. Since $R_Y(f, g)$ is not identically zero, because f and g are not associated, we have that there exists a neighborhood D of the origin in $\mathbb{C}$ such that $R_Y(f, g)(a) \neq 0$, for all $a \in D \setminus \{0\}$. This implies that $f(a, Y)$ and $g(a, Y)$ have no common roots in $\mathbb{C}$, for all $a \in D \setminus \{0\}$. If we put $U = D \times \mathbb{C}$, the result follows. $\qquad\square$

COROLLARY 3.28 *Let* $f, g \in \mathbb{C}\{X, Y\}$ *irreducible. Then* $C_f = C_g$ *if and only if* f *and* g *are associated.*

Proof: If f and g are associated, then clearly $C_f = C_g$. The converse follows from the Lemma (3.27). □

PROPOSITION 3.29 *Let* $f, g \in \mathbb{C}\{X, Y\}$ *and suppose there exists a neighborhood of the origin in* $\mathbb{C}^2$ *such that* $C_f = C_g$, *then* red(f) *and* red(g) *are associated.*

Proof: Write $f = f_1^{n_1} \cdots f_r^{n_r}$ and $g = g_1^{m_1} \cdots g_s^{m_s}$. We have that

$$C_f = C_{f_1} \cup \cdots \cup C_{f_r}, \quad \text{and} \quad C_g = C_{g_1} \cup \cdots \cup C_{g_s}.$$

Hence from the Corollaries (3.26) (3.27), we have that

$$C_{f_i} \setminus \{(0,0)\}, \quad i = 1, \ldots, r, \quad \text{and} \quad C_{g_j} \setminus \{(0,0)\}, \quad j = 1, \ldots, s,$$

are respectively the connected components of $C_f \setminus \{(0,0)\}$ and $C_g \setminus \{(0,0)\}$. Since $C_f = C_g$, it then follows that $r = s$, and that each C_{f_i} is equal to C_{g_j}, for some j, which, again from the Corollary (3.28), implies that $(f_i) = (g_j)$, showing that red(f) and red(g) are associated.

□

COROLLARY 3.30 *Let* $f, g \in \mathbb{C}\{X, Y\}$ *be reduced, where* f *is defined in* U *and* g *in* V. *Suppose that there exists a bi-analytic map* $\tilde{\Phi} : U \longrightarrow V$ *such that* $\tilde{\Phi}(C_f) = C_g$. *Then* $(f) \sim (g)$.

Proof: Let Φ be the automorphism of $\mathbb{C}\{X, Y\}$ associated to $\tilde{\Phi}$. The condition $\tilde{\Phi}(C_f) = C_g$ implies that $C_{\Phi(f)} = C_g$, which implies by the Proposition that red($\Phi(f)$) and red(g) are associated. Since f is reduced, it follows that $\Phi(f)$ is reduced, and because also g is reduced, it follows that $\Phi(f)$ and g are associated. □

In what follows we are going to give a geometric interpretation for the tangent cone of a curve. The first result will say that the lines of the tangent cone to an analytic curve at the origin are limits of secant lines through the origin, while the second result will say that the tangent cone consists of limits of tangent lines to the curve at regular points near the origin.

Recall that the *projective space* $\mathbb{C}\mathbf{P}^{n-1}$ associated to $\mathbb{C}^n$ is the quotient of $\mathbb{C}^n \setminus \{0\}$ modulo the equivalence relation

$$z \mathcal{R} z' \iff \exists \lambda \in \mathbb{C} \setminus \{0\}; \quad z = \lambda z'.$$

We will use the notation $(X_0 : \ldots : X_{n-1})$ to represent homogeneous coordinates in $\mathbb{C}\mathbf{P}^{n-1}$. It is well known that $\mathbb{C}\mathbf{P}^{n-1}$, with the quotient topology induced by the topology of $\mathbb{C}^n$, is compact.

To the line $a_0 X + b_0 Y = 0$ in $\mathbb{C}^2$ we associate the point $(a_0 : b_0)$ of $\mathbb{C}\mathbf{P}^1$.

PROPOSITION 3.31 *Let* $C_f \subset U$ *be an analytic curve through the origin, where* $U \subset \mathbb{C}^2$ *is a neighborhood of the origin. Let* (P_m) *be a sequence of points in* $C_f \setminus \{0\}$ *that converges to the origin. Then the sequence* $(\overline{P_m})$ *of the complex lines* $\overline{P_m}$ *through the origin and* P_m, *viewed as points in* $\mathbb{C}\mathbf{P}^1$, *converges to* $\overline{P}$, *that represents a line belonging to the tangent cone of* C_f.

Proof: Let

$$f(X,Y) = \sum_{j \geq n} F_j(X,Y),$$

be the Taylor expansion of f, where F_j is the homogeneous polynomial of degree j and $F_n \neq 0$. The tangent cone of C_f is determined by the equation $F_n(X,Y) = 0$. Define $P_m = (a_m, b_m)$ and $P = (a,b)$. Since

$$\mathbb{CP}^1 = \{\overline{P}; a \neq 0\} \cup \{\overline{P}; b \neq 0\},$$

and $\overline{P} \in \mathbb{CP}^1$, we may assume without loss of generality that $a \neq 0$. Since $\{\overline{P}; a \neq 0\}$ is an open set in $\mathbb{CP}^1$, there exists $m_0 \in \mathbb{N}$ such that for $m \geq m_0$ we have $\overline{P_m} \in \{\overline{P}; a \neq 0\}$. Hence

$$0 = f(P_m) = \sum_{j \geq n} F_j(a_m, b_m) = (a_m)^n \left(F_n\left(1, \frac{b_m}{a_m}\right) + \sum_{i \geq 1}(a_m)^i F_{n+i}\left(1, \frac{b_m}{a_m}\right) \right).$$

Therefore, for $m \geq m_0$,

$$F_n\left(1, \frac{b_m}{a_m}\right) + \sum_{i \geq 1}(a_m)^i F_{n+i}\left(1, \frac{b_m}{a_m}\right) = 0, \tag{3.7}$$

hence taking the limit when $m \to \infty$, we have that

$$\left(1, \frac{b_m}{a_m}\right) \to \left(1, \frac{b}{a}\right) \quad \text{and} \quad a_m \to 0,$$

and therefore from (3.7) we get $F_n(1, \frac{b}{a}) = 0$ and consequently $F_n(\overline{P}) = 0$, which conclude the proof of the result.   □

PROPOSITION 3.32 *Let C be an irreducible analytic plane curve, defined in a neighborhood of the origin and regular away from the origin. For any sequence of points (P_m) in $C \setminus \{0\}$ such that $P_m \to (0,0)$, we have that $T_{P_m}C$ tends to the tangent cone of C.*

Proof: After changing coordinates, if necessary, we may assume that the equation of C is a regular in Y and that its tangent line is horizontal (recall the Unitangent Lemma).

We know that C has a parametrization of the form $(T^n, \varphi(T))$ where $n = \text{mult}(C)$ and

$$\varphi(T) = b_r T^r + b_{r+1}T^{r+1} + \cdots \in \mathbb{C}\{T\}.$$

Hence the directional vector of the tangent line of C at the point $P_m = (t_m^n, \varphi(t_m))$ is given by

$$\left(\frac{\partial T^n}{\partial T} : \frac{\partial \varphi}{\partial T}\right)(t_m) = \left(1 : \frac{r}{n}b_r t_m^{r-n} + \cdots\right).$$

From (3.6) we have that $r - n > 0$. So, when $t_m \to 0$, the directional vectors of the tangents tend to the vector $(1,0)$, which is precisely the directional vector of the tangent cone of C.   □

4 Intersection of curves

4.1 THE LOCAL RING OF A PLANE CURVE

Given a ring A and elements $z_1, \ldots, z_n \in A$, we will denote by $\langle z_1, \ldots, z_n \rangle$ the ideal of A generated by the elements $z_1, \ldots, z_n$.

Let K be an arbitrary field and let f be an element in the maximal ideal $\mathcal{M} = \langle X, Y \rangle$ of $K[[X, Y]]$.

We define the *coordinate ring* of the curve (f) as being the K-algebra

$$\mathcal{O}_f = \frac{K[[X, Y]]}{\langle f \rangle}.$$

If $h \in K[[X, Y]]$ and $B \subset K[[X, Y]]$, we will denote by $\overline{h}$ the residual class of h in $\mathcal{O}_f$, and by $\overline{B}$ the set of the residual classes of the elements of B. We will denote the residual class $\overline{Y}$ by y and $\overline{X}$ by x, respectively. In many situations we will identify x with X itself, since this will cause no confusion.

The ring $\mathcal{O}_f$ is a local ring with maximal ideal

$$\mathcal{M}_f = \overline{\mathcal{M}}.$$

When f is irreducible, $\mathcal{O}_f$ is an integral domain and in this case, the field of fractions of $\mathcal{O}_f$ will be denoted by K_f.

The next result will tell us that the ring $\mathcal{O}_f$ is an important invariant of the equivalence classes of algebroid plane curves. When two local K-algebras $\mathcal{O}_f$ and $\mathcal{O}_g$ are isomorphic we will write $\mathcal{O}_f \simeq \mathcal{O}_g$.

THEOREM 4.1 *Let (f) and (g) be two algebroid plane curves. We have that $(f) \sim (g)$ if and only if $\mathcal{O}_f \simeq \mathcal{O}_g$.*

Proof: We will denote by $\overline{h}$ and by $\overline{\overline{h}}$, respectively, the residual classes in $\mathcal{O}_f$ and in $\mathcal{O}_g$ of an element $h \in K[[X, Y]]$.

Suppose initially that $(f) \sim (g)$. Then there exist an automorphism Φ and a unit u of $K[[X, Y]]$ such that

$$\Phi(f) = ug.$$

It is clear that the surjection $\pi \circ \Phi$,

$$
\begin{array}{ccccc}
K[[X,Y]] & \xrightarrow{\Phi} & K[[X,Y]] & \xrightarrow{\pi} & \mathcal{O}_g, \\
h & \mapsto & \Phi(h) & \mapsto & \overline{\Phi(h)}
\end{array}
$$

where π is the canonical surjection, is such that $\mathrm{Ker}(\pi \circ \Phi) = \langle f \rangle$, and therefore, $\mathcal{O}_f \simeq \mathcal{O}_g$.

Conversely, suppose that $\mathcal{O}_f \simeq \mathcal{O}_g$.

If $\mathrm{mult}(f) = \mathrm{mult}(g) = 1$, we have from the Proposition (3.29) that $(f) \sim (g)$, and we are done in this case.

To conclude the proof, suppose that $\mathrm{mult}(g) \geq 2$ and that $\mathcal{O}_f \simeq \mathcal{O}_g$ through an isomorphism $\tilde{\Phi}$. Let $T_1, T_2 \in \mathcal{M}$ be such that

$$
\begin{cases}
\tilde{\Phi}(\overline{X}) = \overline{\overline{T_1}} \\
\tilde{\Phi}(\overline{Y}) = \overline{\overline{T_2}},
\end{cases}
$$

and define the homomorphism

$$
\begin{array}{cccc}
\Phi: & K[[X,Y]] & \longrightarrow & K[[X,Y]]. \\
& X & \mapsto & T_1 \\
& Y & \mapsto & T_2
\end{array}
$$

Since $\tilde{\Phi}$ is an isomorphism, there exist $R(X,Y), S(X,Y) \in K[[X,Y]]$ such that

$$
\overline{\overline{X}} = \tilde{\Phi}(R(\overline{X},\overline{Y})) = R(\overline{\overline{T_1}},\overline{\overline{T_2}}),
$$

and

$$
\overline{\overline{Y}} = \tilde{\Phi}(S(\overline{X},\overline{Y})) = S(\overline{\overline{T_1}},\overline{\overline{T_2}}).
$$

Now, since $\mathrm{mult}(g) > 1$, it follows that

$$
\begin{aligned}
X - R(T_1, T_2) &\in \langle g \rangle \subset \mathcal{M}^2, \\
Y - S(T_1, T_2) &\in \langle g \rangle \subset \mathcal{M}^2.
\end{aligned}
$$

This implies that

$$
\begin{aligned}
T_1 &= aX + bY + \cdots \\
T_2 &= cX + dY + \cdots
\end{aligned}
$$

for some $a, b, c, d \in K$ such that $ad - bc \neq 0$.

Hence, Φ is an automorphism of $K[[X,Y]]$ such that

$$
\overline{\overline{0}} = \tilde{\Phi}(\overline{f}) = \overline{\overline{f}}(T_1, T_2),
$$

which implies that $\Phi(f) \in \langle g \rangle$ and therefore

$$
\Phi(f) = hg. \tag{4.1}
$$

In particular, (4.1) gives

$$
\mathrm{mult}(f) = \mathrm{mult}(\Phi(f)) = \mathrm{mult}(h) + \mathrm{mult}(g) \geq 2.
$$

Now, observing that Φ^{-1} induces the transformation $\tilde{\Phi}^{-1}$ and inverting the roles of f and g, we get, in the same way as above, that

$$\Phi^{-1}(g) = h'f.$$

It follows from the above equality that $g = \Phi(h')\Phi(f)$, which with (4.1) imply that

$$\Phi(f) = h\Phi(h')\Phi(f).$$

This implies that h is a unit. This, together with (4.1), give $(f) \sim (g)$. $\qquad\square$

COROLLARY 4.2 *If $\mathcal{O}_f \simeq \mathcal{O}_g$, then* $\mathrm{mult}(f) = \mathrm{mult}(g)$.

When f has a well behaved equation with respect to the indeterminate Y, the K-algebra $\mathcal{O}_f$ has also a structure of a $K[[X]]$-module as we show next.

PROPOSITION 4.3 *Let $f \in K[[X, Y]]$ be regular in Y of some order n. Then $\mathcal{O}_f$ is a free $K[[X]]$-module of rank n generated by the residual classes y^i of Y^i, $i = 0, \dots, n-1$, in $\mathcal{O}_f$. In other words,*

$$\mathcal{O}_f = K[[X]] \oplus K[[X]]y \oplus \cdots \oplus K[[X]]y^{n-1}.$$

Proof: From the Division Theorem (Theorem (2.4), we may write

$$g = qf + r, \quad q \in K[[X, Y]], \quad r \in K[[X]][Y] \quad \text{and} \quad \deg_Y r < n$$

recall that we defined the degree of the zero polynomial to be $-\infty$.

Since $g(X, \alpha) = 0$, where $\alpha = \varphi(X^{\frac{1}{n}}) \in K[[X]]^*$, it follows that $r(X, \alpha) = 0$. This implies that r is divisible by the minimal polynomial of α which is of degree n, hence $r = 0$ and therefore $g = fq \in \langle f \rangle$. $\qquad\square$

From the above proposition, we deduce the following fundamental property of curves, represented by irreducible power series regular in Y of order n, with a given Puiseux parametrization $(T^n, \varphi(T))$. Namely, H_φ induces an injective homomorphism of K-algebras

$$H_\varphi : \mathcal{O}_f \longrightarrow K[[T]],$$

which allows us to identify $\mathcal{O}_f$ with the subalgebra of $K[[T]]$

$$A_\varphi := H_\varphi(\mathcal{O}_f) = K[[T^n, \varphi(T)]].$$

If $\psi(T) = \varphi(\zeta T)$, where ζ is an nth root of unity, then we have that $A_\psi \simeq A_\varphi$ through the automorphism

$$\begin{array}{rccc} h_\zeta : & K[[T]] & \longrightarrow & K[[T]]. \\ & P(T) & \mapsto & P(\zeta T) \end{array}$$

We also have from the Proposition (4.3) that

$$A_\varphi = K[[T^n]] \oplus K[[T^n]]\varphi \oplus K[[T^n]]\varphi^2 \oplus \cdots \oplus K[[T^n]]\varphi^{n-1},$$

and from Theorem (3.10.i) that the field of fractions of A_φ is $K((T))$.

An important remark to make here is that when f is regular in X, there are similar results to those we obtained in the case f regular in Y, just by interchanging the roles of X and Y.

THEOREM 4.4 *Let $f \in K[[X]][Y]$ be an irreducible pseudo-polynomial of degree n and let φ be an element of $K[[X^{\frac{1}{n}}]]$ such that $f(X, \varphi) = 0$. If $D_Y(f)(X)$ is the discriminant of $f(X, Y)$ in $K[[X]]$, then*

$$D_Y(f)(T^n)K[[T]] \subset A_\varphi.$$

Proof: Consider the diagram

$$\begin{array}{ccccc} K[[T]] & \subset & K((T)) & = & K((T^n))[\varphi] \\ \cup & & & & | \\ K[[T^n]] & & \subset & & K((T^n)) \end{array}$$

Recall that $K((T^n))[\varphi]/K((T^n))$ is a galoisian extension with Galois group U_n, the multiplicative group of the nth roots of unity (Lemma (3.6) and Theorem (3.10)). Let $\beta \in K[[T]]$. As an element of $K((T)) = K((T^n))[\varphi]$, we may write β as

$$\beta = \sum_{i=0}^{n-1} a_i(T^n)\varphi^i; \quad a_i(T^n) \in K((T^n)). \tag{4.2}$$

Let us denote by ζ a generator of the cyclic group U_n and use notation as in Section 3.3. Put

$$\beta_j = \zeta^j * \beta \quad \text{and} \quad \varphi_j = \zeta^j * \varphi.$$

If we apply the ζ^j to equality (4.2) we get the system

$$\beta_j = \sum_{i=0}^{n-1} a_i(T^n)\varphi_j^i; \quad j = 0, \ldots, n-1, \tag{4.3}$$

whose determinant is the Vandermonde determinant:

$$\det \begin{pmatrix} \varphi_0^0 & \cdots & \varphi_0^{n-1} \\ \varphi_1^0 & \cdots & \varphi_1^{n-1} \\ \vdots & \ddots & \vdots \\ \varphi_{n-1}^0 & \cdots & \varphi_{n-1}^{n-1} \end{pmatrix} = \prod_{r<s}(\varphi_r - \varphi_s) = \Delta.$$

From the Proposition (2.24), we have that $\Delta^2 = \pm D_Y(f)(T^n)$, and therefore $\Delta^2 \in K[[T^n]]$.

By Cramer's rule, the system (4.3) gives

$$a_i(T^n) = \Delta^{-1}M_i,$$

where

$$M_i = \det \begin{pmatrix} \varphi_0^0 & \cdots & \beta_0 & \cdots & \varphi_0^{n-1} \\ \varphi_1^0 & \cdots & \beta_1 & \cdots & \varphi_1^{n-1} \\ \vdots & \ddots & \vdots & \ddots & \vdots \\ \varphi_{n-1}^0 & \cdots & \beta_{n-1} & \cdots & \varphi_{n-1}^{n-1} \end{pmatrix} \in K[[T]],$$

is the determinant of the system (4.3), where we have replaced the ith column by the first members of (4.3).

From this it follows that

$$\Delta^2 a_i(T^n) = \Delta M_i \in K[[T]],$$

and since

$$\zeta^j * \Delta^2 a_i(T^n) = \Delta^2 a_i(T^n), \quad \forall j = 0, \ldots, n-1,$$

it follows that $\Delta^2 a_i(T^n) \in K[[T^n]]$. Since $\beta = \sum_{i=0}^{n-1} a_i(T^n)\varphi^i$, we have that

$$D_Y(f)(T^n)\beta = \pm \sum_{i=0}^{n-1} \Delta^2 a_i(T^n)\varphi^i \in \sum_{i=0}^{n-1} K[[T^n]]\varphi^i = A_\varphi.$$

This implies that

$$K[[T]] \subset \frac{1}{D_Y(f)(T^n)} A_\varphi,$$

and therefore,

$$D_Y(f)(T^n)K[[T]] \subset A_\varphi.$$

$\square$

COROLLARY 4.5 *With the same notation as in Theorem (4.2), we have that*

$$\dim_K \frac{K[[T]]}{A_\varphi} < \infty$$

Proof: The theorem gives the following inequality

$$\dim_K \frac{K[[T]]}{A_\varphi} < \dim_K \frac{K[[T]]}{\langle D_Y(f)(T^n)\rangle} = \text{mult}(D_Y(f)(T^n)) < \infty.$$

$\square$

The above corollary says that $\mathcal{O}_f$ realized as a subring A_φ of $K[[T]]$, by means of a Puiseux parametrization $(T^n, \varphi(T))$ of the irreducible curve (f), has a conductor; that is, there exists a natural number c such that for all $h(T) \in K[[T]]$ with $\text{mult}(h(T)) \geq c$, then $h(T) \in A_\varphi$. The least integer with the above property will be called the *conductor* of (f). We will meet this concept again in Chapter 6, where it will play a fundamental role.

We will define the *valuation associated to* f as being the function

$$\begin{array}{rccc} v_f : & \mathcal{O}_f \setminus \{0\} & \longrightarrow & \mathbb{N}, \\ & \overline{g} & \mapsto & \text{mult}(H_\varphi(\overline{g})) \end{array}$$

and put $v_f(0) = \infty$.

It is clear that v_f may be computed by means of any primitive parametrization $(\psi_1(T), \psi_2(T))$ of (f), since

$$v_f(g) = \text{mult}(g(T^n, \varphi(T))) = \text{mult}(\rho(g(T^n, \varphi(T)))) = \text{mult}(g(\psi_1(T), \psi_2(T))),$$

where ρ is the automorphism of $K[[T]]$ such that

$$(\rho(\psi_1(T)), \rho(\psi_2(T))) = (T^n, \varphi(T)).$$

We will list below some properties of the function v_f, which are easy to verify.

For all $\overline{g}, \overline{h} \in \mathcal{O}_f$, one has

i) $v_f(\overline{g}\overline{h}) = v_f(\overline{g}) + v_f(\overline{h})$,

ii) $v_f(\overline{1}) = 0$,

iii) $v_f(\overline{g} + \overline{h}) \geq \min\{v_f(\overline{g}), v_f(\overline{h})\}$, with equality if $v_f(\overline{g}) \neq v_f(\overline{h})$.

PROBLEMS

1.1) Let (f) be a regular curve. Show that $\mathcal{O}_f \simeq K[[X]]$.

1.2) Show that $(Y^2 - X^5)$ is not equivalent to $(Y^3 - X^5)$.

1.3) Given $f = Y^3 - X^5$, find a Puiseux parametrization $(T^3, \varphi(T))$ for (f). If $g = Y^3 X^3 + Y^5 X^2 - X^8$, compute $v_f(\overline{g})$

4.2   COMPLEMENTS OF LINEAR ALGEBRA

In this section we will present some results about quotients and exact sequences of vector spaces which are not found in ordinary texts of Linear Algebra.

Let U, V, W be vector spaces over a field K. Consider a diagram of the form

$$W \xrightarrow{\psi} V \xrightarrow{\varphi} U, \qquad (4.4)$$

where ψ and φ are homomorphisms of vector spaces. We will say that this diagram forms a sequence. A sequence may be formed with more than two homomorphisms.

DEFINITION 4.6 *We will say that (4.4) is an exact sequence if*

$$\mathrm{Im}(\psi) = \ker(\varphi).$$

An exact sequence of the form

$$0 \longrightarrow V \xrightarrow{\psi} U,$$

means that ψ is injective, while a sequence of the type

$$W \xrightarrow{\varphi} V \longrightarrow 0,$$

means that φ is surjective.

Suppose that W is a subspace of the vector space V. Then

$$\frac{V}{W} = \{\overline{v} = v + W;\ v \in V\},$$

is a vector space. Moreover, if $\dim_K V < \infty$, then

$$\dim_K \frac{V}{W} = \dim_K V - \dim_K W. \qquad (4.5)$$

The above equality follows from the fact that, if $v_1, \ldots v_m, v_{m+1}, \ldots, v_n$, is a basis for V, where $v_1, \ldots v_m$, form a basis for W, then $\overline{v}_{m+1}, \ldots, \overline{v}_n$ is a basis for $\frac{V}{W}$.

PROPOSITION 4.7 *Suppose that we have an exact sequence of vector spaces*

$$0 \longrightarrow W \xrightarrow{\psi} V \xrightarrow{\varphi} U \longrightarrow 0.$$

Then V is finite dimensional if and only if U and W are finite dimensional. In this case,

$$\dim_K V = \dim_K W + \dim_K U.$$

Proof: Since the sequence is exact we have that

$$\dim_K W = \dim_K \psi(W) = \dim_K \ker(\varphi),$$

and from the isomorphism theorem, it follows that

$$\frac{V}{\ker(\varphi)} \simeq U,$$

which in view of (4.5) implies that

$$\dim_K V = \dim_K \ker(\varphi) + \dim_K U = \dim_K W + \dim_K U.$$

$\square$

PROPOSITION 4.8 *Let $W \subset V \subset U$ be a chain of vector spaces such that $\dim_K \frac{U}{W} < \infty$. Then,*

$$\dim_K \frac{U}{W} = \dim_K \frac{V}{W} + \dim_K \frac{U}{V}.$$

Proof: This follows from the Proposition (4.8) and from the exact sequence:

$$0 \longrightarrow \frac{V}{W} \longrightarrow \frac{U}{W} \longrightarrow \frac{U}{V} \longrightarrow 0.$$

$\square$

PROPOSITION 4.9 (Noether Isomorphism Theorem) *Let V and W be subspaces of a vector space U. Then*

$$\frac{V}{V \cap W} \simeq \frac{V + W}{W}.$$

Proof: The map:

$$V + W \quad \longrightarrow \quad \frac{V}{V \cap W}$$
$$v + w \quad \longmapsto \quad \overline{v}$$

is well defined, surjective and its kernel is W.

$\square$

THEOREM 4.10 *Let V be a vector space and let $L : V \longrightarrow V$ be a linear injective map. Let W be a vector subspace of V such that $L(W) \subset W$. We have that*
(i)

$$\frac{V}{W} \simeq \frac{L(V)}{L(W)}.$$

(ii)   *If $\frac{V}{W}$ and $\frac{V}{L(V)}$ are finite dimensional, then*

$$\dim_K \frac{W}{L(W)} = \dim_K \frac{V}{L(V)}.$$

Proof: (i) Consider the sequence,

$$0 \longrightarrow W \xrightarrow{\ j\ } V \xrightarrow{\ \overline{L}\ } \frac{L(V)}{L(W)} \longrightarrow 0,$$

where j is the inclusion homomorphism, and $\overline{L}$ is the homomorphism induced by L. Since L is injective, we have that

$$\ker(\overline{L}) = \{v \in V;\ \ L(v) \in L(W)\} = W,$$

which implies that the above sequence is exact.
(ii) Consider the following chains of inclusions:

$$W \subset W + L(V) \subset V,$$

and

$$L(W) \subset W \cap L(V) \subset L(V).$$

Since from (i) and from the hypothesis of (ii), we have that

$$\dim_K \frac{L(V)}{L(W)} = \dim_K \frac{V}{W} < \infty,$$

we get, from the Proposition (4.9), the following equalities:

$$\begin{array}{ccccc}
\dim_K \frac{V}{W} & = & \dim_K \frac{W+L(V)}{W} & + & \dim_K \frac{V}{W+L(V)} \\
\| & & \| & & \\
\dim_K \frac{L(V)}{L(W)} & = & \dim_K \frac{L(V)}{W\cap L(V)} & + & \dim_K \frac{W\cap L(V)}{L(W)},
\end{array} \qquad (4.6)$$

where the equality

$$\dim_K \frac{W + L(V)}{W} = \dim_K \frac{L(V)}{W \cap L(V)},$$

follows from the Proposition (4.10).
Consequently, we get

$$\dim_K \frac{V}{W + L(V)} = \dim_K \frac{W \cap L(V)}{L(W)} < \infty. \qquad (4.7)$$

Since from hypothesis, $\dim_K \frac{V}{L(V)} < \infty$, from the chain

$$L(V) \subset W + L(V) \subset V,$$

we get

$$\dim_K \frac{V}{L(V)} = \dim_K \frac{V}{W + L(V)} + \dim_K \frac{W + L(V)}{L(V)}. \qquad (4.8)$$

This implies, in presence of the Proposition (4.10) and of (4.7), that

$$\dim_K \frac{W}{W \cap L(V)} = \dim_K \frac{W + L(V)}{L(V)} < \infty. \qquad (4.9)$$

From (4.7), (4.9), and from the chain below,

$$L(W) \subset W \cap L(V) \subset W,$$

we get that

$$\dim_K \frac{W}{L(W)} = \dim_K \frac{W}{W \cap L(V)} + \dim_K \frac{W \cap L(V)}{L(W)} < \infty.$$

The result then follows from this last equality together with (4.7), (4.9) and (4.8).

$\square$

4.3 INTERSECTION INDICES

In this section we will introduce a method to express numerically the order or the degree of contact of two algebroid plane curves. This will be done through the fundamental notion of *intersection index*.

Let K be an arbitrary field. We will denote by $\mathcal{M}$, as usual, the maximal ideal of $K[[X,Y]]$.

PROPOSITION 4.11 *Let $f, g \in \mathcal{M}$. The following conditions are equivalent.*

i) *f and g are relatively prime.*

ii) *The dimension of $\dfrac{K[[X,Y]]}{\langle f, g \rangle}$, as a K-vector space, is finite.*

Proof: The above conditions are preserved by automorphisms of $K[[X,Y]]$ and by multiplication of f and g by units in $K[[X,Y]]$. Therefore, we may assume that f and g are pseudo-polynomials.

(i) $\Rightarrow$ (ii)   Since f and g are relatively prime in $K[[X,Y]]$, they are relatively prime in $K[[X]][Y]$, and therefore $R_Y(f,g) \neq 0$. So $R_Y(f,g) = X^r u$, where $r \geq 0$ and $u(0) \neq 0$. Since $R_Y(f,g) \in \langle f, g \rangle$ (Proposition (2.20)), it follows that $R_Y(f,g)(0) = 0$, hence $r \geq 1$ and we have $X^r \in \langle f, g \rangle$

Suppose that $\deg_Y(f) = n$, then from the Division Theorem, for all $h \in K[[X,Y]]$, there exist elements $q \in K[[X,Y]]$ and $a_0(X), \ldots, a_{n-1}(X) \in K[[X]]$, such that

$$h = fq + a_0(X) + \cdots + a_{n-1}(X)Y^{n-1}.$$

From this it follows that the image $\overline{h}$ of h in $K[[X,Y]]/\langle f, g \rangle$ is in the K-vector space generated by $\overline{X}^i \overline{Y}^j$, $0 \leq i \leq r - 1$ and $0 \leq j \leq n - 1$. Hence (ii) is true.

(ii) $\Rightarrow$ (i)   Suppose that f and g are not relatively prime in $K[[X,Y]]$. There exists $h \in K[[X]][Y]$ non-invertible and pseudo-polynomial such that

$$f = hf_1 \quad \text{e} \quad g = hg_2.$$

Hence $\langle f, g \rangle \subset \langle h \rangle$. We have that $\overline{1}, \overline{X}, \overline{X}^2, \ldots$ are elements linearly independent over K in $K[[X,Y]]/\langle h \rangle$, because otherwise, there would exist a polynomial $P(X) \in$

$K[X]$ such that $P(X) = h_1 h$, $h_1 \in K[[X, Y]]$, which is not possible because h is a pseudo-polynomial.

The result now follows because

$$\dim_K \frac{K[[X, Y]]}{\langle f, g \rangle} \geq \dim_K \frac{K[[X, Y]]}{\langle h \rangle} = \infty.$$

$\square$

DEFINITION 4.12 *Let $f, g \in \mathcal{M}$. The* intersection index *of f and g is the integer (including ∞)*

$$\mathrm{I}(f, g) = \dim_K \frac{K[[X, Y]]}{\langle f, g \rangle}.$$

In what follows we will use the notation z' to represent the image of $z \in K[[X, Y]]$ in $\mathcal{O}_f$.

So, it is immediate to verify that

$$\mathrm{I}(f, g) = \dim_K \frac{\mathcal{O}_f}{\langle g' \rangle}.$$

DEFINITION 4.13 *We will say that two algebroid curves (f) and (g) are* transversal, *if (f) and (g) are regular and their tangent lines are distinct.*

THEOREM 4.14 *Let $f, g, h \in \mathcal{M}$, Φ an automorphism of $K[[X, Y]]$ and u and v units in $K[[X, Y]]$. The intersection index has the following properties:*

i) $\mathrm{I}(f, g) < \infty$ *if and only if f and g are relatively prime in $K[[X, Y]]$.*

ii) $\mathrm{I}(f, g) = \mathrm{I}(g, f)$.

iii) $\mathrm{I}(\Phi(f), \Phi(g)) = \mathrm{I}(uf, vg) = \mathrm{I}(f, g)$.

iv) $\mathrm{I}(f, gh) = \mathrm{I}(f, g) + \mathrm{I}(f, h)$.

v) $\mathrm{I}(f, g) = 1$ *if and only if (f) and (g) are transversal.*

vi) $\mathrm{I}(f, g - hf) = \mathrm{I}(f, g)$.

Proof: The first assertion follows immediately from Proposition (4.12). The assertions (ii), (iii) and (vi) follow immediately from the definition of intersection index.

We are now going to prove (iv). It is sufficient to prove that the following sequence of K-vector spaces is exact,

$$0 \longrightarrow \frac{\mathcal{O}_f}{\langle h' \rangle} \xrightarrow{\psi} \frac{\mathcal{O}_f}{\langle g' h' \rangle} \xrightarrow{\varphi} \frac{\mathcal{O}_f}{\langle g' \rangle} \longrightarrow 0,$$

where the map φ is induced by the projection $\tilde{\varphi}$:

$$\mathcal{O}_f \xrightarrow{\tilde{\varphi}} \frac{\mathcal{O}_f}{\langle g' \rangle} \longrightarrow 0.$$

Notice that

$$\ker(\varphi) = \frac{\langle g' \rangle}{\langle g' h' \rangle}.$$

The map ψ is defined by

$$\psi(\overline{z'}) = \overline{\overline{g'z'}},$$

where the bar represents the residual class modulo $\langle h' \rangle$, whereas the double bar represents the residual class modulo $\langle h'g' \rangle$.

The map ψ is clearly a homomorphism of K-vector spaces. To prove that ψ is injective, suppose that $\psi(\overline{z'}) = 0$ and notice that

$$0 = \psi(\overline{z'}) = \overline{\overline{g'z'}} \implies g'z' \in \langle g'h' \rangle.$$

Hence there exists $\alpha' \in \mathcal{O}_f$ such that

$$g'z' = \alpha'g'h',$$

which in turn implies that there exists $\beta \in K[[X,Y]]$ such that

$$g(z - \alpha h) = gz - \alpha gh = \beta f.$$

Since f and g have no common factors, it follows that f divides $z - \alpha h$, which shows that $\overline{z'} = 0$.

To conclude the proof of (iv), observe that

$$\mathrm{Im}(\psi) = \frac{\langle g' \rangle}{\langle g'h' \rangle} = \ker(\varphi).$$

Finally, we will prove property (v). Suppose that f and g are regular with distinct tangents. Then after a linear change of coordinates we may assume that $f = X + f_1$ and $g = Y + g_1$, where $f_1, g_1 \in \mathcal{M}^2$.

We may then write

$$f = Xu + Yf_2, \quad e \quad g = Yv + Xg_2,$$

where u and v are units and $f_2, g_2 \in \mathcal{M}$.

We then have

$$Y(v - u^{-1}g_2f_2) = Yv - u^{-1}Yg_2f_2 = g - u^{-1}g_2f \in \langle f, g \rangle.$$

Since $(v - u^{-1}g_2f_2)(0,0) = v(0,0) \neq 0$, it follows that $v - u^{-1}g_2f_2$ is a unit and therefore $Y \in \langle f, g \rangle$. In a similar way one shows that $X \in \langle f, g \rangle$. It then follows that $\langle f, g \rangle = \langle X, Y \rangle$, and therefore,

$$\mathrm{I}(f,g) = \dim_K \frac{K[[X,Y]]}{\langle f, g \rangle} = \dim_K \frac{K[[X,Y]]}{\langle X, Y \rangle} = 1.$$

Conversely, if $\mathrm{mult}(f) \geq 2$ or $\mathrm{mult}(g) \geq 2$, or also if (f) and (g) are regular and have the same tangent line we may, after a linear change of coordinates, suppose that $f = Yf_1 + f_2$ and $g = Yg_1 + g_2$, with $f_2, g_2 \in \mathcal{M}^2$ and $f_1, g_1 \in K[[X,Y]]$.

Hence $\langle f, g \rangle \subset \langle Y \rangle + \mathcal{M}^2$ and therefore,

$$\dim_K \frac{K[[X,Y]]}{\langle f, g \rangle} \geq \dim_K \frac{K[[X,Y]]}{\langle Y \rangle + \mathcal{M}^2} = \dim_K \frac{K[[X]]}{\langle X^2 \rangle} = 2.$$

$\square$

The properties listed in the Theorem (4.15) determine uniquely the intersection index of algebroid plane curves. This is the content of the next result.

THEOREM 4.15 *Let* I′ *be a map*

$$
\begin{array}{rcl}
\text{I}': & \mathcal{M} \times \mathcal{M} & \longrightarrow & \mathbb{N} \cup \{\infty\} \\
& (f, g) & \longmapsto & \text{I}'(f, g)
\end{array}
$$

having properties (i) *to* (vi), *of the Theorem* (4.15). *Then* I′ = I.

Proof: The proof will be done by induction on the value of $\text{I}'(f, g)$.

We may assume that $f, g \in \mathcal{M} \setminus \{0\}$ are relatively prime in $K[[X, Y]]$, because otherwise we would have, from (i), that $\text{I}'(f, g) = \text{I}(f, g) = \infty$.

Moreover, if $\text{I}'(f, g) = 1$ then from (v) it follows that $\text{I}(f, g) = 1$.

Suppose by induction hypothesis that:

$$
\text{I}'(f, g) \leq r - 1 \Longrightarrow \text{I}'(f, g) = \text{I}(f, g).
$$

Let $f, g \in \mathcal{M} \setminus \{0\}$ such that $\text{I}'(f, g) = r$. Let Φ be an automorphism of $K[[X, Y]]$ such that $\Phi(f)$ and $\Phi(g)$ are associated respectively to pseudo-polynomials p and q. From property (iii) we have that

$$
\text{I}'(f, g) = \text{I}'(p, q); \quad \text{I}(f, g) = \text{I}(p, q).
$$

Suppose that $\deg_Y(p) = n$ and $\deg_Y(q) = m$. From property (ii) we may assume that $n \leq m$. Define

$$
q_1 = q - Y^{m-n} p = X q_2,
$$

where $q_2 \in K[[X, Y]]$.

If q_2 is a unit, then from properties (iii),(iv) and (vi), we have that

$$
\text{I}'(p, q) = \text{I}'(p, q_1) = \text{I}'(p, X q_2) = \text{I}'(p, X) = \text{I}'(Y^n, X) = n.
$$

In the same way, one shows that $\text{I}(p, q) = n$, and therefore, $\text{I}'(p, q) = \text{I}(p, q)$.

Suppose that q_2 is not a unit, then from properties (vi) and (iv), and by the inductive assumption we have that

$$
r = \text{I}'(p, q) = \text{I}'(p, q_1) = \text{I}'(p, X) + \text{I}'(p, q_2) = \text{I}(p, X) + \text{I}(p, q_2) = \text{I}(p, q),
$$

since $\text{I}'(p, X) \leq r - 1$ and $\text{I}'(p, q_2) \leq r - 1$.

This implies that

$$
\text{I}'(f, g) = \text{I}(f, g).
$$

$\square$

The above theorem is an executable algorithm to compute $\text{I}(f, g)$ for all pair of power series f and g in $\mathcal{M}$ which are pseudo-polynomials.

EXAMPLE 4.16 We will find the intersection index of $Y^7 - X^2$ and $Y^5 - X^3$.

$$
\text{I}(Y^7 - X^2, Y^5 - X^3) = \text{I}(Y^7 - X^2 - Y^2(Y^5 - X^3), Y^5 - X^3) =
$$

$$
\text{I}(X^2(XY^2 - 1), Y^5 - X^3) = \text{I}(X^2, Y^5 - X^3) = 2 \cdot \text{I}(X, Y^5 - X^3) = 2 \cdot \text{I}(X, Y^5) =
$$

$$
10 \cdot \text{I}(X, Y) = 10.
$$

THEOREM 4.17 *Let f and g be pseudo-polynomials in $K[[X]][Y]$. Let*

$$f = f_1 \cdots f_r,$$

be the decomposition of f into irreducible factors, which we may suppose that are pseudo-polynomials. Then

$$I(f,g) = \sum_{i=1}^{r} v_{f_i}(g) = \text{mult}(R_Y(f,g)).$$

Proof: If f and g have common components, we have nothing to prove since the above three terms are infinite. Suppose now that f and g have no common factors.

Since from property (iv) of Theorem (4.15), and from Problem 2.4.2, we have that

$$I(f,g) = \sum_{i=1}^{r} I(f_i,g),$$

and

$$R_Y(f,g) = \prod_{i=1}^{r} R_Y(f_i,g),$$

It will be sufficient to prove that if f is irreducible, then

$$I(f,g) = v_f(g) = \text{mult}(R_Y(f,g)).$$

Consider the K-linear map

$$L: \quad \begin{array}{ccc} K[[T]] & \longrightarrow & K[[T]] \\ h(T) & \mapsto & h(T)g(T^n, \varphi(T)) \end{array}$$

where $n = \deg_Y(f)$ and $(T^n, \varphi(T))$ is a Puiseux parametrization of f. Since f is not a component of g, it follows from Proposition (4.4) that $g(T^n, \varphi(T)) \neq 0$, and therefore L is injective.

Let W be the K-vector subspace $A_\varphi = K[[T^n, \varphi(T)]]$ ($\simeq \mathcal{O}_f$) of $V = K[[T]]$.

We have that

$$\frac{\mathcal{O}_f}{\langle g' \rangle} \simeq \frac{A_\varphi}{g(T^n, \varphi(T))R_\varphi} = \frac{W}{L(W)}.$$

On the other hand,

$$\frac{V}{L(V)} = \frac{K[[T]]}{\langle g(T^n, \varphi(T)) \rangle}.$$

From the Corollary (4.6) we have that

$$\dim_K \frac{V}{W} < \infty,$$

and since

$$\dim_K \frac{V}{L(V)} = \text{mult}(g(T^n, \varphi(T))) < \infty,$$

it follows from the Theorem (4.11) that

$$I(f,g) = \dim_K \frac{\mathcal{O}_f}{\langle g' \rangle} = \dim_K \frac{W}{L(W)} = \dim_K \frac{V}{L(V)} = v_f(g).$$

On the other hand, if ζ is a primitive nth root of unity, from the Proposition (2.22),

$$R_Y(f,g) = \prod_{i=1}^{n} g(X, \varphi(\zeta^i X^{\frac{1}{n}})).$$

Hence

$$\mathrm{mult}(R_Y(f,g)) = n.\mathrm{mult}(g(X, \varphi(X^{\frac{1}{n}}))) = \mathrm{mult}(g(T^n, \varphi(T))).$$

$\square$

THEOREM 4.18 *Let $f, g \in \mathcal{M}$. We have that*

$$\mathrm{I}(f,g) \geq \mathrm{mult}(f) \cdot \mathrm{mult}(g),$$

with equality if and only if (f) and (g) have no common tangents.

Proof: Let $f = f_1 \cdots f_r$ and $g = g_1 \cdots g_s$, be respectively the decompositions of f and g into irreducible factors.

From the Theorem (4.15) (ii) and (iv), we have that

$$\mathrm{I}(f,g) = \sum_{i,j} \mathrm{I}(f_i, g_j). \tag{4.10}$$

On the other hand, since $\mathrm{mult}(h_1 h_2) = \mathrm{mult}(h_1) + \mathrm{mult}(h_2)$, we have that

$$\mathrm{mult}(f) \cdot \mathrm{mult}(g) = \sum_{i,j} \mathrm{mult}(f_i) \cdot \mathrm{mult}(g_j). \tag{4.11}$$

Therefore, from (4.10) and (4.11), it follows that it is sufficient to prove the theorem for f and g irreducible.

Suppose that coordinates have been chosen in such a way that f and g are associated to pseudo-polynomials in $K[[X]][Y]$ and that (f) has Y as its tangent line.

Let $(T^n, \varphi(T))$, where $n = \mathrm{mult}(f)$, be a Puiseux parametrization of (f). Since Y is the tangent line of (f), recall from (3.6) that $\mathrm{mult}(\varphi(T)) > n$.

Suppose now that

$$g(X, Y) = (aX + bY)^m + g_{m+1}(X, Y) + \cdots,$$

where $g_{m+1}(X, Y) + \cdots \in \mathcal{M}^{m+1}$. Then we have that

$$\mathrm{I}(f,g) = \mathrm{mult}((aT^n + b\varphi(T))^m + g_{m+1}(T^n, \varphi(T)) + \cdots) \geq nm,$$

with equality if and only if $a \neq 0$; that is, the tangent line of (g) is not Y, or equivalently, if the tangent lines of (f) and g are distinct. $\square$

The intersection index of two curves is easily and effectively computable, either by means of the Theorem (4.15), or by means of the Theorem (4.18), when either both curves are given in cartesian form or one of them in cartesian form and the other in parametric form. In Chapter 8 we will show how to compute the intersection index when both curves are given in parametric form.

PROBLEMS

3.1)   Compute the intersection indices of the following pairs of curves:
a) $(Y^2 - X^3)$ and $(Y^2 - X^5)$.
b) $(Y^2 - X^3)$ and $(X^2 - Y^3)$.
c) $((X^2 + Y^2)^2 - (X^2 - Y^2))$ and $((X^2 + Y^2)^2 + (X^2 - Y^2))$.
d)   Compute the intersection indices of the curves of Section 3.3, taken by pairs.

3.2)   Let $f = Y^4 - 2X^3Y^2 - 4X^5Y - X^6 - X^7$. Denote by $f_Y^{(i)}$ the partial derivative of order i of f with respect to the indeterminate Y. Compute $\mathrm{I}(f, f_Y^{(1)})$, $\mathrm{I}(f, f_Y^{(2)})$ and $\mathrm{I}(f, f_Y^{(3)})$.

5 Resolution of singularities

In this chapter we will be concerned with the process of resolution of singularities of irreducible plane algebroid curves. In the analytic context, this process consists in transforming a germ of an analytic branch into a non-singular germ, by means of a finite sequence of certain birational transformations. This technique was introduced by Max Noether, the father of Algebraic Geometry, in the nineteenth century.

In order to motivate our study, we will introduce the subject in the more geometric context of germs of analytic plane curves, and then develop it in the context of formal geometry.

5.1 QUADRATIC TRANSFORMS IN $\mathbb{C}^2$

DEFINITION 5.1 *A* quadratic transformation *or a* blowing-up *centered at the origin of $\mathbb{C}^2$ is a map which in some coordinate system of $\mathbb{C}^2$ is of the form:*

$$T: \quad \mathbb{C}^2 \quad \longrightarrow \quad \mathbb{C}^2$$
$$(X_1, Y_1) \quad \mapsto \quad (X, Y) = (X_1, X_1 Y_1)$$

The line $E : X_1 = 0$ is called the *exceptional divisor* of the blowing-up and it is equal to $T^{-1}(0,0)$.

Let $L : Y - aX = 0$ be a line through the origin of $\mathbb{C}^2$, then $T^{-1}(L)$ is the union of the exceptional divisor E and the line $Y_1 = a$, which cuts E in the point of coordinates $(0, a)$. Hence, the points of the exceptional divisor correspond to the directions through the origin of $\mathbb{C}^2$, excepting the direction of the Y axis.

The transformation T induces an analytic automorphism of $\mathbb{C}^2 \setminus E$.

Let C_f be a germ of an analytic curve defined by an element $f \in \mathcal{M} \subset \mathbb{C}\{X, Y\}$. Let us see what is the set $T^{-1}(C_f)$. Suppose that the expansion of f in homogeneous polynomials is given by

$$f(X, Y) = F_n(X, Y) + F_{n+1}(X, Y) + \cdots$$

So, the equation (5.1) of $T^{-1}(C_f)$ is given by the series

$$\begin{aligned} f(T(X_1, Y_1)) &= F_n(X_1, X_1 Y_1) + F_{n+1}(X_1, X_1 Y_1) + \cdots \\ &= X_1^n (F_n(1, Y_1) + X_1 F_{n+1}(1, Y_1) + \cdots), \end{aligned}$$

which will be called the *total transform* of f, whose associated curve $T^{-1}(C_f)$ will be called the total transform of C_f.

The curve $C_{f^{(1)}}$, determined by the equation $f^{(1)}(X_1, Y_1) = 0$, where

$$f^{(1)}(X_1, Y_1) = F_n(1, Y_1) + X_1 F_{n+1}(1, Y_1) + \cdots + X_1^{j-n} F_j(1, Y_1) + \cdots,$$

is called the *strict transform* of C_f. Since $T^{-1}(0,0)$ is the exceptional divisor, with equation $E : X_1 = 0$, we have that

$$T^{-1}(C_f) = E \cup C_{f^{(1)}}.$$

EXAMPLE 5.2 Consider the curve C_f where $f = Y^2 - a^2 X^2 - X^3 = 0$, with $a \neq 0$. Since the curve is defined by a polynomial, we will not have convergence problems and therefore it is globally defined in all $\mathbb{C}^2$.

The total transform of C_f is

$$T^{-1}(C_f) : \ X_1^2 Y_1^2 - a^2 X_1^2 - X_1^3 = X_1^2(Y_1^2 - a^2 - X_1) = 0,$$

whereas the strict transform of C_f is given by $C_{f^{(1)}} : Y_1^2 - a^2 - X_1 = 0$. Notice that the tangent lines of C_f at the origin, given by

$$Y^2 - a^2 X^2 = (Y - aX)(Y + aX) = 0,$$

have as strict transforms two horizontal lines which pass through the points $P_1 = (0, a)$ and $P_2 = (0, -a)$, which are precisely the points of intersection of $C_{f^{(1)}}$ with E (see the figure below).

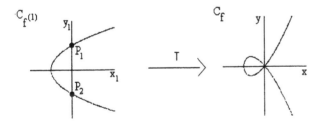

Figure 1

Notice that after one blowing-up, the curve C_f is transformed into a smooth curve $C_{f^{(1)}}$ and its two branches through the origin are transformed into portions of the curve passing through the points P_1 and P_2.

EXAMPLE 5.3 Consider the curve C_f, where $f = Y^2 - X^3 = 0$. Here again we have a curve defined globally in $\mathbb{C}^2$.

The total transform of C_f is

$$T^{-1}(C_f) : \ X_1^2 Y_1^2 - X_1^3 = X_1^2(Y_1^2 - X_1) = 0,$$

whereas the strict transform of C_f is given by $C_{f^{(1)}} : Y_1^2 - X_1 = 0$, which has the vertical line $X_1 = 0$ as a tangent line in the origin.

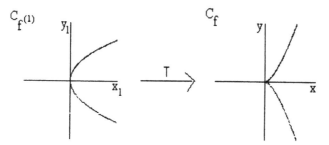

Figure 2

Observe that also here after one blowing-up we succeeded to transform C_f into a smooth curve $C_{f^{(1)}}$, and since C_f has only one branch through the origin, it is transformed into the unique branch of $C_{f^{(1)}}$ at the origin.

EXAMPLE 5.4 Consider the curve C_f where $f = Y^2 - X^5$. Here again we have a curve defined globally in $\mathbb{C}^2$.

The total transform of C_f is

$$T^{-1}(C_f): \quad X_1^2 Y_1^2 - X_1^5 = X_1^2(Y_1^2 - X_1^3) = 0,$$

whereas the strict transform of C_f is given by $C_{f^{(1)}} : Y_1^2 - X_1^3 = 0$, which has as a tangent line the line $Y_1 = 0$. Notice that the curve $C_{f^{(1)}}$ is the curve C_f of the previous example, and therefore with one more blowing-up we transform it into the parabola $Y_2^2 - X_2 = 0$.

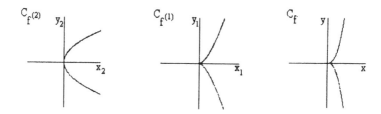

Figure 3

Some of the facts pointed out in the above examples always happen. For example, we will show that after a finite number of blowing-ups, it is possible to transform any algebroid irreducible plane curve into an algebroid regular curve. This procedure will be called the resolution of the singularities of the curve, which will allow to study a singularity transforming it into several simpler singularities.

5.2 RESOLUTION OF SINGULARITIES OF PLANE CURVES

From now on, K will be an arbitrary algebraically closed field.

DEFINITION 5.5 *The* quadratic transformation σ *from the ring* $K[[X,Y]]$ *into the ring* $K[[X_1,Y_1]]$ *is the homomorphism of K-algebras defined by*

$$\begin{array}{cccc}
\sigma: & K[[X,Y]] & \longrightarrow & K[[X_1,Y_1]] \\
& X & \mapsto & X_1 \\
& Y & \mapsto & X_1 Y_1
\end{array}$$

or by

$$\begin{array}{cccc}
\tau: & K[[X,Y]] & \longrightarrow & K[[X_1,Y_1]] \\
& X & \mapsto & X_1 Y_1 \\
& Y & \mapsto & Y_1
\end{array}$$

These transformations are not invertible, but they are birational; that is, they define isomorphisms between $K((X,Y))$ and $K((X_1,Y_1))$, respectively the fields of fractions of $K[[X,Y]]$ and $K[[X_1,Y_1]]$.

The homomorphism σ transforms the ideal $\langle X,Y \rangle$ into the ideal

$$\langle X_1, X_1 Y_1 \rangle = \langle X_1 \rangle.$$

When we are working with the ring $\mathbb{C}\{X,Y\}$, this geometrically corresponds to the condition $T^{-1}(0,0) = E$ of the previous section.

The transform by σ of

$$f(X,Y) = F_n(X,Y) + F_{n+1}(X,Y) + \cdots \in K[[X,Y]],$$

is the element of $K[[X_1,Y_1]]$ defined by

$$\begin{aligned}
\sigma(f) = f(X_1, X_1 Y_1) &= F_n(X_1, X_1 Y_1) + F_{n+1}(X_1, X_1 Y_1) + \cdots \\
&= X_1^n \left(F_n(1,Y_1) + X_1 F_{n+1}(1,Y_1) + \cdots \right).
\end{aligned}$$

The series

$$\sigma^*(f) = \frac{1}{X_1^n}\sigma(f) = \frac{1}{X_1^n}f(X_1, X_1 Y_1),$$

where $n = \text{mult}(f)$, will be called the *strict transform* by σ of (f), and will be denoted also by $f^{(1)}$.

PROPOSITION 5.6 *Let* $f,g \in K[[X,Y]]$.

  i) $\sigma^*(f)$ *is invertible in* $K[[X_1,Y_1]]$ *if, and only if, f is regular in X; that is, the initial form of f is of the form* $F_n = (cX^n + \cdots)$, *where* $c \in K \setminus \{0\}$ *and n is a non-negative integer (the case $n = 0$ corresponds to f invertible in $K[[X,Y]]$).*

 ii) $\sigma^*(fg) = \sigma^*(f)\sigma^*(g)$.

iii) $\text{mult}(\sigma^*(f)) \leq \text{mult}(f)$.

iv) *If f is a Weierstrass polynomial in $K[[X]][Y]$ of degree n with tangent cone (Y^n), then $\sigma^*(f)$ is a pseudo-polynomial of degree n in Y.*

v) *If f is irreducible, then either $\sigma^*(f)$ is irreducible or a unit.*

Proof: If we write $f = F_n(X, Y) + F_{n+1}(X, Y) + \cdots$, then we have $\sigma^*(f) = F_n(1, Y_1) + X_1 F_{n+1}(1, Y_1) + \cdots$.

(i) So, $\sigma^*(f)$ is a unit if and only if $F_n(1, 0)$ is a non-zero constant. This in turn is equivalent to say that $F_n(X, Y) = (cX^n + \cdots)$, where $c \in K^*$.

(ii) This follows easily from the definition of σ^*.

(iii) In $F_n(1, Y_1)$ there are monomials of degree less or equal than $n = \operatorname{mult}(f)$. The result follows by observing that these monomials cannot cancel with other monomials of $\sigma^*(f)$, because the latter are all multiple of X_1.

(iv) Let $f = Y^n + a_1(X)Y^{n-1} + \cdots + a_n(X)$ be a Weierstrass polynomial with tangent cone (Y^n); that is $\operatorname{mult}(a_i(X)) > i$. Since,

$$\sigma^*(f) = Y_1^n + \frac{a_1(X_1)}{X_1}Y_1^{n-1} + \cdots + \frac{a_n(X_1)}{X_1^n},$$

and

$$\operatorname{mult}\left(\frac{a_i(X_1)}{X_1^i}\right) > 0,$$

the result follows.

(v) Since f is irreducible, from the Unitangent Lemma we must have either f regular in X or regular in Y. When f is regular in Y but not regular in X, then the tangent cone of (f) is of the form (Y^n).

If f is regular in X, then from (i) we get that $\sigma^*(f)$ is a unit. So it is sufficient to show that if we assume f irreducible with tangent cone (Y^n) and $\sigma^*(f)$ reducible, then we get a contradiction. Let's assume that f is so.

From (i) and (ii) we have that if f and g are associated, then $\sigma^*(f)$ and $\sigma^*(g)$ are associated. Hence we may assume that f is a Weierstrass polynomial in Y. From (iv) we have that $\sigma^*(f)$ is a pseudo-polynomial of degree n. Since $\sigma^*(f)$ is reducible in $K[[X, Y]]$, from the Lemmas (2.10) and (2.11), it factors into a product of irreducible pseudo-polynomials. Let h be an irreducible pseudo-polynomial of degree $0 < r < n$ that divides $\sigma^*(f)$, hence $X^r h(X, Y/X)$ would be a non-trivial factor of f, a contradiction.   □

Remark that if $f \in K[[X, Y]]$ is irreducible and such that $\sigma^*(f)$ is not a unit, then f is regular in Y with tangent cone (Y^n); that is, f may be written as

$$f = a_0(X)Y^n + a_1(X)Y^{n-1} + \cdots + a_n(X) + Y^{n+1}h(X, Y),$$

with $a_i(X) \in K[[X]]$, $\operatorname{mult}(a_i(X)) > i$, $a_0(0) \neq 0$ and $h \in K[[X, Y]]$.

PROPOSITION 5.7 *Let $f \in K[[X, Y]]$ be an irreducible power series with tangent cone (Y^n). Suppose that $\operatorname{I}(f, Y) = m$. Then*

(i) $\operatorname{I}(\sigma^*(f), Y_1) = m - n$ *and* $\operatorname{I}(\sigma^*(f), X_1) = n$.

ii) *If $m - n \geq n$, then $\operatorname{mult}(\sigma^*(f)) = \operatorname{mult}(f) = n$. Moreover, if $m - n > n$, then $(\sigma^*(f))$ has tangent cone (Y_1^n); and if $m - n = n$, then neither (X_1) nor (Y_1) are the tangent lines of $(\sigma^*(f))$.*

iii) *If $m - n < n$, then* $\text{mult}(\sigma^*(f)) = m - n < \text{mult}(f)$ *and* $\sigma^*(f)$ *has tangent cone* (X_1^{m-n}).

Proof: Since f is regular of order n in Y, we may write it as follows:

$$f = a_0(X)Y^n + a_1(X)Y^{n-1} + \cdots + a_n(X) + Y^{n+1}h(X,Y),$$

where $a_0(0) \neq 0$, $h(X,Y) \in K[[X,Y]]$, and $\text{mult}(a_n(X)) = \text{I}(f,Y) = m$. From the Lemma (3.17) we also have that

$$\text{mult}(a_i(X)) \geq i\frac{m}{n}.$$

Now,

$$\sigma^*(f) = b_0(X_1)Y_1^n + b_1(X_1)Y_1^{n-1} + \cdots + b_n(X_1) + X_1Y_1^{n+1}h(X_1, X_1Y_1),$$

where

$$b_i(X_1) = \frac{a_i(X_1)}{X_1^i}.$$

So, we have that $b_0(X_1) = a_0(X_1)$ and

$$\text{mult}(b_i(X_1)) = \text{mult}(a_i(X_1)) - i \geq i\frac{m}{n} - i = i\frac{m-n}{n}. \tag{5.1}$$

(i) From (3.6) we have that $m > n$ because the tangent cone of (f) is (Y^n). Since $b_0(X_1)$ is a unit and $i\frac{m-n}{n} > 0$, for $i \geq 1$, it follows from (5.1) that

$$\text{I}(\sigma^*(f), X_1) = \text{I}(b_0(X_1)Y_1^n, X_1) = \text{I}(Y_1^n, X_1) = n.$$

On the other hand,

$$\text{I}(\sigma^*(f), Y_1) = \text{I}(b_n(X_1), Y_1) = \text{mult}(b_n(X_1)) = \text{mult}(a_n(X_1)) - n = m - n.$$

(ii) If $m - n \geq n$, then from (5.1) it follows that $\text{mult}(b_i(X_1)) \geq i$ and since $b_0(0) \neq 0$, we have that $\text{mult}(\sigma^*(f)) = n = \text{mult}(f)$.

Now, if $m - n > n$, then from (5.1), $\text{mult}(b_i(X_1)) > i$, showing that (Y_1^n) is the tangent cone of $\sigma^*(f)$.

On the other hand, if $m - n = n$, then the tangent line of $\sigma^*(f)$ has as summands at least the two terms of lower order in $b_0(X_1)Y^n$ and in $b_n(X_1)$. This shows that neither (Y_1) nor (X_1) are the tangent lines of $(\sigma^*(f))$.

(iii) Suppose that $n < m < 2n$. Since

$$\text{mult}(b_i(X_1)Y_1^{n-i}) \geq i\frac{m-2n}{n} + n > n\frac{m-2n}{n} + n = m - n,$$

we have that $\text{mult}(\sigma^*(f)) = m - n$ and that (X_1^{m-n}) is the tangent cone of $(\sigma^*(f))$.
□

Similarly, we may define the notion of strict transform $\tau^*(f)$ of f, and prove results analogous to those of the above propositions, interchanging the roles of X and Y.

The following result will be applied often.

LEMMA 5.8 *Let $f \in K[[X,Y]]$ be irreducible of multiplicity n and regular in Y. There exists an automorphism Φ of $K[[X,Y]]$ such that $\Phi(f)$ is irreducible of multiplicity n, regular in Y and such that $I(\Phi(f),Y)$ is not a multiple of n.*

Proof: Since f is irreducible and regular of order n in Y, we may write

$$f = a_0(X)Y^n + a_1(X)Y^{n-1} + \cdots + a_n(X) + Y^{n+1}h(X,Y),$$

where $a_0(0) \neq 0$, $h(X,Y) \in K[[X,Y]]$, and $I(f,Y) = \mathrm{mult}(a_n(X))$. From Lemma (3.17) we also have that

$$\mathrm{mult}(a_i(X)) \geq i\frac{I(f,Y)}{n}.$$

If n doesn't divide $I(f,Y)$, then just take $\Phi = \mathrm{Id}$.

Suppose now that $I(f,Y) = nr$, for some integer $r \geq 1$. Let c be a parameter in K and define

$$\Phi_1(X,Y) = (X, Y + cX^r).$$

We then have

$$\Phi_1(f) = \begin{array}{l} a_0(X)(Y + cX^r)^n + a_1(X)(Y + cX^r)^{n-1} + \cdots + a_n(X) + \\ (Y + cX^r)^{n+1}h(X, Y + cX^r), \end{array}$$

and since

$$\mathrm{mult}(a_i(X)) \geq i\frac{I(f,Y)}{n} = ir,$$

it follows that we may write

$$\Phi_1(f) = b_0(c,X)Y^n + b_1(c,X)Y^{n-1} + \cdots + b_n(c,X) + P(c)X^{nr} + Y^{n+1}h'(c,X,Y),$$

with $b_0(c,0) = a_0(0) \neq 0$,

$$\mathrm{mult}(b_i(c,X)) \begin{cases} \geq ir, & \text{if } 1 \leq i \leq n-1 \\ > nr, & \text{if } i = n, \end{cases}$$

and $P(c)$ is a polynomial of degree n in c.

Let $c_0 \in K$ be such that $P(c_0) = 0$, then

$$\Phi_1(f) = b_0(X)Y^n + b_1(X)Y^{n-1} + \cdots + b_n(X) + Y^{n+1}h''(X,Y),$$

where $b_i(X) = b_i(c_0,X)$, for $0 = 1, \ldots, n$, and $h''(X,Y) = h'(c_0,X,Y)$. So, $\Phi_1(f)$ is irreducible, regular in Y and such that $\mathrm{mult}(b_n(X)) > nr$.

If n doesn't divide $\mathrm{mult}(b_n(X))$, we are done, since $I(\Phi_1(f),Y)$ is equal to $\mathrm{mult}(b_n(X))$. Otherwise, we repeat the above process obtaining an automorphism Φ_2 such that

$$\Phi_2(f) = c_0(X)Y^n + c_1(X)Y^{n-1} + \cdots + c_n(X) + Y^{n+1}h''(X,Y),$$

with $c_0(0) \neq 0$ and $\mathrm{mult}(c_n(X)) > \mathrm{mult}(b_n(X))$.

This process must stop at some stage because otherwise we would get an automorphism Φ such that

$$\Phi(f) = a_0'(X)Y^n + a_1'(X)Y^{n-1} + \cdots + a_{n-1}'(X)Y + Y^{n+1}\tilde{h}(X,Y),$$

which is reducible; a contradiction. $\square$

PROPOSITION 5.9 *Let (f) be irreducible with tangent cone (Y^n). Let $f^{(1)} = \sigma^*(f)$ and $f^{(i)} = \sigma^*(f^{(i-1)})$. If n doesn't divide $m := \mathrm{I}(f, Y)$, then*

$$\left[\frac{m}{n}\right] = \min\left\{i;\ \mathrm{mult}(f^{(i)}) \neq \mathrm{mult}(f)\right\}.$$

Proof: Since n doesn't divide m, by the euclidian algorithm we have $m = nq + r$, with $0 < r < n$ and $q = \left[\dfrac{m}{n}\right]$. Now, from the Proposition (5.7) we have that

$$\mathrm{mult}(f^{(q)}) \neq \mathrm{mult}(f^{(q-1)}) = \mathrm{mult}(f).$$

□

We are now ready to describe the process of resolution of singularities for an irreducible algebroid plane curve.

Assume that $f \in K[[X, Y]]$ is irreducible of multiplicity n, regular in Y and such that n doesn't divide $m = \mathrm{I}(f, Y)$. We use the notation $f^{(1)} = \sigma^*(f)$ and $f^{(i)} = \sigma^*(f^{(i-1)})$, for $i = 1, \ldots, q = [m/n]$. Put $n' = \mathrm{mult}(f^{(q)})$, then we have that $n' < n$. If $n' = 1$ we stop. Otherwise, from Proposition (5.7.iii) we have that $f^{(q)}$ is power series regular of order $n' > 1$ in the indeterminate X_q. Let $m' = \mathrm{I}(f^{(q)}, X_q)$ and put $q' = [m'/n']$. If m' is a multiple of n', then by the Lemma (5.8) we may change coordinates by an automorphism Φ of $K[[X, Y]]$ in order that m' is no more divisible by n'. We still denote $\Phi(f^{(q)})$ by $f^{(q)}$. We then define

$$f^{(q+1)} = \tau^*(f^{(q)}),$$

and

$$f^{(i)} = \tau^*(f^{(i-1)}), \quad i = q+2, \ldots, q+q'.$$

Now, either $n'' = \mathrm{mult}(f^{(q+q')}) = 1$, and we stop; or from the analog of Proposition (5.7.iii) we have that $f^{(q+q')}$ is a regular power series of order $n'' > 1$ in the indeterminate $Y_{q+q'}$, to which we continue to apply the above process, using σ this time. Since the multiplicities are decreasing at some point we get a regular curve $(f^{(N)})$.

The sequence $f, f^{(1)}, \ldots, f^{(N)}$ is called the *canonical resolution* of (f). We have proved the following result:

THEOREM 5.10 *The canonical resolution of an irreducible algebroid plane curve leads after finitely many steps to a non-singular algebroid plane curve.*

The canonical resolution of (f) determines the numerical sequence

$$\mathrm{mult}(f) \geq \mathrm{mult}(f^{(1)}) \geq \cdots \geq \mathrm{mult}(f^{(i)}) \geq \cdots \geq mult(f^{(N)}) = 1,$$

of great importance in this theory, called the *multiplicity sequence* of (f).

We say that two plane branches are *equiresoluble* if their multiplicity sequences are equal.

EXAMPLE 5.11 We will determine the canonical resolution of $f = Y^4 - X^7$. In the present case, we have that $f^{(1)} = \sigma^*(f) = Y_1^4 - X_1^3$. Therefore, applying τ^* to $f^{(1)}$, we get $f^{(2)} = Y_2 - X_2^3$, which is not singular. The multiplicity sequence in this case is $(4, 3, 1)$ (see figure below).

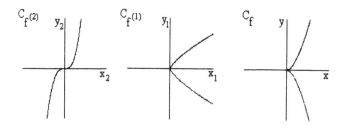

Figure 4

PROBLEMS

2.1)   Resolve the singularities of the curves given below and determine their multiplicity sequences.

a) $Y^5 - X^8$.

b) $Y^4 - 2X^3Y^2 - 4X^5Y - X^6 - X^7$.

5.3   NOETHER'S FORMULA

In this section we will present a useful formula due to Max Noether that relates the intersection index of two curves with that of their strict transforms.

Let $f \in K[[X, Y]]$ be irreducible of multiplicity n and regular in Y. We know from Section 3.4 that f has a Puiseux parametrization $(T^n, \varphi(T))$ with $m = \text{mult}(\varphi(T)) > n$. According to the Lemma (5.8), performing a change of coordinates, if necessary, we may assume that m is not a multiple of n.

Consider

$$f^{(1)}(X_1, Y_1) := \sigma^*(f) = X_1^{-n} f(X_1, X_1 Y_1).$$

Since $f^{(1)}(X_1, Y_1)$ is regular in Y_1 of order n, and

$$f^{(1)}\left(T^n, \frac{\varphi(T)}{T^n}\right) = (T^n)^{-n} f(T^n, \varphi(T)) = 0,$$

it follows that

$$(T^n, \psi(T)) = \left(T^n, \frac{\varphi(T)}{T^n}\right)$$

is a Puiseux parametrization of $f^{(1)} = \sigma^*(f)$. Hence, for $g \in K[[X_1, Y_1]]$, we always have

$$I(\sigma^*(f), g) = \text{mult}\left(g(T^n, \psi(T))\right).$$

Recall now that the local ring $\mathcal{O}_f$ injects into $K[[T]]$ by means of the homomorphism

$$H_\varphi : \quad \begin{array}{ccc} \mathcal{O}_f & \xrightarrow{\sim} & A_\varphi \subset K[[T]]. \\ \overline{g} & \mapsto & g(T^n, \varphi(T)) \end{array}$$

For the same reason, the local ring $\mathcal{O}_{f^{(1)}}$ injects into $K[[T]]$ by means of the homomorphism

$$H_\psi : \begin{array}{ccc} \mathcal{O}_{f^{(1)}} & \xrightarrow{\sim} & A_\psi \subset K[[T]]. \\ \overline{\dfrac{1}{g}} & \mapsto & g(T^n, \psi(T)) \end{array}$$

We will see in the next proposition that the local ring of an irreducible algebroid plane curve injects naturally into the local ring of its blow-up.

PROPOSITION 5.12 *Let* $f \in K[[X, Y]]$ *be irreducible with tangent cone* (Y^n). *Then there is a natural injective homomorphism* ϕ *which makes the diagram below commutative*

$$\begin{array}{ccc} \mathcal{O}_f & \xrightarrow{\phi} & \mathcal{O}_{f^{(1)}} \\ \| & & \| \\ A_\varphi & \xrightarrow{\text{Id}} & A_\psi \end{array}$$

Proof: Let

$$\phi : \begin{array}{ccc} \dfrac{\mathcal{O}_f}{g(X,Y)} & \longrightarrow & \dfrac{\mathcal{O}_{f^{(1)}}}{g(X_1, X_1 Y_1)}, \\ \overline{g(X,Y)} & \mapsto & \overline{g(X_1, X_1 Y_1)} \end{array}$$

where the overlines have the obvious meaning. Suppose that the multiplicity of f is n. The map ϕ is well defined and injective because, from the Proposition (4.3), it may be viewed as a homomorphism of $K[[X]]$-module

$$\phi : \begin{array}{ccc} K[[X]] \oplus \cdots \oplus K[[X]]y^{n-1} & \longrightarrow & K[[X]] \oplus \cdots \oplus K[[X]]y_1^{n-1}, \\ g(X, y) & \mapsto & g(X, Xy_1) \end{array}$$

which is clearly well defined and injective.

Now, let $\overline{g(X,Y)} \in \mathcal{O}_f$, then

$$H_\psi \phi(g) = H_\psi(g(X, Xy_1)) = g(T^n, T^n \psi(T)) = g(T^n, \varphi(T)).$$

On the other hand,

$$\text{Id}(H_\varphi(g)) = g(T^n, \varphi(T)),$$

which proves the assertion. $\qquad\qquad\qquad\qquad\qquad\qquad\qquad\qquad\qquad\qquad\qquad$ □

PROPOSITION 5.13 *Let* (f) *and* (g) *be two algebroid plane curves with* f *irreducible. Then we have*

$$\text{I}(f, g) = \text{mult}(f) \cdot \text{mult}(g) + \text{I}(f^{(1)}, g^{(1)}).$$

Proof: Denote by n and n' respectively the multiplicities of f and g. We may change coordinates in such a way that f becomes regular in Y. Let $(T^n, \varphi(T))$ be a Puiseux parametrization of (f), then $(T^n, \frac{\varphi(T)}{T^n})$ is a primitive parametrization of $f^{(1)}$. So we have

$$\begin{array}{rl} \text{I}(f^{(1)}, g^{(1)}) = & \text{mult}\left(\frac{1}{(T^n)^{n'}} g\left(T^n, T^n \frac{\varphi(T)}{T^n}\right)\right) = -n \cdot n' + v_f(g) = \\ & -\text{mult}(f).\text{mult}(g) + \text{I}(f, g), \end{array}$$

which proves the result. $\qquad\qquad\qquad\qquad\qquad\qquad\qquad\qquad\qquad\qquad\qquad\qquad\qquad$ □

Let f and g be two algebroid irreducible plane curves. With a finite sequence, say of length N, of blowing-ups of either type σ, or τ, we are led necessarily to the situation in which $f^{(N)}$ and $g^{(N)}$ have distinct tangents. This observation, together with the Proposition (5.13) and the Theorem (4.19), proves a classical formula due to Max Noether, whose statement we give below.

THEOREM 5.14 (Noether's formula) *Let f and g be two irreducible algebroid plane curves. We have that*

$$I(f,g) = \sum_{i=0}^{N} \mathrm{mult}(f^{(i)})\mathrm{mult}(g^{(i)}),$$

where $f^{(0)} = f$ and $g^{(0)} = g$.

6   Semigroups of plane branches

In this chapter we will introduce the semigroup of values of an algebroid irreducible plane curve. This is an invariant under the equivalence of curves and was classically recognized as the classifying topological invariant in the case of germs of complex analytic irreducible plane curves (see for example [5]). We will relate this semigroup to the characteristic integers, to the Puiseux pairs and to other numerical invariant of the curve. This will give us tools to compute such semigroups.

6.1   SEMIGROUPS OF THE NATURALS

Let $G \neq \{0\}$ be a subset of $\mathbb{N}$, containing the element 0. We say that G is a *semigroup* in $\mathbb{N}$ if it is closed under addition. In this section we will study semigroups under a strictly arithmetical point of view.

The element $\min(G \setminus \{0\})$ will be called the *multiplicity* of G, and will be denoted by $\mathrm{mult}(G)$.

If $x_0, \ldots, x_r \in \mathbb{N}$, then the set

$$\langle x_0, \ldots, x_r \rangle = \{\lambda_0 x_0 + \cdots + \lambda_r x_r; \quad \lambda_1, \ldots, \lambda_r \in \mathbb{N}\}$$

is a semigroup in $\mathbb{N}$, called the semigroup generated by $x_0, \ldots, x_r$. The elements $x_0, \ldots, x_r$ are called *generators* for G.

For example, we have

$$\langle 3 \rangle = \{0, 3, 6, 9, \ldots\},$$

and

$$\langle 3, 5 \rangle = \{0, 3, 5, 6, 8, 9, 10, 11, \ldots\}.$$

PROPOSITION 6.1 *Given any semigroup G in $\mathbb{N}$, there exists a unique finite set of elements $v_0, \ldots, v_g$ in G such that*

  i) $v_0 < \cdots < v_g$, *and* $v_i \not\equiv v_j \mod v_0$, *for* $i \neq j$,

 ii) $G = \langle v_0, \ldots, v_g \rangle$,

iii) $\{v_0, \ldots, v_g\}$ *is contained in any set of generators of G.*

Proof: The elements $v_0, \ldots, v_g$ are defined inductively as follows. Put $v_0 = \text{mult}(G)$ and define
$$v_1 = \min\left(G \setminus \langle v_0 \rangle\right).$$

It is clear that $v_0 \not\equiv v_1 \mod v_0$, because otherwise v_1 would be in $\langle v_0 \rangle$, which is a contradiction.

Define for $i \geq 2$,
$$v_i = \min\left(G \setminus \langle v_0, \ldots, v_{i-1} \rangle\right).$$

It is clear that $v_i \not\equiv v_j \mod v_0$, for $j < i$, because otherwise v_i would be in $\langle v_0, \ldots, v_{i-1} \rangle$, which is a contradiction.

Since $v_i \not\equiv v_j \mod v_0$, for $i \neq j$, then for some $g < v_0$, we must have
$$G = \langle v_0, \ldots, v_g \rangle.$$

It is now easy to check that $v_0, \ldots, v_g$ satisfy conditions (i) and (iii) in the statement of the proposition.   □

The set $\{v_0, \ldots, v_g\}$ of the Proposition (5.6) will be called the *minimal system of generators* of G, and the integer g will be called the *genus* of the semigroup G (notice that $g \leq \text{mult}(G) - 1$).

Given a semigroup in $\mathbb{N}$, the elements of $\mathbb{N} \setminus G$ are called the *gaps* of G.

A semigroup may have finitely or infinitely many gaps. For example the semigroup $\langle 3 \rangle$ has infinitely many gaps, while the gaps of $\langle 3, 5 \rangle$ are the integers $1, 2, 4$ and 7.

When the number of gaps of G is finite, there exists a unique element $c \in G$, called the *conductor* of G, such that
a) $c - 1 \notin G$
b)   if $z \in \mathbb{N}$ and $z \geq c$, then $z \in G$.

If $n > 1$, then obviously the semigroup $\langle n \rangle$ has no conductor, since no element $m \in \mathbb{N}$ with $m \not\equiv 0 \mod n$ is in $\langle n \rangle$.

PROPOSITION 6.2 *Let G be a semigroup in $\mathbb{N}$. The following assertions are equivalent*

   i) *G has a conductor,*

  ii) *The elements of G have GCD equal to one,*

 iii) *There exist two consecutive integers in G.*

Proof: (i) $\Longrightarrow$ (ii)   Since $G \subset \langle \text{GCD}(G) \rangle$, it is clear that, if G has a conductor, then $\text{GCD}(G) = 1$.
(ii) $\Longrightarrow$ (iii)   Let $v_0, \ldots, v_g$ be the minimal system of generators of G. Then clearly $\text{GCD}(G) = 1$ implies $\text{GCD}(v_0, \ldots, v_g) = 1$. So, in this case, there exist integers $\lambda_0, \ldots, \lambda_g$ such that
$$\lambda_0 v_0 + \cdots + \lambda_g v_g = 1.$$

Transfering to the second member the terms with the λ_i's negative, the result follows immediately.
(iii) $\Longrightarrow$ (i)   Let a and $a + 1$ be two elements in G, then the set
$$\{0, a + 1, 2(a + 1), \ldots, (a - 1)(a + 1)\}$$

is a complete residue system, modulo a. So, any integer $n \geq (a-1)(a+1)$ may be written as

$$n = i(a+1) + ja, \quad i = 0, \ldots, (a-1), \quad j \geq 0,$$

which implies that G has a conductor $c \leq (a-1)(a+1)$. $\square$

Notice that in the proof of the proposition we got the estimate $c \leq (a-1)(a+1)$ for the conductor of G, when a and $a+1$ are elements of G.

Let $G = \langle x_0, \ldots, x_r \rangle$ be a semigroup in $\mathbb{N}$ with conductor c. We define below two sequences of natural numbers associated to the set of generators $x_0, \ldots, x_r$ of G.

Put $e_0 = x_0$ and $n_0 = 1$. For $i = 1, \ldots, r$, define

$$e_i = \mathrm{GCD}(x_0, \ldots, x_i) \quad \text{and} \quad n_i = \frac{e_{i-1}}{e_i}.$$

From the definition of the e_i's it is clear that $e_i | e_{i-1}$, for all $i = 1, \ldots, r$, and also $e_r = \mathrm{GCD}(x_0, \ldots, x_r) = 1$.

PROBLEMS

1.1) Let $x_0, x_1, \cdots, x_r$ be a set of generators of a semigroup with conductor. Let the n_i's and e_i's be as defined in this section. Show that $x_0 = n_1 n_2 \cdots n_i e_i$. In particular, $x_0 = n_1 \cdots n_r$. Show also that $e_i = n_{i+1} \ldots n_r$.

6.2   SEMIGROUPS OF VALUES

From now on, f will denote an irreducible element in $\mathcal{M} \subset K[[X, Y]]$. Hence, the ring $\mathcal{O}_f = K[[X, Y]]/\langle f \rangle$ will be a domain. Let $v = v_f$ the valuation associated to f, as defined in Chapter 4.

To simplify notation, we will denote the residual class of Y modulo $\langle f \rangle$ by y and that of X by x or also by X, since this will not cause any confusion.

DEFINITION 6.3 *The* semigroup of values *associated to the curve* (f) *is the set*

$$S(f) = \{\mathrm{I}(f, g); g \in K[[X, Y]] \setminus \langle f \rangle\} \subset \mathbb{N}.$$

A straightforward verification, by means of the Theorem (4.15), shows that the set $S(f)$ is effectively a semigroup in $\mathbb{N}$ and that two equivalent algebroid plane curves have the same semigroup of values.

The semigroup of values $S(f)$ may also be viewed as follows. Let f be an irreducible power series in $K[[X, Y]]$, regular in Y with the Puiseux parametrization $(T^n, \varphi(T))$. Let H_φ the homomorphism we defined in Section 4.1. Then one has

$$S(f) = \{v_f(\overline{g}); \overline{g} \in \mathcal{O}_f \setminus \{0\}\} = \{\mathrm{mult}_T(H_\varphi(\overline{g})); \; \overline{g} \in \mathcal{O}_f \setminus \{0\}\} =$$

$$\{\mathrm{mult}_T(h); \; h \in A_\varphi \setminus \{0\}\}.$$

Two plane branches (f) and (g) will be called *equisingular* if $S(f) = S(g)$. It is clear that if (f) and (g) are equivalent, then they are equisingular.

The semigroup of values of a plane branch has a conductor. In fact, since equivalent branches have same semigroup, we may assume that the branch is represented by a Weierstrass polynomial f, so from the Theorem (4.5) we have that any element in $K[[T]]$ of multiplicity greater than $\text{mult}_T(D_Y(f)(T^n))$ is already in A_φ, hence its multiplicity is in $S(f)$.

Alternatively, this also follows from the observation we made in Section 4.1, before the Theorem (4.5), that the field of fractions of A_φ is $K((T))$. Indeed, there exist $h_1, h_2 \in A_\varphi$ such that $T = h_1/h_2$, so $\text{mult}(h_1) = \text{mult}(h_2) + 1$, which in view of the Proposition (5.7), implies that $S(f)$ has a conductor.

EXAMPLE 6.4 If $f = Y^2 - X^3$, we have that

$$S(f) = \{0, 2, 3, 4, 5, \ldots\}.$$

Indeed, $v(x) = \mathrm{I}(f, X) = 2$ and $v(y) = \mathrm{I}(f, Y) = 3$. Since any natural number $l \geq 2$ may be written as $l = r \cdot 2 + s \cdot 3$, with $r, s \in \mathbb{N}$, it follows that $l = v(x^r y^s) \in S(f)$. The conductor of $S(f)$ is 2.

EXAMPLE 6.5 If $f = Y^2 - X^5$, we have that

$$S(f) = \{0, 2, 4, 5, \ldots\}.$$

Indeed, $v(x) = \mathrm{I}(f, X) = 2$, $v(y) = \mathrm{I}(f, Y) = 5$. Now, it is easy to write any natural number $l \geq 4$ as $l = r \cdot 2 + s \cdot 5$, with $r, s \in \mathbb{N}$, and therefore $l = v(x^r y^s) \in S(f)$. The conductor in this case is 4.

EXAMPLE 6.6 Let $f = Y^4 - X^7$. Since $X = T^4$ and $Y = T^7$ is a Puiseux parametrization of (f), then given $g = \sum_{i,j} a_{i,j} X^i Y^j \in K[[X, Y]]$, we have that

$$v_f(\overline{g}) = \text{mult}(\sum_{i,j} a_{i,j} T^{4i+7j}) = 4r + 7s,$$

for some natural numbers r and s. This shows that $S(f)$ is generated by 4 and 7. Therefore,

$$S(f) = \{0, 4, 7, 8, 11, 12, 14, 15, 16, 18, 19 \ldots\},$$

with conductor 18.

How one can, in general, determine the semigroup $S(f)$? This is not a priori an easy question to answer and an important part of this chapter will be devoted to it.

The least element of $S(f) \setminus \{0\}$ is the integer $\text{mult}(f)$ since from the Theorem (4.19), we have that $v(\overline{g}) = \mathrm{I}(f, g) \geq \text{mult}(f)$, for all $g \in \mathcal{M} \subset K[[X, Y]]$, and if (L) is a line not belonging to the tangent cone of (f), we have that $v(\overline{L}) = \text{mult}(f)$.

To study the semigroup of values of a plane branch (f), we will assume that $\text{char}(K) = 0$ and that coordinates have been chosen in order that f is regular in Y and the tangent cone of (f) is (Y^n). So, f is of the form

$$f = a_0(X)Y^n + a_1(X)Y^{n-1} + \cdots + a_n(X) + Y^{n+1}h(X, Y),$$

with $a_0(0) \neq 0$ and $\text{mult}(a_i) > i$. In particular, we have that $a_n(X) = cX^m + \cdots$, with $m > n$. We also will choose coordinates such that m is not a multiple of n (see Lemma (5.8)).

Let

$$\begin{cases} X = T^n \\ Y = \varphi(T) = \sum_{i \geq m} b_i T^i, \quad b_m \neq 0, \quad m > n, \end{cases}$$

with m and n as above, be a Puiseux parametrization of (f).

REMARK 6.7 In the above situation we have that

$$m = \min(S(f) \setminus \langle n \rangle).$$

Indeed, $m \in S(f)$, because $v(y) = m$. On the other hand, if $g = \sum_{\alpha,\beta} a_{\alpha,\beta} X^\alpha Y^\beta \in K[[X,Y]]$, then

$$I(f,g) = \text{mult}(\sum_{\alpha,\beta} a_{\alpha,\beta} T^{n\alpha} \varphi(T)^\beta).$$

From this, it is clear that $\min(S(f) \setminus \langle n \rangle) \geq m$, and since $I(f,Y) = m$, the equality follows.

So, if $v_0, v_1, \ldots, v_g$ represent the minimal system of generators of $S(f)$, we have, in this case, that $v_0 = n$ and $v_1 = m$.

If we denote by $\beta_0, \beta_1, \ldots \beta_\gamma$ the characteristic exponents of the above Puiseux parametrization of (f), we have that $\beta_0 = n$ and $\beta_1 = m$. We will still denote by η_j and ε_j the integers corresponding to the characteristic exponents as defined in Section 3.4.

Recall that we denoted the group of the rth roots of 1 in K by U_r. For $j = 0, \ldots \gamma$, we put

$$G_j = U_{\varepsilon_j} = \{\zeta \in K \mid \zeta^{\varepsilon_j} = 1\},$$

hence

$$U_n = G_0 \supset G_1 \supset \cdots \supset G_\gamma = \{1\}.$$

LEMMA 6.8 *If $\zeta \in G_j \setminus G_{j+1}$, then $\zeta^{\beta_{j+1}} \neq 1$.*

Proof: Indeed, if the we had $\zeta^{\beta_{j+1}} = 1$, then we would have $\zeta^{\varepsilon_j} = \zeta^{\beta_{j+1}} = 1$, and since $\varepsilon_{j+1} = \text{GCD}(\varepsilon_j, \beta_{j+1})$, we would have $\zeta^{\varepsilon_{j+1}} = 1$, a contradiction, since $\zeta \notin G_{j+1}$. $\square$

LEMMA 6.9 *Let $\zeta \in G_{k-1} \setminus G_k$. If $l \geq k$, then $\zeta G_l \subset G_{k-1} \setminus G_k$ (see figure below).*

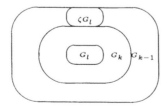

Figure 1

Proof: Indeed, from the hypothesis, we have that $\zeta^{\varepsilon_{k-1}} = 1$, but $\zeta^{\varepsilon_k} \neq 1$. Let ρ be any element of G_l. Since ε_l divides ε_{k-1}, it follows that $(\zeta\rho)^{\varepsilon_{k-1}} = 1$, and therefore, $\zeta\rho \in G_{k-1}$. We now want to prove that $\zeta\rho \notin G_k$. Suppose by absurdity that $\zeta\rho \in G_k$. Since ε_l divides ε_k, we have that

$$1 = (\zeta\rho)^{\varepsilon_k} = \zeta^{\varepsilon_k}\rho^{\varepsilon_k} = \zeta^{\varepsilon_k},$$

which is a contradiction. $\square$

For every $j = 2, \ldots, \gamma$, we define

$$P_j(T) = \sum_{i=\beta_1}^{\beta_j-1} b_i T^i.$$

For commodity, we define $\beta_{\gamma+1} = \infty$. So,

$$P_{\gamma+1} = \sum_{i=\beta_1}^{\infty} b_i T^i = y.$$

Consider now, for $j = 1, \ldots, \gamma$, $P_j(X^{\frac{1}{n}}) \in K((X^{\frac{1}{n}}))$. We want to study the following tower of extensions:

$$
\begin{array}{c}
K((X^{\frac{1}{n}})) \\
| \\
K((X))(P_j(X^{\frac{1}{n}})) \\
| \\
K((X))
\end{array}
$$

LEMMA 6.10 Let $\zeta, \rho \in G_0 = U_n$ and $j \in \{2, \ldots, \gamma\}$. We have that

$$P_j(\zeta T) = P_j(\rho T) \Longleftrightarrow \zeta\rho^{-1} \in G_{j-1}.$$

Proof: Suppose that

$$\sum_{i=\beta_1}^{\beta_j-1} b_i \zeta^i T^i = \sum_{i=\beta_1}^{\beta_j-1} b_i \rho^i T^i.$$

Hence, $\zeta^i = \rho^i$ for $i = n, \beta_1, \ldots, \beta_{j-1}$ and therefore, $(\zeta\rho^{-1})^i = 1$, for $i = \mathrm{GCD}(n, \beta_1, \ldots, \beta_{j-1}) = \varepsilon_{j-1}$. This implies that $\zeta\rho^{-1} \in G_{j-1}$.

Conversely, suppose that $\zeta^{\varepsilon_{j-1}} = \rho^{\varepsilon_{j-1}}$. From Section 3.4 we know that ε_{j-1} divides all the exponents i such that $\beta_{j-1} \leq i < \beta_j$, for which $b_i \neq 0$ and since ε_{j-1} divides ε_ℓ for all $\ell \leq j-1$, it follows that ε_{j-1} divides all the exponents i with $\beta_1 \leq i < \beta_j$ for which $b_i \neq 0$. Therefore,

$$b_i \zeta^i = b_i \rho^i, \quad \forall i = \beta_1, \beta_1 + 1, \ldots, \beta_j - 1,$$

proving thus that $P_j(\zeta T) = P_j(\rho T)$. $\square$

The above result gives immediately the following corollary.

COROLLARY 6.11 *The Galois group of the extension* $K((X^{\frac{1}{n}}))/K((X))(P_j(X^{\frac{1}{n}}))$ *is* G_{j-1}.

Since the Galois group of the extension $K((X^{\frac{1}{n}}))/K((X))$ is $G_0 = U_n$, hence abelian, it follows readily from the above corollary that the Galois group of $K((X))(P_j(X^{\frac{1}{n}}))/K((X))$ is G_0/G_{j-1}.

Denoting by $[\zeta]$ the residual class of ζ in $G_0 = U_n$, modulo the subgroup G_{j-1}, we define

$$P_j([\zeta]T) := P_j(\zeta T),$$

which makes sense by the Lemma (6.10). It then follows that the conjugates of $P_j(X^{\frac{1}{n}})$ in $K((X))^*$ over $K((X))$, are the $P_j([\zeta]X^{\frac{1}{n}})$, where $[\zeta] \in G_0/G_{j-1}$. Therefore,

$$f_j(X,Y) = \prod_{[\zeta] \in G_0/G_{j-1}} \left(Y - P_j([\zeta]X^{\frac{1}{n}})\right) \in K[[X]][Y].$$

The polynomial $f_j(X,Y)$ is of degree $n/\varepsilon_{j-1} = \eta_1\eta_2\cdots\eta_{j-1}$ in Y (see Problem 3.4.1) and it is the minimal polynomial of $P_j(X^{\frac{1}{n}})$ over $K((X))$.

Consequently, we have that

$$f_j = f_j(x,y) \in \mathcal{O}_f.$$

We define

$$v_j = \begin{cases} \beta_0 = n, & \text{if } j = 0 \\ \beta_1, & \text{if } j = 1 \\ v(f_j), & \text{if } j = 2,\ldots,\gamma. \end{cases}$$

Therefore, the v_j's are elements in $S(f)$. The following theorem, due to Zariski [15] will give us important information about these integers, relating them to the characteristic exponents $\beta_0,\ldots,\beta_\gamma$ of (f).

To state the Theorem we will need the following notation. Define for $k = 0,\ldots,n$,

$$M_k = K[[X]] + K[[X]]y + K[[X]]y^2 + \cdots + K[[X]]y^k \subset \mathcal{O}_f,$$

where $n = \text{mult}(f)$ and y is the residual class of Y in $\mathcal{O}_f$. From the Division Theorem it follows easily that $M_{n-1} = \mathcal{O}_f$.

THEOREM 6.12 i) *For* $j = 2,\ldots,\gamma$, *we have that*

$$v_j = \sum_{k=1}^{j-1} \frac{\varepsilon_{k-1} - \varepsilon_k}{\varepsilon_{j-1}} \beta_k + \beta_j \tag{6.1}$$

ii) *If* $h \in M_k$ *with* $k < \eta_1\cdots\eta_j$, *then* $v(h) \in \sum_{i=0}^{j} v_i \mathbb{N}$. *In particular,*

$$S(f) = \langle v_0,\ldots,v_\gamma \rangle.$$

iii) *The GCD of* ε_{j-1} *and* v_j *is* ε_j, *for* $j = 1,\ldots,\gamma$. *Moreover,* v_j *is the smallest non-zero element of* $S(f)$ *which is not divisible by* ε_{j-1}.

iv) *The genus of* $S(f)$ *is* γ *and the integers* $v_0,\ldots,v_\gamma$ *are a minimal system of generators.*

Proof: (i) From the Lemma (6.9) it follows that if $j > k$, then each set $G_{k-1} \setminus G_k$ is partitioned into the disjoint union of $\dfrac{\varepsilon_{k-1} - \varepsilon_k}{\varepsilon_{j-1}}$ cosets of G_{j-1}. Moreover, using the definitions and the Lemma (6.10), we have that

$$v(y - P_j([\zeta]T)) = \beta_k, \quad \forall \zeta \in G_{k-1} \setminus G_k.$$

It then follows that

$$v(f_j) = \sum_{k=1}^{j-1} \frac{\varepsilon_{k-1} - \varepsilon_k}{\varepsilon_{j-1}} \beta_k + \beta_j.$$

(ii) We will prove this result by induction over j.
Case $j = 1$: let $h = a_0(x) + y a_1(x) + \cdots + y^k a_k(x)$, with $k < \eta_1$. For every $i = 0, \ldots, \eta_1 - 1$, we have that

$$v(y^i a_i(x)) \equiv i \beta_1 \mod n.$$

If i and l are two distinct non-negative integers less than η_1, we have that

$$(i - l)\beta_1 \not\equiv 0 \mod n,$$

because otherwise,

$$(i - l)\frac{\beta_1}{\varepsilon_1} = \lambda \frac{n}{\varepsilon_1} = \lambda \eta_1,$$

which, since $\dfrac{\beta_1}{\varepsilon_1}$ and $\eta_1 = \frac{n}{\varepsilon_1}$ are relatively prime, would imply that η_1 divides $i - l$. This is a contradiction because $0 < |i - l| < \eta_1$.
Therefore,

$$v(y^i a_i(x)) \neq v(y^l a_l(x)), \quad \text{if } i, l < \eta_1 \text{ with } i \neq l.$$

It then follows that $v(h) = v(y^i a_i(x))$ for some $i = 0, \ldots, k$. This shows that

$$v(h) \in n\mathbb{N} + \beta_1\mathbb{N} = v_0\mathbb{N} + v_1\mathbb{N}.$$

Suppose now that the result is true for $j - 1$. Let $h \in M_k \subset \mathcal{O}_f$, and suppose that $k < \eta_1 \cdots \eta_j$. Suppose that $h = \overline{h(X, Y)}$, where $h(X, Y)$ is a polynomial in $K[[X]][Y]$ of degree $\leq k$. Expanding $h(X, Y)$ with respect to the powers of $f_j(X, Y)$, we may write

$$h(X, Y) = A_0(X, Y) + A_1(X, Y)f_j(X, Y) + \cdots + A_s(X, Y)f_j(X, Y)^s,$$

where $s < \eta_j$ and $A_0(X, Y), \ldots, A_s(X, Y) \in K[[X]][Y]$ are of degrees less than $\deg_Y f_j(X, Y) = \eta_1 \cdots \eta_{j-1}$.
Therefore,

$$h = A_0(x, y) + A_1(x, y)f_j + \cdots A_s(x, y)(f_j)^s,$$

where $A_i(x, y) \in M_{k'}$, with $k' < \eta_1 \cdots \eta_{j-1}$, for $i = 0, \ldots, s$. From the inductive hypothesis this implies, for $i = 0, \ldots, s$, that

$$v(A_i(x, y)) \in \sum_{\nu=0}^{j-1} v_\nu \mathbb{N}. \tag{6.2}$$

In particular, from (6.2), it follows that

$$v(A_i(x,y)) \equiv 0 \mod \varepsilon_{j-1}.$$

From this and again from (6.2), it follows that

$$v(A_i(x,y)f_j^i) \equiv iv_j \equiv i\beta_j, \mod \varepsilon_{j-1}.$$

If $i, l < \eta_j$ and $i \neq l$, we have that $(i-l)\beta_j \not\equiv \mod \varepsilon_{j-1}$; because otherwise, we get

$$(i-l)\frac{\beta_j}{\varepsilon_j} = \lambda\frac{\varepsilon_{j-1}}{\varepsilon_j} = \lambda\eta_j,$$

which implies that η_j divides $i - l$, because $\dfrac{\beta_j}{\varepsilon_j}$ and $\eta_j = \dfrac{\varepsilon_{j-1}}{\varepsilon_j}$ are relatively prime.
This is a contradiction because $0 < |i - l| < \eta_j$.

This shows that the $v(A_i(x,y)f_j^i)$ are distinct for $i = 0, \ldots, s$.

Therefore, using (6.2), we see that there exists i, with $i = 0, \ldots, s$, such that

$$v(h) = v(A_i(x,y)f_j^i) = v(A_i(x,y)) + iv_j \in \sum_{\nu=0}^{j} v_\nu \mathbf{N},$$

which concludes the proof of (ii).

(iii)   Observe that from (6.1) we clearly have that $\mathrm{GCD}(v_j, \varepsilon_{j-1}) = \varepsilon_j$. It also follows easily from (6.1) that $v_0 < v_1 < \cdots < v_\gamma$.

Since, from (ii), $v_0 < v_1 < \cdots < v_\gamma$, generate $S(f)$, and that ε_{i-1} divides $v_0, \ldots, v_{i-1}$, we have that v_i is the least non-zero element of $S(f)$ which is not divisible by ε_{i-1}.

(iv) It follows from (ii) and (iii) above that $v_0, v_1, \ldots, v_\gamma$, is the minimal system of generators of $S(f)$, because $v_0 = n = \min(S(f) \setminus \{0\})$, $v_1 = \beta_1 = \min(S(f) \setminus n\mathbf{N})$, and for all $i = 2, \ldots, \gamma$,

$$v_i = \min\left(S(f) \setminus \sum_{\nu=0}^{i-1} v_\nu \mathbf{N}\right).$$

In particular this shows that γ is equal to the genus g of $S(f)$. $\square$

From this we see that for all $i = 0, \ldots, g$ one has

$$e_i = \varepsilon_i \quad \text{and} \quad \eta_i = n_i,$$

where the e_i's and the n_i's are the integers we associated in Section 1 to any sequence of integers $v_0, \ldots, v_g$. In particular, since $\eta_i > 1$ for $i = 1, \ldots, g$ (see Problem 3.4.2), it follows that $n_i > 1$, for $i = 1, \ldots, g$.

By the above observation, and using the relations in Problem 1.1, formula (6.1) may be written as follows:

$$v_j = \begin{aligned}&(n_1 - 1)n_2 \cdots n_{j-1}\beta_1 + (n_2 - 1)n_3 \cdots n_{j-1}\beta_2 + \cdots + \\ &(n_{j-1} - 1)\beta_{j-1} + \beta_j.\end{aligned} \qquad (6.3)$$

From this formula we get easily the following formula:

$$v_i = n_{i-1}v_{i-1} - \beta_{i-1} + \beta_i, \quad i = 2, \ldots, g. \qquad (6.4)$$

Because the β_i's are increasing, it follows immediately from (6.4) that

$$v_{i+1} > n_i v_i, \quad i = 0, \ldots, g-1. \tag{6.5}$$

A semigroup S with a minimal system of generators $v_0, v_1, \ldots, v_g$ satisfying condition (6.5) will be called a *strongly increasing semigroup*.

So, we have shown that the semigroup of values of a plane branch is a strongly increasing semigroup.

THEOREM 6.13 *Let f be an irreducible power series in $K[[X, Y]]$ with a given Puiseux parametrization. Then $S(f)$ and the characteristic integers of (f) determine each other.*

Proof: Indeed, given the characteristic integers $(\beta_0, \beta_1, \ldots, \beta_g)$ of (f), we may determine by the formulas (6.4) the integers $v_0, \ldots, v_g$ and therefore the semigroup $S(f)$ that they generate.

Conversely, given $S(f)$, then its minimal system of generators $v_0, \ldots, v_g$ is well determined so also are the integers $n = v_0$, and the integers n_i, defined inductively by

$$n_i = \frac{e_{i-1}}{e_i}, \quad e_i = \mathrm{GCD}(e_{i-1}, v_i), \quad e_0 = n.$$

Now, the β_i's may be determined recursively by the formulas (6.4).   □

Next we show that the condition on $v_0, \ldots, v_g$ to be strongly increasing, together with the condition $n_i > 1$, for all $i = 1, \ldots, g$ are sufficient to guarantee that there exists a curve (f) such that $S(f)$ has $v_0, \ldots, v_g$ as its minimal system of generators.

THEOREM 6.14 *Let S be a given semigroup with a conductor and with minimal system of generators $v_0, \ldots, v_g$ such that $n_i > 1$ and $v_i > n_{i-1}v_{i-1}$ for all $i = 1, \ldots, g$. Then there exists a plane algebroid irreducible curve (f) such that $S(f) = S$.*

Proof: Put $n = \beta_0 = v_0$, and $\beta_1 = v_0$. By the equations (6.4) we get an increasing sequence of integers $\beta_0, \beta_1, \beta_2, \ldots, \beta_g$, because of the conditions $v_i > n_{i-1}v_{i-1}$. The conditions $n_i > 1$, $i = 1, \ldots, g$, imply that $\mathrm{GCD}(\beta_0, \ldots, \beta_{i-1}) \neq \mathrm{GCD}(\beta_0, \ldots, \beta_i)$.

Now, let

$$\varphi(T) = T^{\beta_1} + T^{\beta_2} + \cdots + T^{\beta_g},$$

and take

$$f(X, Y) = \prod_{i=1}^{n} \left(Y - \varphi(\zeta^i X^{\frac{1}{n}}) \right),$$

where ζ is an nth primitive root of 1.

It is easy to see that $f(X, Y) \in K[[X]][Y]$ determines an irreducible algebroid plane curve (f) with characteristic exponents $\beta_0, \ldots, \beta_g$, and therefore with semigroup S.   □

PROBLEMS

2.1)   Determine the minimal system of generators of the semigroups of the branches given in Problem 3.4.3.

2.2)   Determine $S(f)$ in the following cases:
a) $f = Y^5 - X^8$.
b) $f = Y^n - X^{n+1}$.
c) $f = Y^7 - X^9$.

2.3)   Show that one has, for $i = 1, \ldots, g$, the formulas:

$$\beta_i = v_i - (n_{i-1} - 1)v_{i-1} - \cdots - (n_2 - 1)v_2 - (n_1 - 1)v_1.$$

[Hint: Use formulas (6.4)]

2.4)   Let (f) be an irreducible algebroid plane curve and let $v_0, \ldots, v_g$ be the minimal set of generators of $S(f)$. Suppose that $h \in K[[X, Y]]$ is such that $\mathrm{I}(f, h) = v_i$, for some $i = 0, \ldots, g$, show that h is an irreducible power series.

6.3   SEMIGROUPS AND CARTESIAN EQUATIONS

Up to now we have shown how to determine the semigroup $S(f)$ through a Newton-Puiseux parametrization of (f). In this section we will show how to determine from f itself the characteristic integers of the curve (f), and therefore, by formulas (6.4) we will be able to recover the generators of $S(f)$.

Let (f) be an irreducible algebroid plane curve regular in Y and parametrized by

$$X = T^n, \quad Y = \varphi(T) = \sum_{i \geq m} b_i T^i, \quad b_m \neq 0$$

LEMMA 6.15   If $\zeta \in G_k \setminus G_{k+1}$, then $\mathrm{mult}(\varphi(T) - \varphi(\zeta T)) = \beta_{k+1}$.

Proof: Since

$$\varphi(T) = \sum_{i=m}^{\beta_{k+1}-1} b_i T^i + \sum_{i \geq \beta_{k+1}} b_i T^i,$$

and when $\zeta \in G_k \setminus G_{k+1}$ we have

$$\sum_{i=m}^{\beta_{k+1}-1} b_i(\zeta T)^i = \sum_{i=m}^{\beta_{k+1}-1} b_i T^i,$$

then

$$\varphi(T) - \varphi(\zeta T) = (1 - \zeta^{k+1})b_{\beta_{k+1}} T^{\beta_{k+1}} + \cdots.$$

Since $\zeta^{k+1} \neq 1$, the result follows immediately.   □

We will use the notation: $f_Y^{(j)} := \dfrac{\partial^j f}{\partial Y^j}$.

THEOREM 6.16   Let $f \in K[[X, Y]]$ be irreducible of multiplicity n and regular in Y. Let j, k be integers such that $1 \leq j \leq n$ and $0 \leq k \leq g - 1$. If $e_{k+1} \leq j \leq e_k$, then

$$\mathrm{I}(f, f_Y^{(j)}) = (n - e_1)\beta_1 + \cdots + (e_{k-1} - e_k)\beta_k + (e_k - j)\beta_{k+1}.$$

Proof: If u is a unit in $K[[X,Y]]$ then

$$I(uf, (uf)_Y) = I(uf, u_Y f + u f_Y) = I(uf, u f_Y) = I(f, f_Y).$$

So, we may assume that f is a Weierstrass polynomial. If $(T^n, \varphi(T))$ is a Puiseux parametrization of (f), then we have

$$f(X,Y) = \prod_{\zeta \in G_0} (Y - \varphi(\zeta X^{\frac{1}{n}})).$$

Differentiating the above expression, we get

$$f_Y^{(j)}(X,Y) = j! \sum_{\substack{\Lambda \subset G_0 \\ \# \Lambda = n-j}} \left[\prod_{\zeta \in \Lambda} (Y - \varphi(\zeta X^{\frac{1}{n}})) \right].$$

Then

$$I(f, f_Y^{(j)}) = \operatorname{mult} \left(\sum_{\substack{\Lambda \subset G_0 \\ \# \Lambda = n-j}} \left[\prod_{\zeta \in \Lambda} (\varphi(T) - \varphi(\zeta T)) \right] \right).$$

We know by the Lemma (6.15) that the multiplicity of $\prod_{\zeta \in \Lambda}(\varphi(T) - \varphi(\zeta T))$ depends upon the set $G_i \setminus G_{i+1}$, where ζ is located, and it will be smaller as i is smaller. Since $e_{k+1} \leq j \leq e_k$, it follows that $n - e_k \leq n - j \leq n - e_{k+1}$, and therefore smallest multiplicities in the above expression will occur when $\Lambda = (G_0 \setminus G_k) \cup J$, where $J \subset G_k \setminus G_{k+1}$ (see figure below).

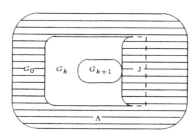

Figure 2

Therefore, we must evaluate the multiplicity of the following sum:

$$\sum_{\substack{J \subset G_k \setminus G_{k+1} \\ \# J = e_k - j}} \left[\prod_{\zeta \in (G_0 \setminus G_k) \cup J} (\varphi(T) - \varphi(\zeta T)) \right] =$$

$$\prod_{\zeta \in (G_0 \setminus G_k)} (\varphi(T) - \varphi(\zeta T)). \sum_{\substack{J \subset G_k \setminus G_{k+1} \\ \# J = e_k - j}} \left[\prod_{\eta \in J} (\varphi(T) - \varphi(\eta T)) \right].$$

From the Lemma (6.15), the multiplicity of the first factor is

$$(n - e_1)\beta_1 + (e_1 - e_2)\beta_2 + \cdots + (e_{k-1} - e_k)\beta_k.$$

On the other hand, each summand in the second factor has multiplicity

$$(e_k - j)\beta_{k+1},$$

and leading coefficient

$$b_{\beta_{k+1}}^{e_k - j} \prod_{\eta \in J} (1 - \eta^{\beta_{k+1}}).$$

To finish the proof, we must show that

$$\sum_{\substack{J \subset G_k \setminus G_{k+1} \\ \#J = e_k - j}} \left(\prod_{\eta \in J} (1 - \eta^{\beta_{k+1}}) \right), \tag{6.6}$$

is non-zero.

But, since $1 - \eta^{\beta_{k+1}} = 0$, for $\eta \in G_{k+1}$, because e_{k+1} divides β_{k+1}, it follows that (6.6) is equal to

$$\sum_{\substack{J \subset G_k \\ \#J = e_k - j}} \left(\prod_{\eta \in J} (1 - \eta^{\beta_{k+1}}) \right). \tag{6.7}$$

Observe that when η varies in G_k the expression $\eta^{\beta_{k+1}}$ assumes $n_{k+1} = \dfrac{e_k}{e_{k+1}}$ distinct values, each one repeated e_{k+1} times. With this, we may conclude that, modulo the sign, the expression (6.7) is the coefficient of Z^j of the polynomial

$$g(Z) = [(1 - Z)^{n_{k+1}} - 1]^{e_{k+1}},$$

which is also, modulo the sign, equal to

$$\sum_{i_1 + \cdots + i_{e_{k+1}} = j} \binom{n_{k+1}}{i_1} \cdots \binom{n_{k+1}}{i_{e_{k+1}}} \neq 0.$$

$\square$

COROLLARY 6.17 *If $e_{k+1} < j \leq e_k$, then*

$$\beta_{k+1} = \mathrm{I}(f, f_Y^{(j-1)}) - \mathrm{I}(f, f_Y^{(j)}). \tag{6.8}$$

Remark that if f is regular in X, then all results in this section are valid interchanging the roles of X and Y.

The above corollary allows to compute the characteristic integers of an irreducible algebroid plane curve, given in cartesian form. For this, if f is regular in Y then it is sufficient to compute the differences (6.8).

More precisely, put in (6.8) $j = 2 > e_g = 1$. Then we have

$$\mathrm{I}(f, f_Y^{(1)}) - \mathrm{I}(f, f_Y^{(2)}) = \beta_g.$$

Next, compute the difference (6.8) for $j = 3, \ldots$, until the last value of j for which it remains unchanged. This value of j is precisely e_{g-1}. Next, compute the difference (6.8) for $j = e_{g-1} + 1$, getting β_{g-1}. Next, compute the difference (6.8) for $j = e_{g-1} + 2, \ldots$, until the last value of j for which this difference remains unchanged. This new value of j is e_{g-2}. Proceeding in this way we get all the characteristic integers of (f).

If instead f is not regular in Y then it is necessarily regular in X. Hence we may use the same method as above interchanging the roles of X and Y.

The results of this section were essentially taken from [7].

PROBLEMS

3.1)   Determine the characteristic integers of the curve (f), where

$$f = Y^4 - 2X^3Y^2 - 4X^5Y + X^6 - X^7.$$

[Hint: Use the computations you did in Problem 4.3.2 and then apply the above procedure.]

6.4 APÉRY SEQUENCE OF THE SEMIGROUP OF VALUES

In his 1947 note [2], M. Apéry introduced the important notion of Apéry sequence associated to a semigroup of the naturals. As we will see, several notions and invariants may be expressed in terms of these sequences.

Let S be a semigroup with conductor c and let $p \in S \setminus \{0\}$. We define the *Apéry sequence* $a_0, \ldots, a_{p-1}$ of S, with respect to p, inductively by the conditions: $a_0 = 0$ and

$$a_j = \min \left(S \setminus \bigcup_{i=0}^{j-1} (a_i + p\mathbb{N}) \right), \quad 1 \le j \le p-1,$$

where

$$a_i + p\mathbb{N} = \{a_i + \lambda p \mid \lambda \in \mathbb{N}\}.$$

If $a \in \mathbb{N}$, we will denote by $[a]$ the residual class of a, modulo p, in $\mathbb{N}$.

It is easy to verify from the above definition that the following properties hold:

(a) $a_0 = 0 < a_1 < \cdots < a_{p-1}$.

(b) $a_i \not\equiv a_j$, mod p, $0 \le i < j \le p-1$.

(c) $a_i = \min([a_i] \cap S)$.

(d) $S = \bigcup_{j=0}^{p-1} (a_j + p\mathbb{N})$.

(e) $c = a_{p-1} - (p-1)$.

The property (d) implies that $\{p, a_1, \ldots, a_{p-1}\}$ generates S over $\mathbb{N}$. Property (e) follows on one hand, from the fact that $c - 1 = a_{p-1} - p$ doesn't belong to S, because of (c). On the other hand, for all $r \geq 0$, we have that $c + r = a_{p-1} - (p-1) + r \in S$, because otherwise we would have for some $i = 0, \ldots, p - 1$ that $c + r = a_i + \lambda p$, with $\lambda < 0$. This would imply that $a_{p-1} - a_i = \lambda p + p - 1 < 0$, a contradiction.

According to (c) and (d) above, the elements of S are of the form $a_i + \lambda p$ for some $i = 0, \ldots, p - 1$ and $\lambda \geq 0$; while the gaps of S are of the form $a_i + \lambda p$ for some $i = 0, \ldots, p - 1$ and $\lambda < 0$.

When we take $p = n = \min{(S \setminus \{0\})}$, then the *Apéry sequence* with respect to n will be called simply the *Apéry sequence of* S and the set

$$\mathcal{A} = \{a_0, \ldots, a_{n-1}\}$$

will be called the *Apéry set* of S.

Let $f \in K[[X, Y]]$ be an irreducible power series, regular of order n in Y. Recall from the Proposition (4.3) that

$$\mathcal{O}_f = K[[X]] \oplus K[[X]]y \oplus \cdots \oplus K[[X]]y^{n-1}.$$

Recall also that in Chapter 3 we defined, for $k = 0, 1, \ldots, n - 1$,

$$M_k = K[[X]] \oplus K[[X]]y \oplus \cdots \oplus K[[X]]y^k.$$

So,

$$K[[X]] = M_0 \subset M_1 \subset M_2 \subset \cdots \subset M_{n-1} = \mathcal{O}_f.$$

The next result, concerning the spaces M_k, will play an important role in what follows.

PROPOSITION 6.18 (Apéry) *Let $f \in K[[X, Y]]$ be irreducible, regular of order n in Y and such that n doesn't divide $m = \mathrm{I}(f, Y)$. Put $M_{-1} = \{0\}$ amd $y_0 = 1$. Then for every $k = 0, 1, \ldots, n - 1$, there exists an element $y_k \in y^k + M_{k-1}$ such that $v(y_k) \notin v(M_{k-1})$.*

Proof: Let $(T^n, \varphi(T))$ be a Puiseux parametrization of (f), where $\varphi(T) = a_m T^m + \cdots$, with m not a multiple of n. Observe that $y_0 \in y^0 + M_{-1}$ and

$$v(y_0) = 0 \notin v(M_{-1}) = \{\infty\}.$$

Since m is not a multiple of n, defining $y_1 = y$, we have that

$$v(y_1) = v(y) = m \notin v(M_0) = n\mathbb{N}.$$

Notice that

$$M_1 = K[[X]] + K[[X]]y_1.$$

Consider the element $\varphi_0(X) + \varphi_1(X)y_1 \in M_1$ such that $y_2 = y^2 - \varphi_0(X) - \varphi_1(X)y_1$, when viewed as a series in T, doesn't contain any term in T of order multiple of n or of order $v(y_1)$ plus a multiple of n. Notice that $v(y_2) \neq \infty$ because otherwise $y_2 \in M_1$, and consequently, $y^2 \in M_1$, which is absurd. We then have that $y_2 \in y^2 + M_1$ and $v(y_2) \notin v(M_1)$. Notice that

$$M_2 = K[[X]] + K[[X]]y_1 + K[[X]]y_2.$$

Consider the element $\varphi_0(X) + \varphi_1(X)y_1 + \varphi_2(X)y_2 \in M_2$ such that $y_3 = y^3 - \varphi_0(X) - \varphi_1(X)y_1 - \varphi_2(X)y_2$, when viewed as a series in T, doesn't contain any term in T either of order multiple of n or of order $v(y_1)$ plus a multiple of n, or of order $v(y_2)$ plus a multiple of n. Notice that $v(y_3) \neq \infty$ because otherwise, $y_3 \in M_2$ and consequently, $y^3 \in M_2$, which is absurd. We then have that $y_3 \in y^3 + M_2$ and $v(y_3) \notin v(M_2)$. Notice that

$$M_3 = K[[X]] + K[[X]]y_1 + K[[X]]y_2 + K[[X]]y_3.$$

We continue this way until we prove the result. $\square$

Remark that during the proof of the above proposition we have shown that

$$\mathcal{O}_f = K[[X]] + K[[X]]y_1 + \cdots + K[[X]]y_{n-1}. \tag{6.9}$$

We then have that the y_k, $k = 0, 1, \ldots, n - 1$, found in the above proposition generate $\mathcal{O}_f$ as a $K[[X]]$-module.

LEMMA 6.19 (Azevedo) *Suppose that, for $k = 0, \ldots, n-1$, we have elements $y_k \in y^k + M_{k-1}$ with $v(y_k) \notin v(M_{k-1})$, where $y_0 = 1$ and $M_{-1} = \{0\}$. Then for all i, j with $0 \leq i, j \leq n - 1$, and $i + j \leq n - 1$, we have that*

$$v(y_i) + v(y_j) \leq v(y_{i+j}).$$

Proof: Write $y_i = a + y^i$ and $y_j = b + y^j$ with $a \in M_{i-1}$ and $b \in M_{j-1}$. Then

$$y_i y_j = ab + ay^j + by^i + y^i y^j = c + y^{i+j},$$

where $c = ab + ay^j + by^i \in M_{i+j-1}$. Since $y^{i+j} = y_{i+j} - d$, for some $d \in M_{i+j-1}$, it follows that $y_i y_j = c - d + y_{i+j}$. Define $e = c - d \in M_{i+j-1}$. If $v(y_{i+j}) < v(y_i y_j)$, then $v(y_{i+j}) = v(e) \in v(M_{i+j-1})$, which is a contradiction because $v(y_{i+j}) \notin v(M_{i+j-1})$.
$\square$

REMARK 6.20 We have that $v(y_i) < v(y_j)$, whenever $0 \leq i < j \leq n - 1$.
    Indeed, since for $i \geq 1$, $v(y_i) \notin v(M_{i-1})$, we have that $v(y_i) \neq 0$, hence it follows that $v(y_j) \geq v(y_i) + v(y_{j-i}) > v(y_i)$.

PROPOSITION 6.21 *Let $f \in K[[X, Y]]$ be irreducible and regular in Y of order n. Put $y_0 = 1$ and let $y_1, \ldots, y_{n-1}$ be elements of $\mathcal{O}_f$ such that $y_k \in y^k + M_{k-1}$, and $v(y_k) \notin v(M_{k-1})$. Denoting by $[r]$ the residual class of the integer r, modulo n, then for all $k = 0, \ldots, n - 1$, we have that*

  (i) $v(M_k) = \cup_{i=0}^{k}\{v(y_i)\} + n\mathbb{N}$,

  (ii) $v(y_i) \not\equiv v(y_j) \mod n$, *for all $i, j = 0, \ldots, n - 1$, with $i \neq j$.*

  (iii) $v(y_k) = \min\left([v(y_k)] \cap S(f)\right)$, *for all $k = 0, \ldots, n - 1$.*

Proof: If $n = 1$, we have nothing to prove since in this case, $S(f) = \mathbb{N}$, and assertions (i), (ii) e (iii) are trivially satisfied. We then may assume that $n > 1$.

(i) We are going to prove this part by induction on k. For $k = 0$, the assertion is easily proved since every element of $K[[X]]$ is the product of a power of x by a unit in this ring, hence its value will be a positive multiple of n. Suppose then that for some k such that $1 \leq k \leq n - 1$, we have

$$v(M_{k-1}) = \{v(y_i) + \lambda n; \ 0 \leq i \leq k - 1, \lambda \geq 0\}.$$

We will prove the assertion for k. Since $M_k = M_{k-1} + K[[X]]y_k$, we may write any element β of M_k in the form $\beta = \alpha + a(X)y_k$, with $\alpha \in M_{k-1}$ and $a(X) \in K[[X]]$. We initially prove that $v(\alpha) \neq v(a(X)y_k)$. Indeed, if the contrary was true, using the inductive hypothesis, we would have, for some $i \leq k - 1$, a relation of the following type

$$v(y_i) + \lambda n = \mu n + v(y_k),$$

and therefore, in view of the Remark (6.20), $v(y_k) = v(y_i) + (\lambda - \mu)n$, with $\lambda > \mu$. Hence, again from the inductive hypothesis, $v(y_k) \in v(M_{k-1})$, which is a contradiction. Hence, from the inductive hypothesis, $v(\alpha) \neq v(a(X)y_k)$. It follows from this that either, $v(\beta) = v(\alpha) \in v(M_{k-1})$, or $v(\beta) = v(a(X)y_k) = v(y_k) + \mu n$, which proves the result.

(ii) Suppose by absurdity that $v(y_i) \equiv v(y_j) \mod n$, for some pair of integers (i, j) with $i \neq j$. Suppose that $v(y_j) = v(y_i) + \lambda n$, and without loss of generality that $j > i$. It then follows from the Remark (6.20) that $\lambda > 0$, hence from (i), $v(y_j) \in v(M_i)$, which is a contradiction.

(iii) It follows from (ii) that each residual class of integers modulo n contains exactly one of the integers $v(y_i)$, $i = 0, \ldots, n - 1$. On the other hand, the only elements in the residual class of $v(y_i)$ are from (i) the integers of the form $v(y_i) + \lambda n$, with $\lambda \geq 0$. Therefore,

$$v(y_k) = \min\left([v(y_k)] \cap S(f)\right); \quad \forall k, \ k = 0, \ldots, n - 1.$$

$\square$

Conditions (ii) and (iii) in the above proposition say that $a_i = v(y_i)$, $i = 0, \ldots, n - 1$, is the Apéry sequence of $S(f)$.

COROLLARY 6.22 *Let i and j be two distinct integers, $i, j = 0, \ldots, n - 1$, and let $\alpha_i(X), \alpha_j(X) \in K[[X]] \setminus \{0\}$. Then $v(\alpha_i(x)y_i) \not\equiv v(\alpha_j(x)y_j) \mod n$.*

Proof: We have already observed before that $v(\alpha_i(X)) = \lambda_i n$ for some natural number λ_i. If we had $v(\alpha_i(X)y_i) = v(\alpha_j(X)y_j)$, then assuming $j > i$, we would have $0 < v(y_j) - v(y_i) = (\lambda_i - \lambda_j)n$. It would follow that $\lambda_i > \lambda_j$, hence $v(y_j) \in v(M_i) \subset v(M_{j-1})$, a contradiction. $\square$

COROLLARY 6.23 *We have*

$$\mathcal{O}_f = K[[X]] \oplus K[[X]]y_1 \oplus \cdots \oplus K[[X]]y_{n-1}.$$

Proof: In view of (6.9), it is sufficient to prove that $y_0, \ldots, y_{n-1}$ are independent over $K[[X]]$. In fact, if we had a non-trivial relation $\alpha_0(X) + \alpha_1(X)y_1 + \cdots +$

$\alpha_{n-1}(X)y_{n-1} = 0$, from the Corollary (6.11) above, it would exist $i \in \{0, \ldots, n-1\}$ such that $\alpha_i \neq 0$ and

$$\infty = v(0) = v(\alpha_0(X) + \alpha_1(X)y_1 + \cdots + \alpha_{n-1}(X)y_{n-1}) = v(\alpha_i(X)y_i),$$

which is a contradiction.   □

COROLLARY 6.24 *Let* $1 = z_0, z_1, \ldots, z_{n-1} \in \mathcal{O}_f$ *be such that*

(i) $z_k \in y^k + M_{k-1}, \forall k = 0, \ldots, n-1$, *and*

(ii) $v(z_0), v(z_1), \ldots, v(z_{n-1})$ *are pairwise non-congruent modulo* n.

Then $v(z_k) = v(y_k)$, *for all* $k = 0, \ldots, n-1$.

Proof: From Proposition (6.21.iii), it is sufficient to show that $v(z_k) \notin v(M_{k-1})$ for all $k = 1, \ldots, n-1$, since (i) and this condition determine uniquely the $v(y_k)$. Since from the Proposition (6.21.i), $v(M_{k-1})$ intersects only k residual classes modulo n, and since $v(z_i) \in v(M_{k-1})$, for all $i = 0, \ldots, k-1$, it follows that $v(z_k) \notin v(M_{k-1})$, which proves the assertion.   □

Since the Apéry set of a semigroup is a set of generators, it is clear that the Apéry set and the minimal set of generators determine each other. In the next chapter we will see how, in the case of semigroups of values of branches, we can obtain the Apéry sequence from the minimal set of generators of the semigroup and how to recognize which elements of the Apéry sequence form the minimal system of generators of the semigroup.

PROBLEMS

4.1)   Determine the Apéry sequences for the following semigroups:
a) $\langle 5, 8 \rangle$.
b) $\langle n, n+1 \rangle$.
c) $\langle 6, 9, 19 \rangle$.

6.5   SEMIGROUPS AND BLOWING-UPS

In this section we will relate the semigroup of a singularity of a plane branch with the multiplicity sequence obtained from the desingularization process that we discussed in Chapter 5.

Let $f \in K[[X, Y]]$ be an irreducible power series of multiplicity n and regular in Y. Then from the Proposition (5.12) it follows that if $f^{(1)} = \sigma^*(f)$, then $S(f) \subset S(f^{(1)})$.

PROPOSITION 6.25 *Let* $f \in K[[X, Y]]$ *be an irreducible power series regular in* Y *and with tangent cone* (Y^n). *Let* $f^{(1)} = \sigma^*(f)$ *and let* $a_0 < a_1 < \cdots < a_{n-1}$ *and* $a'_0 < a'_1 < \cdots < a'_{n-1}$ *be respectively the Apéry sequences of* $S(f)$ *and* $S(f^{(1)})$, *with respect to the integer* n. *Then*

$$a_j = a'_j + jn, \quad \forall j = 0, \ldots, n-1.$$

Proof: In the above conditions, from the Proposition (5.7), we have that $f^{(1)} = \sigma^*(f)$ is regular of order n in Y_1 and is such that $v_{f^{(1)}}(x) = n$.

Since $a'_0, a'_1, \ldots, a'_{n-1}$ form a complete residue system modulo n, the set $\{a'_0, a'_1 + n, \ldots, a'_{n-1} + (n-1)n\}$ is also a complete residue system modulo n. From the Corollary (6.24), it will be sufficient to show that for every $k = 1, \ldots, n-1$, there exists $y_k \in M_{k-1} + y^k$, such that $v(y_k) = a'_k + kn$.

For every $k = 0, \ldots, n-1$, define M'_{k-1} in an analogous way as we did for M_k with $f^{(1)}$ in the place of f. Let $y'_k \in (\frac{y}{X})^k + M'_{k-1}$, be such that $v(y'_k) = a'_k$, whose existence in $\mathcal{O}_{f^{(1)}}$ is secured by the Propositions (6.18) and (6.21) and put $y'_0 = 1$. Multiplying y'_k by X^k, for $k = 0, \ldots, n-1$, we get $y_k = X^k y'_k \in M_{k-1} + y^k$. Hence $v(y_k) = v(y'_k) + v(X^k) = a'_k + kn$, finishing the proof. $\qquad\square$

When the power series is regular in X, we may apply a similar argument using τ instead of σ, showing that the result may applied to the sequence of semigroups occurring in the canonical resolution of (f).

COROLLARY 6.26 *Let c and $c^{(1)}$ be respectively the conductors of $S(f)$ and $S(f^{(1)})$. Then $c = c^{(1)} + n(n-1)$.*

Proof: We have only to use the above relations among the Apéry sequences of $S(f)$ and $S(f^{(1)})$ with respect to n and the relations $c = a_{n-1} - (n-1)$ and $c^{(1)} = a'_{n-1} - (n-1)$. $\qquad\square$

Notice that the equality $c^{(1)} = c - n(n-1)$, we have just proved, gives us another proof that after finitely many blowing-ups we get at the end a non-singular algebroid curve.

COROLLARY 6.27 *$S(f^{(1)})$ and $\mathrm{mult}(f)$ determine $S(f)$ and conversely.*

COROLLARY 6.28 *Two plane branches are equisingular if, and only if, they are equiresoluble.*

Proof: Consider the canonical resolution of a branch (f):

$$f, f^{(1)}, f^{(2)}, \ldots f^{(N)}.$$

Since $f^{(N)}$ is non-singular, its associated semigroup is $\mathbb{N}$. Since the Apéry sequence of $\mathbb{N}$ is given by $a_i^{(N)} = i$, then from Proposition (6.25) we have that the Apéry sequence of $S(f^{(N-1)})$ is determined if we know the multiplicity of $f^{(N-1)}$ and vice-versa. Using an inductive reasoning, we see that the multiplicity sequence $n^{(i)}$ of the $f^{(i)}$ determines and is determined by the semigroup $S(f)$. $\qquad\square$

Corollary (6.11), used repeatedly gives us the following formula for the conductor:

$$c = \sum_{i=0}^{r} n^{(i)}(n^{(i)} - 1), \qquad (6.10)$$

where $n^{(0)} = \mathrm{mult}(f)$.

EXAMPLE 6.29 In the case of the curve $h = Y^4 - X^7$, we saw in Chapter 5 that the multiplicity sequence in the canonical resolution is $(4, 3, 1)$, which gives $c = 4(4-1) + 3(3-1) = 18$, in accordance with the result in the Example (6.6).

PROBLEMS

5.1)   Determine the sequence of semigroups and their respective conductors in the canonical resolution of the curves in Problem 2.1, Chapter 5.

7   Semigroups of the naturals

In this chapter we will study semigroups of natural numbers from a purely arithmetical point of view. This will show that most of the properties of semigroups of values of plane branches are consequence of the very basic inequality (6.5) we proved in Section 6.2 involving their minimal system of generators.

7.1   SEMI GROUPS WITH CONDUCTORS

Given a semigroup $G = \langle x_0, \ldots, x_r \rangle$ in $\mathbb{N}$, then any element in G may be represented in several ways in the form

$$\lambda_0 x_0 + \cdots + \lambda_r x_r; \quad \lambda_0, \ldots, \lambda_r \in \mathbb{N}.$$

For example, if

$$G = \langle x_0, x_1 \rangle = \langle 3, 5 \rangle,$$

then we have

$$45 = 15x_0 = 9x_1 = 5x_0 + 6x_1.$$

However, when $G = \langle x_0, \ldots, x_r \rangle$ is a semigroup with a conductor, we will show in the Proposition (7.2) below that the elements of G may be represented uniquely as a combination of special type of the elements $x_0, \ldots, x_r$.

Let n_i and e_i, $i = 0, \ldots, g$, be the integers associated to $x_0, \ldots, x_g$ as defined in the Section 6.1.

The following lemma will be fundamental.

LEMMA 7.1 *Let* $x_0, x_1, \ldots, x_r \in \mathbb{N}$ *with* $\mathrm{GCD}(x_0, x_1, \ldots, x_r) = 1$ *and let* e_i *and* n_i, $i = 0, \ldots, r$, *be their associated integers. For every* $m \in \mathbb{N}$ *there is a unique solution for the congruence*

$$m \equiv \sum_{i=1}^{r} s_i x_i \mod x_0; \quad with \ 0 \leq s_i < n_i, \ i = 1, \ldots, r.$$

Proof: By induction on r. Case $r = 1$: Since $e_1 = \text{GCD}(x_0, x_1) = 1$, it follows that $n_1 = \dfrac{e_0}{e_1} = \dfrac{x_0}{1} = x_0$, and $\lambda x_0 + \mu x_1 = 1$ for some integers λ and μ. Then

$$m = m\lambda x_0 + m\mu x_1.$$

Dividing $m\mu$ by x_0, we get q and s_1 with $0 \le s_1 < x_0 = n_1$, such that

$$\mu m = q x_0 + s_1.$$

Putting together the above two displayed equalities, we get

$$m \equiv s_1 x_1 \mod x_0, \quad \text{with } 0 \le s_1 < n_1.$$

Let us now suppose the result true for $r \ge 1$. Let $x_0, \ldots, x_{r+1}$ be positive integers satisfying the hypothesis of the Lemma. Consider the sequence $x_0' = \dfrac{x_0}{e_r}$, $x_1' = \dfrac{x_1}{e_r}, \ldots, x_r' = \dfrac{x_r}{e_r}$, which satisfies the inductive assumption; that is, for every integer m' there exist s_i and λ such that

$$m' = \sum_{i=1}^{r} s_i x_i' + \lambda x_0', \quad \text{with } 0 \le s_i < n_i', \; i = 1, \ldots, r,$$

where

$$n_i' = \frac{e_{i-1}'}{e_i'} = \frac{\text{GCD}(x_0', \ldots, x_{i-1}')}{\text{GCD}(x_0', \ldots, x_i')} = \frac{\text{GCD}(x_0, \ldots, x_{i-1})}{\text{GCD}(x_0, \ldots, x_i)} = \frac{e_{i-1}}{e_i} = n_i.$$

By hypothesis, there exist $\lambda_0, \ldots, \lambda_{r+1}$ such that

$$\lambda_0 x_0 + \lambda_1 x_1 + \cdots + \lambda_{r+1} x_{r+1} = 1,$$

and therefore, for every integer m, we have

$$m = \lambda_0' x_0 + \lambda_1' x_1 + \cdots + \lambda_r' x_r + \lambda_{r+1}' x_{r+1}.$$

Dividing λ_{r+1}' by e_r, we get integers q and s_{r+1}, with $0 \le s_{r+1} < e_r = n_{r+1}$, such that $\lambda_{r+1}' = q e_r + s_{r+1}$. Therefore,

$$m = \lambda_0' x_0 + \lambda_1' x_1 + \cdots + \lambda_r' x_r + (q e_r + s_{r+1}) x_{r+1}.$$

Since e_r divides $x_0, \ldots, x_r$, there exists an integer m' such that

$$m = m' e_r + s_{r+1} x_{r+1}.$$

Now, the result follows by writing m' in terms of the the x_i', as above, and using the fact that $x_i = x_i' e_r$, $i = 1, \ldots, r$.

Uniqueness   Since we have proved the existence of the solution of the congruences, the uniqueness will follow from the fact that

$$\#\left\{ \sum_{i=1}^{r} s_i x_i; \; 0 \le s_i < n_i, \; i = 1, \ldots, r \right\} = n_1 n_2 \cdots n_r = x_0.$$

$\square$

PROPOSITION 7.2 *Let* $x_0, x_1, \ldots, x_r$ *be natural relatively prime numbers.*

i) *Every natural number m has a unique representation as*

$$m = \sum_{i=0}^{r} s_i x_i; \quad 0 \le s_i < n_i, \ i = 1, \ldots, r, \ s_0 \in \mathbb{Z}.$$

ii) *If $m > \sum_{i=1}^{r}(n_i - 1)x_i - x_0$, then in the above representation $s_0 \ge 0$.*

Proof: Part (i) is a consequence of the Lemma (7.1). To prove (ii), suppose that $m > \sum_{i=1}^{r}(n_i - 1)x_i - x_0$. From (i) we have a unique representation

$$m = \sum_{i=0}^{r} s_i x_i, \quad 0 \le s_i < n_i, \ i = 1, \ldots, r, \ s_0 \in \mathbb{Z}.$$

So, from the hypothesis we get

$$\sum_{i=0}^{r} s_i x_i = m > \sum_{i=1}^{r}(n_i - 1)x_i - x_0 \ge \sum_{i=1}^{r} s_i x_i - x_0,$$

which implies that $s_0 x_0 > -x_0$, giving $s_0 \ge 0$. $\qquad\square$

COROLLARY 7.3 *Let G be a semigroup in $\mathbb{N}$, with conductor c, and let $x_0, \ldots, x_r$ be any set of generators of G. Then we have*

$$c \le \sum_{i=1}^{r}(n_i - 1)x_i - x_0 + 1.$$

Notice that in the representation of the Proposition (7.2), we have that if $s_0 \ge 0$, then $m \in G$. The converse is not always true; that is, we may have $m \in G$ with $s_0 < 0$, as one can see in the following example.

EXAMPLE 7.4 Let

$$G = \langle 8, 10, 11 \rangle = \{0, 8, 10, 11, 16, 18, 19, 20, 21, 22, 24, 26, 27, \ldots\}.$$

We have that the conductor of G is $c = 26$. Also, $e_0 = 8$, $e_1 = 2$, $e_2 = 1$, $n_1 = 4$ and $n_2 = 2$. The element $22 \in G$, in the above representation, is written as

$$22 = 3 \cdot x_1 + 0 \cdot x_2 - x_0.$$

There is an important class of semigroups in $\mathbb{N}$, where this converse, we mentioned above, is true. To describe this class of semigroups, let us define the following notion.

Let $x_0, \cdots, x_r$ be a sequence of relatively prime integers, and let the n_i's be the associated integers. We say that this sequence is *nice* if, for all $i = 1, \ldots, r$, we have

$$n_i x_i \in \langle x_0, \ldots, x_{i-1} \rangle.$$

PROPOSITION 7.5 *Let $x_0, \ldots, x_r$ be a nice sequence of integers. If $G = \langle x_0, \ldots, x_r \rangle$, then*

i) *every element $m \in G$ is uniquely representable in the form*

$$m = \sum_{i=0}^{r} s_i x_i, \quad 0 \le s_i < n_i, \; i = 1, \ldots, r, \; s_0 \in \mathbb{N},$$

ii) *the conductor c of G is given by*

$$c = \sum_{i=1}^{r} (n_i - 1)x_i - x_0 + 1.$$

Proof: (i) Let $m = \lambda_0 x_0 + \cdots + \lambda_r x_r \in G$, with $\lambda_i \in \mathbb{N}$. Write $\lambda_r = q_r n_r + s_r$ with $0 \le s_r < n_r$. Then we have

$$m = s_r x_r + q_r n_r x_r + \lambda_0 x_0 + \cdots + \lambda_{r-1} x_{r-1}, \quad 0 \le s_r < n_r.$$

Since the sequence is nice, we get that

$$q_r n_r x_r + \lambda_0 x_0 + \cdots + \lambda_{r-1} x_{r-1} \in \langle x_0, \ldots x_{r-1} \rangle.$$

Now, repeat this procedure to the coefficient of x_{r-1} of the above element written as a combination of $x_0, \ldots, x_{r-1}$ with natural coefficients, and so on.
(ii) From the Corollary (7.3), we know that

$$c \le \sum_{i=1}^{r} (n_i - 1)x_i - x_0 + 1.$$

Now, since from (i) we have that $\sum_{i=1}^{r}(n_i - 1)x_i - x_0 \notin G$, the result follows. $\square$

In reality, it can be proved that assertions (i) and (ii) in the Proposition (7.5) are equivalent to each other, and also equivalent to the fact that $x_0, \ldots, x_r$ is a nice sequence (see for example [1]).

A semigroup G in $\mathbb{N}$ with conductor c will be called *symmetric* if,

$$\forall z \in \mathbb{Z}, \quad z \in G \Longleftrightarrow c - 1 - z \notin G.$$

PROPOSITION 7.6 *Let G be a semigroup in $\mathbb{N}$ with conductor c. The following assertions are equivalent.*

i) *G is symmetric,*

ii) *$2|G \cap [0, c)| = c$,*

iii) *$2|\mathbb{N} \setminus G| = c$,*

iv) *$|\mathbb{N} \setminus G| = |G \cap [0, c)|$.*

Proof: Since $(\mathbb{N} \setminus G) \cup (G \cap [0,c)) = [0,c)$, it follows that (ii),(iii) and (iv) are equivalent.

Consider now the bijection

$$\varphi : [0, c-1] \longrightarrow [0, c-1]$$
$$z \longmapsto c-1-z$$

Since G is a semigroup and $c-1 \notin G$, we have that $\varphi(G \cap [0,c)) \subset \mathbb{N} \setminus G$. Now, G symmetric is equivalent to the fact that $\varphi(G \cap [0, c-1]) = \mathbb{N} \setminus G$. This proves that (i) is equivalent to (iv). $\qquad\square$

So, if G is symmetric, c is even and there as many gaps as non-gaps of G in the interval $[0,c)$.

PROPOSITION 7.7 *Every semigroup generated by a nice sequence is symmetric.*

Proof: Let $x_0, \ldots, x_r$ be a nice sequence. We know from the Proposition (7.5.ii), that the conductor c of $G = \langle x_0, \ldots, x_r \rangle$ is equal to $\sum_{i=1}^{r}(n_i - 1)x_i - x_0 + 1$. Now from the Proposition (7.2.i), we also know that any $z \in \mathbb{N}$ may be written uniquely as

$$z = \sum_{i=0}^{r} s_i x_i; \quad 0 \leq s_i < n_i, \quad i = 1, \ldots, r; \quad s_0 \in \mathbb{Z},$$

and, from the Proposition (7.5.i), $z \in G$ if and only if $s_0 \geq 0$. Now, consider

$$c - 1 - z = \sum_{i=1}^{r}(n_i - 1 - s_i)x_i - (s_0 + 1)x_0.$$

Hence, we have that $s_0 \geq 0$ if and only if $-(s_0 + 1) \leq -1$, which proves that G is symmetric. $\qquad\square$

PROBLEMS

1.1) Let $1 < x_0 < x_1$ be two relatively prime integers. Show that x_0, x_1 forms a nice sequence. Conclude that the semigroup they generate is symmetric and its conductor is $c = (v_0 - 1)(v_1 - 1)$.

7.2 STRONGLY INCREASING SEMIGROUPS

In this section we will explore some properties of strongly increasing semigroups, which will be used intensively in what follows.

Let $x_0, x_1, \cdots, x_r$ be a sequence of natural relatively prime numbers and let e_i and n_i, $i = 0, \ldots, r$, be their associated numbers. We say that the sequence is *strongly increasing* if

$$x_i > n_{i-1}x_{i-1}, \quad \forall i = 1, \ldots, r.$$

It follows immediately from the above definition that a strongly increasing sequence is an increasing sequence.

LEMMA 7.8 *Let $x_0 < x_1 < \cdots < x_r$ be a strongly increasing sequence of natural numbers. Then*

$$x_{i+1} > \sum_{j=0}^{i}(n_j - 1)x_j, \quad i = 0, \ldots, r - 1.$$

Proof: By induction on i. For $i = 0$, this is obvious. Suppose the inequality true for x_i, then

$$x_{i+1} > n_i x_i = (n_i - 1)x_i + x_i > (n_i - 1)x_i + \sum_{j=0}^{i-1}(n_j - 1)x_j = \sum_{j=0}^{i}(n_j - 1)x_j.$$

$\square$

PROPOSITION 7.9 *Every strongly increasing sequence of natural numbers is nice. In particular, it generates a symmetric semigroup.*

Proof: Let $x_0 < x_1 < \cdots < x_r$ be a strongly increasing sequence of natural numbers. We will show that this sequence is nice, which in particular, by the Proposition (7.7), will imply that the semigroup generated by it is symmetric.

Consider the sequence $x_0' = \dfrac{x_0}{e_i}$, $x_1' = \dfrac{x_1}{e_i}, \ldots, x_i' = \dfrac{x_i}{e_i}$. Then it is easy to verify that

$$n_i' = \frac{\text{GCD}(x_0', \ldots, x_{i-1}')}{\text{GCD}(x_0', \ldots, x_i')} = \frac{\text{GCD}(x_0, \ldots, x_{i-1})}{\text{GCD}(x_0, \ldots, x_i)} = n_i.$$

From the Lemma (7.8), we have that

$$n_i x_i \geq x_i > \sum_{j=0}^{i-1}(n_j - 1)x_j - x_0,$$

which implies that

$$n_i' x_i' > \sum_{j=0}^{i-1}(n_j' - 1)x_j' - x_0',$$

and therefore, from the corollary (7.3) it follows that

$$n_i' x_i' \in \langle x_0', x_1', \ldots, x_{i-1}' \rangle,$$

which implies clearly that

$$n_i x_i \in \langle x_0, x_1, \ldots, x_{i-1} \rangle.$$

$\square$

Recall from Section 6.2 that a semigroup G is called strongly increasing if its minimal system of generators is strongly increasing.

Let $x, y \in \mathbb{N}^r$. We will say that x is smaller than y in the reverse lexicographical order, writing $x \prec y$, if the last non-zero coordinate of $y - x$ is positive. This establishes a total order relation in $\mathbb{N}^r$.

Let $x_0, \ldots, x_r$ be a sequence of positive relatively prime integers. Consider the integers $n_0, \ldots, n_r$ associated to $x_0, \ldots, x_r$ and define the set

$$E(x_0, \ldots, x_r) = \{(s_1, \ldots, s_r) \in \mathbb{N}^r; \ 0 \le s_i < n_i, \ i = 1, \ldots, r\} \subset \mathbb{N}^r.$$

We will always consider $E(x_0, \ldots, x_r)$ equipped with the reverse lexicographical order and $\mathbb{N}$ with its usual order.

LEMMA 7.10 *If $x_0, \ldots, x_r$ is a strongly increasing sequence, then the map*

$$\xi: \begin{array}{ccc} E(x_0, \ldots, x_r) & \longrightarrow & \mathbb{N}, \\ (s_1, \ldots, s_r) & \mapsto & \sum_{i=1}^r s_i x_i \end{array}$$

preserves orders.

Proof: Clearly, it is enough to prove, for all $j = 1, \ldots, r-1$, and all s_i with $s_i < n_i$, $i = 1, \ldots, j$, that

$$\sum_{i=1}^j s_i x_i < x_{j+1}.$$

The above inequality is proved as follows:

$$\begin{array}{rl} \sum_{i=1}^j s_i x_i \le & (n_1 - 1)x_1 + \cdots + (n_j - 1)x_j < \\ & n_1 x_1 + (n_2 - 1)x_2 + \cdots + (n_j - 1)x_j < \\ & n_2 x_2 + (n_3 - 1)x_3 + \cdots + (n_j - 1)x_j < \\ & \cdots < n_j x_j < x_{j+1}. \end{array}$$

$\square$

PROBLEMS

2.1) Give an example of a nice sequence which is not strongly increasing.

2.2) Determine the conductors of the semigroups of the branches in Problem 3.4.3. [Hint: Use the computations you did in Problem 6.2.1.]

7.3 APÉRY SEQUENCES

Recall from Section 6.4 the definitions of the Apéry sequence and the Apéry set of a semigroup S.

PROPOSITION 7.11 *Let $S = \langle x_0, \ldots, x_r \rangle$, where $x_0, \ldots, x_r$ is a nice sequence. We have that*

$$\mathcal{A} = \left\{ \sum_{i=1}^r s_i x_i; \ 0 \le s_i < n_i, \ i = 1, \ldots, r \right\},$$

where $n_1, \ldots, n_r$ are the integers associated to $x_0, \ldots, x_r$ and $\mathcal{A}$ is the Apéry set of S.

Proof: We know from the Proposition (7.5) that every $m \in S$ may be written in a unique way as

$$\sum_{i=1}^{r} s_i x_i + s_0 x_0, \ \ 0 \le s_i < n_i, \ \ i = 1, \dots, r \ \text{ and } \ s_0 \in \mathbb{N}.$$

This implies that each sum $\sum_{i=1}^{r} s_i x_i$ is minimal in its congruence class modulo $n = x_0$, so we have

$$\left\{ \sum_{i=1}^{r} s_i x_i; \ \ 0 \le s_i < n_i, \ \ i = 1, \dots, r \right\} \subset \mathcal{A}.$$

Now, since

$$\# \left\{ \sum_{i=1}^{r} s_i x_i; \ \ 0 \le s_i < n_i, \ \ i = 1, \dots, r \right\} = n_1 \cdots n_r = n,$$

the result follows. $\qquad\qquad\qquad\qquad\qquad\qquad\qquad\qquad\qquad\qquad\qquad\square$

LEMMA 7.12 *Let* $x_0, \dots, x_r$ *be a sequence of positive integers such that* $GCD(x_0, \dots, x_r) = 1$. *The map*

$$\begin{array}{rccc} \lambda : & E(x_0, \dots, x_r) & \longrightarrow & \mathbb{N} \\ & (s_1, \dots, s_r) & \mapsto & \sum_{i=1}^{r} s_i n_0 \dots n_{i-1} \end{array}$$

is an order preserving map that is a bijection between $E(x_0, \dots, x_r)$ *and* $\{0, 1, \dots, n - 1\}$, *where* $n = x_0 = n_1 \dots n_r$.

Proof: The inequalities

$$\sum_{i=1}^{j} s_i n_0 \dots n_{i-1} \le (n_1 - 1)n_0 + (n_2 - 1)n_0 n_1 + \cdots + (n_j - 1)n_0 \dots n_{j-1} =$$

$$n_0 n_1 \dots n_j - n_0 < n_0 n_1 \dots n_j,$$

show that λ preserves orders and that its image is contained in the set $\{0, 1, \dots, n - 1\}$. Since λ is order preserving, it follows that it is injective and because both sets $E(x_0, \dots, x_r)$ and $\{0, 1, \dots, n - 1\}$ have the same cardinality n, it follows that λ is a bijection onto this last set. $\qquad\qquad\qquad\qquad\qquad\qquad\qquad\qquad\square$

The next result will relate the Apéry sequence and the minimal set of generators of a strongly increasing semigroup. More precisely, since we know that

$$\{a_0, \dots, a_{n-1}\} = \left\{ \sum_{i=1}^{r} s_i x_i; \ \ 0 \le s_i < n_i, \ \ i = 1, \dots, r \right\},$$

but not necessary in the same order, we want to identify which a_j corresponds to a given sum $\sum_{i=1}^{r} s_i x_i$, and which element of the Apéry sequence corresponds to a given generator v_i. This will be easy to answer when the semigroup is strongly increasing.

THEOREM 7.13 *Let $v_0, \ldots, v_g$ be a strongly increasing sequence of integers and let $a_0, \ldots, a_{n-1}$, where $n = v_0$, be the Apéry sequence of $G = \langle v_0, \ldots, v_g \rangle$. Then one has:*

i) *If $(s_1, \ldots, s_g) \in E(v_0, \ldots, v_g)$, then*

$$a_{\sum_{i=1}^r s_i n_0 \ldots n_{i-1}} = \sum_{i=1}^r s_i v_i.$$

ii) *For $i = 1, \ldots, g$, $v_i = a_{\frac{n}{e_{i-1}}}$.*

Proof: The proof of (i) follows immediately from the facts we proved in the Lemmas (7.10) and (7.12) that ξ and λ are maps preserving orders. To prove (ii) just observe that $(s_1, \ldots, s_g) = (0, \ldots, 1, \ldots, 0)$, where 1 is in position i, gives us that $v_i = a_{n_0 \ldots n_{i-1}} = a_{\frac{n}{e_{i-1}}}$. $\qquad\qquad\square$

7.4   APPLICATION TO SEMIGROUPS OF VALUES

Since the semigroup of values $S(f)$ of a plane branch (f) is strongly increasing, from the Proposition (7.9) its minimal system of generators $v_0, \ldots, v_g$ forms a nice sequence. Hence from the Proposition (7.5) the conductor of $S(f)$ is given by

$$c = \sum_{i=1}^g (n_i - 1)v_i - v_0 + 1, \qquad\qquad (7.1)$$

and from the Proposition (7.7), it follows that $S(f)$ is a symmetric semigroup, and hence from the Proposition (7.6), c is an even integer.

PROPOSITION 7.14 *Let (f) be a plane branch and let $v_0, \ldots, v_g$ be the minimal system of generators of $S(f)$. If c is the conductor of $S(f)$, then*

$$c = n_g v_g - \beta_g - v_0 + 1.$$

Proof: Proposition (7.5) combined with Problem 6.2.3 imply that

$$c = n_g v_g - [v_g - (n_{g-1} - 1)v_{g-1} - \cdots - (n_1 - 1)v_1] - v_0 + 1 = n_g v_g - \beta_g - v_0 + 1.$$

$$\square$$

COROLLARY 7.15 *Let (f) be an algebroid irreducible plane curve. If $v_0, \ldots, v_g$ is the minimal system of generators of $S(f)$ and c its conductor, then*

$$c = \sum_{i=1}^g (e_{i-1} - e_i)\beta_i - \beta_0 + 1.$$

Proof: From (6.3) we have that

$$v_g = (n_1 - 1)n_2 \cdots n_{g-1}\beta_1 + (n_2 - 1)n_3 \cdots n_{g-1}\beta_2 + \cdots + (n_{g-1} - 1)\beta_{g-1} + \beta_g,$$

which replaced in the formula of the Proposition (7.11), gives the result, taking into account that $\beta_0 = v_0$ and that $e_i = n_{i+1} \cdots n_g$ (see Problem 6.1.1). $\qquad\square$

COROLLARY 7.16 *Let $f \in K[[X,Y]]$ be an irreducible power series. Let $n = I(f,X)$ and $m = I(f,Y)$. If c is the conductor of $S(f)$, then*

$$I(f, f_Y) = c + n - 1, \quad and \quad I(f, f_X) = c + m - 1.$$

Proof: Suppose that f is regular in Y (the case f regular in X is analogous). From the Theorem (6.16), corresponding to the case $k = g - 1$ and $j = 1$, we have that

$$I(f, f_Y) = (n - e_1)\beta_1 + \cdots + (e_{g-2} - e_{g-1})\beta_{g-1} + (e_{g-1} - 1)\beta_g,$$

which is equal, from the Corollary (7.15), to $c + n - 1$.

Let $(x(T), y(T))$ be a primitive parametrization of (f). From the chain rule we have that $f_X(x(T), y(T))x'(T) + f_Y(x(T), y(T))y'(T) = 0$, hence

$$
\begin{aligned}
I(f, f_X) &= \operatorname{mult}(f_X(x(T), y(T))) \\
&= \operatorname{mult}(f_Y(x(T), y(T))y'(T)) - \operatorname{mult}(x'(T)) \\
&= c + n - 1 + m - 1 - (n - 1) \\
&= c + m - 1.
\end{aligned}
$$

$\square$

In [10], Milnor introduced the following number:

$$\mu_f := \dim_K \frac{K[[X,Y]]}{\langle f_X, f_Y \rangle}.$$

The theorem below, proved by Milnor by topological methods, was reproved algebraically by Risler in [12]. First we need a proposition.

PROPOSITION 7.17 *Let $f \in K[[X,Y]]$ be irreducible and regular in Y. Then*

$$I(f^{(1)}, (f^{(1)})_{Y_1}) = I(nf^{(1)} + X_1(f^{(1)})_{X_1}, (f^{(1)})_{Y_1}).$$

Proof: Suppose that $(f^{(1)})_{Y_1} = p_1^{\alpha_1} \ldots p_r^{\alpha_r}$, where each p_i is an irreducible power series. So,

$$I(f^{(1)}, (f^{(1)})_{Y_1}) = \sum_{i=1}^{r} \alpha_i I(f^{(1)}, p_i),$$

and

$$I(nf^{(1)} + X_1(f^{(1)})_{X_1}, (f^{(1)})_{Y_1}) = \sum_{i=1}^{r} \alpha_i I(f^{(1)} + X_1(f^{(1)})_{X_1}, p_i).$$

So it is sufficient to prove, for (p) any branch of $((f^{(1)})_{Y_1})$, that

$$I(f^{(1)}, p) = I(nf^{(1)} + X_1(f^{(1)})_{X_1}, p). \tag{7.2}$$

Suppose (p) parametrized by power series (φ, ψ) in T. Then

$$\frac{d}{dT}\varphi^n \cdot f^{(1)}(\varphi, \psi) = \varphi^{n-1}\left[nf^{(1)}(\varphi, \psi) + \varphi(f^{(1)})_{X_1}(\varphi, \psi)\right]\frac{d\varphi}{dT}. \tag{7.3}$$

Since mult $\left(\frac{d}{dT}\varphi^n \cdot f^{(1)}(\varphi, \psi)\right) = n \cdot \text{mult}(\varphi) + \text{mult}\left(f^{(1)}(\varphi, \psi)\right) - 1$, and

$$\text{mult}\left(\varphi^{n-1}\left[nf^{(1)}(\varphi, \psi) + \varphi(f^{(1)})_{X_1}(\varphi, \psi)\right]\frac{d\varphi}{dT}\right) =$$
$$(n-1)\text{mult}(\varphi) + \text{mult}\left(nf^{(1)}(\varphi, \psi) + \varphi(f^{(1)})_{X_1}(\varphi, \psi)\right) + \text{mult}(\varphi) - 1,$$

we get (7.2) from (7.3). □

THEOREM 7.18 *Let* (f) *be a plane branch. If* μ_f *is Milnor's number of* (f) *and* c *the conductor of* $S(f)$, *then* $\mu_f = c$.

Proof: Suppose that f is irreducible of multiplicity n and regular in Y (the other case is analogous). From the definition of strict transform we have

$$X_1^n f^{(1)}(X_1, Y_1) = f(X_1, X_1 Y_1). \tag{7.4}$$

Differentiating expression (7.4) in Y_1 we get

$$X_1^n (f^{(1)})_{Y_1} = X_1 f_{Y_1}(X_1, X_1 Y_1),$$

hence

$$(f^{(1)})_{Y_1} = \frac{1}{X_1^{n-1}} f_{Y_1}(X_1, X_1 Y_1) = (f_Y)^{(1)}. \tag{7.5}$$

Differentiating (7.4) in X_1 we get

$$nf^{(1)} + X_1 (f^{(1)})_{X_1} = (f_X)^{(1)} + Y_1(f_Y)^{(1)}. \tag{7.6}$$

So, from (7.6) we get

$$\text{I}(nf^{(1)} + X_1(f^{(1)})_{X_1}, (f_Y)^{(1)}) = \text{I}((f_X)^{(1)}, (f_Y)^{(1)}),$$

hence from the Proposition (7.17) and (7.5) we get

$$\text{I}(f^{(1)}, (f_Y)^{(1)}) = \text{I}((f_X)^{(1)}, (f_Y)^{(1)}), \tag{7.7}$$

Now, using the Proposition (5.13) we get

$$\text{I}(f^{(1)}, (f_Y)^{(1)}) = -n(n-1) + \text{I}(f, f_Y) = -n(n-1) + c + n - 1. \tag{7.8}$$

and

$$\text{I}((f_X)^{(1)}, (f_Y)^{(1)}) = -(n-1)^2 + \text{I}(f_X, f_Y). \tag{7.9}$$

Putting (7.7), (7.8) and (7.9) together we get the result. □

Theorem (7.18) implies in particular that Milnor's number is invariant under the relation of equisingularity of curves which, in turn, implies that it is invariant under equivalence of curves.

We now will relate the semigroup of a plane branch (f) with that of its strict transform $(f^{(1)})$.

THEOREM 7.19 *Suppose that* (f) *is a plane branch with tangent cone* (Y^n) *and that* $I(f, Y)$ *is not divisible by* n. *Let* $v_0, \ldots, v_g$ *is the minimal set of generators of* $S(f)$. *Then the minimal system of generators of* $S(\sigma^*(f))$ *is given by*

$$
\begin{array}{lll}
\text{i)} & v_0 < v_1' < \cdots < v_g' & \text{if } v_0 < v_1', \\
\text{ii)} & v_1' < v_0 < v_2' < \cdots < v_g' & \text{if } v_0 > v_1', \text{ and } v_1' \nmid v_0, \\
\text{iii)} & v_1' < v_2' < \cdots < v_g' & \text{if } v_0 > v_1', \text{ and } v_1' | v_0,
\end{array}
$$

where

$$
v_j' = v_j - \frac{v_0^2}{e_{j-1}}, \quad e_j = \mathrm{GCD}(v_0, \ldots, v_j), \quad j = 1, \ldots, g.
$$

The proof of the above theorem is a consequence of the Theorem (7.13) and will be left as an exercise.

8   Contact among branches

In this chapter we will study the contact among branches. This will provide us the last and till now missing way to compute the intersection index of two branches when both are given in parametric form. The contact formula is classical and has many applications. It was used by O. Zariski in [14] to study the saturation of the local ring of a singular curve and by M. Merle in [M] to describe the behaviour of the generic polar of a plane branch.

8.1   THE CONTACT FORMULA

Let (f) and (g) be two branches defined by two power series regular in Y, both with tangent line (Y) and Puiseux parametrizations:

$$(f): \begin{cases} x = T^n \\ y = \varphi(T) = \sum_{i \geq i_0} b_i T^i \end{cases}$$

$$(h): \begin{cases} x' = T^{n'} \\ y' = \psi(T) = \sum_{j \geq j_0} b'_j T^j, \end{cases} \qquad (8.1)$$

with characteristic integers $(\beta_0, \ldots, \beta_g)$ and $(\beta'_0, \ldots, \beta'_{g'})$ respectively.

Let ζ and ξ denote respectively an nth and n'th roots of unity. We will say that the branches (f) and (h) have *contact order* $\alpha \in \mathbb{Q} \cup \{\infty\}$, if

$$\alpha = \frac{\max_{\zeta, \xi} \mathrm{mult}(\varphi(\zeta T^{n'}) - \psi(\xi T^n))}{nn'}.$$

By changing, if necessary, the above Puiseux parametrizations of (f) and (h) by equivalent ones, we may suppose that

$$\alpha = \frac{\mathrm{mult}(\varphi(T^{n'}) - \psi(T^n))}{nn'}.$$

If (f) and (h) have distinct tangent lines then the contact order is not defined, and if they have same tangent line but distinct from (Y), then we change coordinates in order that (Y) becomes the common tangent line.

From the above definition it follows that if, (f) and (h) have contact order α, then the power series $\varphi(T^{n'})$ and $\psi(T^n)$ coincide up to order $\alpha n n' - 1$.

If $S(f) = \langle v_0, \ldots, v_g \rangle$ and $S(h) = \langle v'_0, \ldots, v'_{g'} \rangle$, we will use, as previously, the notation n_i, e_i, n'_j and e'_j for the associated integers to the v_i and to the v'_j respectively (or which is the same to the β_i and the β'_j, respectively).

REMARK 8.1 Observe that if $\mathrm{mult}(\varphi(T^{n'}) - \psi(T^n))$ is realized by a power of a term in $\varphi(T^{n'})$, then $n n' \alpha = i n'$ for some integer i. Alternatively, if the multiplicity is realized by a power of a term in $\psi(T^n)$, then $n n' \alpha = j n$ for some integer j. So, either $n\alpha$ or $n'\alpha$ is an integer and if, for example, $n\alpha$ is an integer with $\beta_q \leq i = n\alpha < \beta_{q+1}$, for some q (where we put $\beta_{g+1} = \infty$), then e_q divides $n\alpha$. Analogous result is valid in the other case.

PROPOSITION 8.2 *Let (f) and (h) be two branches with contact order α such that $\dfrac{\beta_q}{n} \leq \alpha < \dfrac{\beta_{q+1}}{n}$ for some $q \geq 1$. Then*

$$
\frac{n}{n'} = \frac{e_i}{e'_i} = \frac{\beta_i}{\beta'_i} \quad for \quad
\begin{cases}
0 \leq i \leq q-1, & if \ \alpha = \dfrac{\beta_q}{n}, \\[2ex]
0 \leq i \leq q, & if \ \alpha > \dfrac{\beta_q}{n}.
\end{cases}
$$

Proof: Let us write

$$
\varphi(T^{n'}) = b_{i_0} T^{n' i_0} + b_{i_1} T^{n' i_1} + \cdots + b_{i_r} T^{n' i_r} + b_{i_{r+1}} T^{n' i_{r+1}} + \cdots
$$

and

$$
\psi(T^n) = b'_{j_0} T^{n j_0} + b'_{j_1} T^{n j_1} + \cdots + b'_{j_s} T^{n j_s} + b'_{j_{s+1}} T^{n j_{s+1}} + \cdots
$$

with $b_{i_0} \cdots b_{i_{r+1}} \neq 0$, $b'_{j_0} \cdots b'_{j_{s+1}} \neq 0$ and $n' i_r < \alpha n n' \leq n' i_{r+1}$.

Suppose that $\alpha n n' = \min\{n' i_{r+1}, n j_{s+1}\}$, then $r = s$ and

$$
n' i_\ell = n j_\ell, \quad b_{i_\ell} = b'_{j_\ell}, \quad \ell = 0, \ldots, r.
$$

On the other hand, $b_{i_{r+1}} \neq b'_{j_{r+1}}$ if $n' i_{r+1} = n j_{r+1}$.

Now, since $n' \beta_q \leq \alpha n n' < n' \beta_{q+1}$ and

$$
\mathrm{GCD}(n'n, n' i_0, \ldots, n' i_\ell) = n' \mathrm{GCD}(n, i_0, \ldots, i_\ell) = n \, \mathrm{GCD}(n', j_0, \ldots, j_\ell),
$$

the result follows from the definitions of the e_i, e'_j, β_i and β'_j.   □

COROLLARY 8.3 *Let (f) and (h) be two plane branches with characteristic integers respectively $(\beta_0, \beta_1, \ldots, \beta_g)$ and $(\beta'_0, \beta'_1, \ldots, \beta'_{g'})$. Suppose $S(f) = \langle v_0, \ldots, v_g \rangle$ and $S(f') = \langle v'_0, \ldots, v'_{g'} \rangle$ and that (f) and (h) have contact order α with $\dfrac{\beta_q}{n} \leq \alpha < \dfrac{\beta_{q+1}}{n}$, where $q \geq 1$, then*

$$
\frac{n}{n'} = \frac{e_i}{e'_i} = \frac{\beta_i}{\beta'_i} = \frac{v_i}{v'_i} \quad for \quad
\begin{cases}
0 \leq i \leq q-1, & if \ \alpha = \frac{\beta_q}{n}, \\[2ex]
0 \leq i \leq q, & if \ \alpha > \frac{\beta_q}{n}.
\end{cases}
$$

Proof: From relations (6.4) we get the following:

$$\frac{n}{n'} = \frac{\beta_0}{\beta_0'} = \cdots = \frac{\beta_i}{\beta_i'} \iff \frac{n}{n'} = \frac{v_0}{v_0'} = \cdots = \frac{v_i}{v_i'}.$$

The result now follows from the Proposition (8.2). □

Remark that if $n'e_i = ne_i'$ for $i = 0, \ldots, r$, then $n_i = n_i'$ in the same range.

The above Corollary motivates the following definition:

$$\rho = \max_j \left\{ j \in \mathbb{N}; \ \frac{v_i}{v_i'} = \frac{v_0}{v_0'}, \ \forall i, \ 0 \leq i \leq j \right\}. \tag{8.2}$$

If $S(f) = S(h)$, then $\rho = g = g'$. On the other hand, if $S(f) \neq S(h)$, then $\beta_{\rho+1} n' \neq \beta_{\rho+1}' n$ and the contact order α among (f) and (h) satisfies the inequality:

$$\alpha nn' < \max\{\beta_{\rho+1} n', \beta_{\rho+1}' n\}.$$

Thus, we have

$$\alpha < \max\left\{ \frac{\beta_{\rho+1}}{n}, \frac{\beta_{\rho+1}'}{n'} \right\}, \tag{8.3}$$

which is also valid when $S(f) = S(h)$ if we set, as we already did, $\beta_{g+1} = \infty$.

Recall the following definition:

$$G_i = \{\zeta \in K; \ \zeta^{e_i} = 1\}.$$

THEOREM 8.4 *Let (f) and (h) be branches defined by power series regular in Y, with tangent line (Y), parametrized as in (8.1). Suppose that (f) and (h) have contact order α and let $S(f) = \langle v_0, \ldots, v_g \rangle$. If $\alpha < \dfrac{\beta_1}{n}$, then $I(f, h) = \alpha nn'$. In addition, if we put $n_0 = 1$, then the assertions below are equivalent.*

i) $\dfrac{\beta_q}{n} \leq \alpha < \dfrac{\beta_{q+1}}{n}$, *for some* $q = 1 \ldots, g$.

ii) $\dfrac{I(f, h)}{\text{mult}(h)} = \dfrac{v_q}{n_0 \ldots n_{q-1}} + \dfrac{n\alpha - \beta_q}{n_1 \ldots n_q}.$

Proof: Let $G_0 = \{\zeta \in K; \ \zeta^n = 1\}$. Since f is associated to the Y-polynomial $\prod_{\zeta \in G_0} \left(Y - \varphi(\zeta x^{\frac{1}{n}}) \right)$, it follows that

$$I(f, h) = \text{mult} f(T^{n'}, \psi(T)) = \text{mult} \prod_{\zeta \in G_0} \left(\psi(T) - \varphi(\zeta T^{\frac{n'}{n}}) \right) =$$

$$\frac{1}{n} \text{mult} \left(\prod_{\zeta \in G_0} \left(\psi(T^n) - \varphi(\zeta T^{n'}) \right) \right) = \frac{1}{n} \sum_{\zeta \in G_0} \text{mult} \left(\psi(T^n) - \varphi(\zeta T^{n'}) \right) =$$

$$\frac{1}{n} \sum_{i=1}^{g+1} \sum_{\zeta \in G_{i-1} \setminus G_i} \text{mult} \left(\psi(T^n) - \varphi(\zeta T^{n'}) \right),$$

where $G_{g+1} = \emptyset$. From the above formula it follows that if $\alpha < \dfrac{\beta_1}{n}$, then $I(f,h) = \alpha nn'$.

i) $\Longrightarrow$ ii) Suppose now that $\dfrac{\beta_q}{n} \leq \alpha < \dfrac{\beta_{q+1}}{n}$, for some $q = 1, \ldots, g$. If we put $G_{-1} = \emptyset$, then from the Lemma (6.8), for $\zeta \in G_{i-1} \setminus G_i$, $i = 1, \ldots, q-1$, we have that

$$\mathrm{mult}\left(\psi(T^n) - \varphi(\zeta T^{n'})\right) = n'\beta_i. \tag{8.4}$$

At this point we divide our proof into two cases.

Case 1) $\alpha = \dfrac{\beta_q}{n}$. In this case, when $\zeta \in G_{q-1}$, we have that

$$\mathrm{mult}\left(\psi(T^n) - \varphi(\zeta T^{n'})\right) = \mathrm{mult}\left(\psi(T^n) - \varphi(T^{n'})\right) = n'\beta_q = \alpha nn'. \tag{8.5}$$

The above equation together with (8.4) imply

$$I(f,h) = \frac{1}{n}\left(e_{q-1}\alpha nn' + \sum_{i=1}^{q-1}(e_{i-1} - e_i)n'\beta_i\right),$$

which from formula (6.1) gives

$$\frac{I(f,h)}{\mathrm{mult}(h)} = \alpha e_{q-1} + \frac{1}{n}e_{q-1}(v_q - \beta_q) = \frac{v_q}{n_0 \ldots n_{q-1}}.$$

Case 2) $\dfrac{\beta_q}{n} < \alpha < \dfrac{\beta_{q+1}}{n}$. In this case, for $\zeta \in G_q$, we have that (8.5) holds. This together with (8.4) imply that

$$I(f,h) = \frac{1}{n}\left(e_q\alpha nn' + \sum_{i=1}^{q}(e_{i-1} - e_i)n'\beta_i\right).$$

From this last equality together with (6.1) and (6.4) we may conclude that

$$\frac{I(f,h)}{n'} = e_q\alpha + \frac{1}{n}e_q(v_{q+1} - \beta_{q+1}) = e_q\alpha + \frac{1}{n}e_q(n_q v_q - \beta_q) =$$

$$\frac{v_q}{n_0 \ldots n_{q-1}} + \frac{n\alpha - \beta_q}{n_1 \ldots n_q}.$$

ii) $\Longrightarrow$ i) Suppose that, for some $q = 1, \ldots, g$,

$$\frac{I(f,h)}{\mathrm{mult}(h)} = \frac{v_q}{n_0 \ldots n_{q-1}} + \frac{n\alpha - \beta_q}{n_1 \ldots n_q}.$$

If (f) and (h) have contact of order $\tilde{\alpha}$, with $\dfrac{\beta_{\tilde{q}}}{n} \leq \tilde{\alpha} < \dfrac{\beta_{\tilde{q}+1}}{n}$, we have to prove that $\tilde{q} = q$ and $\tilde{\alpha} = \alpha$.

From the first part of the proof, we have that $\tilde{q} > 0$ and

$$\frac{I(f,h)}{n'} = \frac{v_{\tilde{q}}}{n_0 \ldots n_{\tilde{q}-1}} + \frac{n\tilde{\alpha} - \beta_{\tilde{q}}}{n_1 \ldots n_{\tilde{q}}}.$$

The proof will be complete if we show that if $q > 0$, $q' > 0$ and $q' \neq q''$, then

$$\forall \alpha', \alpha'' \text{ with } \frac{\beta_{q'}}{n} \leq \alpha' < \frac{\beta_{q'+1}}{n} \text{ and } \frac{\beta_{q''}}{n} \leq \alpha'' < \frac{\beta_{q''+1}}{n},$$

we have

$$\frac{v_{q'}}{n_0 \ldots n_{q'-1}} + \frac{n\alpha' - \beta_{q'}}{n_1 \ldots n_{q'}} \neq \frac{v_{q''}}{n_0 \ldots n_{q''-1}} + \frac{n\alpha'' - \beta_{q''}}{n_1 \ldots n_{q''}}.$$

Indeed, if $q'' = q' + 1$, then from (6.4) we have

$$\frac{v_{q'+1}}{n_0 \ldots n_{q'}} + \frac{n\alpha'' - \beta_{q'+1}}{n_1 \ldots n_{q'+1}} \geq \frac{v_{q'+1}}{n_1 \ldots n_{q'}} = \frac{n_{q'} v_{q'} + \beta_{q'+1} - \beta_{q'}}{n_1 \ldots n_{q'}} =$$

$$= \frac{v_{q'}}{n_0 \ldots n_{q'-1}} + \frac{\beta_{q'+1} - \beta_{q'}}{n_1 \ldots n_{q'}} > \frac{v_{q'}}{n_0 \ldots n_{q'-1}} + \frac{n\alpha' - \beta_{q'}}{n_1 \ldots n_{q'}},$$

where the last inequality follows from the condition $\frac{\beta_{q'}}{n} \leq \alpha' < \frac{\beta_{q'+1}}{n}$. The result now follows from these considerations. $\qquad\square$

The Theorem (8.4) will allow us to compute the intersection index of two branches when both are given in parametric form.

PROBLEMS

1.1) Determine the contact order and the intersection index for the following pairs of branches:
a) $(T^4, T^6 + T^8 + T^9)$ and $(T^2, T^3 + T^4 + T^5)$.
b) $(T^4, T^6 + T^8 + T^{11})$ and $(T^2, T^3 + T^4 + T^5)$.

1.2) Prove in details the Corollary (8.3).

8.2   PAIRS OF BRANCHES WITH GIVEN SEMIGROUPS

Let $F = \langle v_0, \ldots, v_g \rangle$ and $H = \langle v_0', \ldots, v_{g'}' \rangle$ be two strongly increasing semigroups. We will put $v_{g+1} = \infty$. Our objective in this section will be to determine the set

$$\Gamma(F, H) = \{ \mathrm{I}(f, h); \;\; S(f) = F \text{ and } S(h) = H \}.$$

This will be done here in characteristic zero, since we use Puiseux parametrization. In arbitrary characteristic this was done by V. Bayer in [4].

Motivated by the Corollary (8.3), we define the integer $\rho(F, H)$ as being the integer ρ in (8.2)

THEOREM 8.5   *With notation as above, we have*

$$\Gamma(F, H) = \{ e_{i-1}' v_i + j e_i' e_i, \; 0 \leq i \leq \rho, \; 0 \leq j < \pi_{i+1} \} \cup \{ I_\infty \}$$

where $e_{-1} = e_{-1}' = 1$,

$$\pi_{i+1} = \min \left\{ \frac{v_{i+1} - n_i v_i}{e_i}, \frac{v_{i+1}' - n_i' v_i'}{e_i'} \right\}, \quad i = 0, \ldots, \rho,$$

and

$$I_\infty = \begin{cases} e_\rho v'_{\rho+1}, & \text{if } v_0 v'_{\rho+1} < v'_0 v_{\rho+1} \\ e'_\rho v_{\rho+1}, & \text{if } v_0 v'_{\rho+1} > v'_0 v_{\rho+1}. \end{cases}$$

Proof: We will prove a stronger result. Namely, for (f) any fixed branch with $S(f) = F$ and when h varies in such a way that $S(h) = H$, then we have

$$\{I(f,h); \ h \in K[[X,Y]], \ S(h) = H\} = \Gamma(F,H).$$

The intersection index $I(f,h)$ will be computed using the formula in the Theorem (8.4).

Recall from the Corollary (8.3) that $\dfrac{\beta'_i}{n'} = \dfrac{\beta_i}{n}$ for $i = 0, \ldots, \rho$.

If for some $i = 0, \ldots \rho - 1$ we have

$$\frac{\beta'_i}{n'} = \frac{\beta_i}{n} \le \alpha < \frac{\beta_{i+1}}{n} = \frac{\beta'_{i+1}}{n'}, \tag{8.6}$$

then from the Theorem (8.4) and the remark after the Corollary (8.3) we have

$$I(f,h) = n'\left(\frac{v_i}{n_1 \ldots n_{i-1}} + \frac{n\alpha - \beta_i}{n_1 \ldots n_i}\right) = e'_{i-1} v_i + e'_i (n\alpha - \beta_i).$$

Since $e'_i(n\alpha - \beta_i) = e_i(n'\alpha - \beta'_i)$, it follows from the Remark (8.1) that $e'_i(n\alpha - \beta_i) = je'_i e_i$ for some integer j. Since $\alpha \in \mathbb{Q}$ varies as in (8.6) then we have j varying in the interval

$$0 \le j = \frac{n\alpha - \beta_i}{e_i} < \frac{\beta_{i+1} - \beta_i}{e_i} = \frac{\beta'_{i+1} - \beta'_i}{e'_i}.$$

Observe that for $i = 0, \ldots, \rho - 1$,

$$\pi_{i+1} = \frac{v_{i+1} - n_i v_i}{e_i} = \frac{\beta_{i+1} - \beta_i}{e_i} = \frac{\beta'_{i+1} - \beta'_i}{e'_i} = \frac{v'_{i+1} - n'_i v'_i}{e'_i},$$

and we get part of $\Gamma(F,H)$.

Now, if

$$\frac{\beta'_\rho}{n'} = \frac{\beta_\rho}{n} \le \alpha < \min\left\{\frac{\beta_{\rho+1}}{n}, \frac{\beta'_{\rho+1}}{n'}\right\},$$

then from the Theorem (8.4),

$$I(f,h) = n'\left(\frac{v_\rho}{n_1 \ldots n_{\rho-1}} + \frac{n\alpha - \beta_\rho}{n_1 \ldots n_\rho}\right) = e'_{\rho-1} v_\rho + je'_\rho e_\rho,$$

where j is an integer such that

$$0 \le j = \frac{n\alpha - \beta_\rho}{e_\rho} < \min\left\{\frac{\beta_{\rho+1} - \beta_\rho}{e_\rho}, \frac{\beta'_{\rho+1} - \beta'_\rho}{e'_\rho}\right\}.$$

Putting

$$\pi_{\rho+1} = \min\left\{\frac{\beta_{\rho+1} - \beta_\rho}{e_\rho}, \frac{\beta'_{\rho+1} - \beta'_\rho}{e'_\rho}\right\} = \min\left\{\frac{v_{\rho+1} - n_\rho v_\rho}{e_\rho}, \frac{v'_{\rho+1} - n'_\rho v'_\rho}{e'_\rho}\right\},$$

we get the rest of $\Gamma(F, H)$, with the exception of I_∞, if $F \neq H$.

The remaining possibility, in view of (8.3), is

$$\min\left\{\frac{\beta_{\rho+1}}{n}, \frac{\beta'_{\rho+1}}{n'}\right\} \leq \alpha < \max\left\{\frac{\beta_{\rho+1}}{n}, \frac{\beta'_{\rho+1}}{n'}\right\}.$$

We will split our analysis into two cases.

Case a) $\dfrac{\beta_{\rho+1}}{n} < \dfrac{\beta'_{\rho+1}}{n'}$. Observe that this condition is equivalent to $\dfrac{v_{\rho+1}}{n} < \dfrac{v'_{\rho+1}}{n'}$.

In such case we must have $\alpha = \dfrac{\beta_{\rho+1}}{n}$, because, if otherwise, $\alpha > \dfrac{\beta_{\rho+1}}{n}$, we would

have $\dfrac{\beta_{\rho+1}}{n} = \dfrac{\beta'_{\rho+1}}{n'}$, contradicting the definition of ρ.

Hence, from the Theorem (8.4), we have

$$I(f, h) = n' \frac{v_{\rho+1}}{n_1 \ldots n_\rho} = e'_\rho v_{\rho+1} = I_\infty.$$

Case b) $\dfrac{\beta_{\rho+1}}{n} > \dfrac{\beta'_{\rho+1}}{n'}$. The proof is analogous to that of case a). $\qquad\square$

Put $\rho = \rho(F, H)$ and let π_{i+1} be as in the Theorem (8.5). Consider the following map

$$\epsilon: \begin{array}{ccc} \{(i, j); \ 0 \leq i \leq \rho, \ 0 \leq j < \pi_{i+1}\} \cup \{\infty\} & \longrightarrow & \Gamma(F, H) \subset \mathbb{N} \\ (i, j) & \mapsto & e'_{i-1} v_i + j e'_i e_i, \\ \infty & \mapsto & I_\infty. \end{array}$$

Concerning the map ϵ we have the following result.

PROPOSITION 8.6 *If we give the lexicographical order to the domain of ϵ and consider ∞ as its biggest element, then ϵ is an order preserving map.*

Proof: Since the v_i form a strongly increasing sequence we have

$$e_i v_{i+1} > e_i n_i v_i = e_{i-1} v_i, \quad \forall i = 0, \ldots g. \tag{8.7}$$

The above inequality with the fact that $\dfrac{e_i}{e'_i} = \dfrac{v_0}{v'_0}$ for all $i = 0, \ldots, \rho$ imply that

$$e'_{i-1} v_i > e'_{i-2} v_{i-1}, \quad i = 1, \ldots, \rho. \tag{8.8}$$

Let i and ℓ be integers such that $0 \leq \ell < i \leq \rho$. To prove that

$$e'_{i-1} v_i + j e_i e'_i > e'_{\ell-1} v_\ell + k e_\ell e'_\ell,$$

for all j and k such that $0 \leq j < \dfrac{v_{i+1} - n_i v_i}{e_i}$ and $0 \leq k < \dfrac{v_{\ell+1} - n_\ell v_\ell}{e_\ell}$, it is enough to show that

$$e'_{i-1} v_i > e'_{\ell-1} v_\ell + \left(\frac{v_{\ell+1} - n_\ell v_\ell}{e_\ell} - 1\right) e_\ell e'_\ell = e'_\ell v_{\ell+1} - e_\ell e'_\ell,$$

which is true because of (8.8).

On the other hand, if $i = \ell$, it is clear that for $j > k$, we have

$$e'_{i-1}v_i + je_ie'_i > e'_{\ell-1}v_\ell + ke_\ell e'_\ell.$$

To conclude the proof we have to show that

$$\mathrm{I}_\infty > e_{\rho-1}v_\rho + (\pi_{\rho+1} - 1)e'_\rho e_\rho.$$

If $F = H$, this is obvious since $\mathrm{I}_\infty = \infty$.

Suppose now that $F \neq H$. We will prove the above inequality only in the case $v_0v'_{\rho+1} < v_{\rho+1}v'_0$, since the other case is similar.

Remark that in the present case we have that $\pi_{\rho+1} = \dfrac{v'_{\rho+1} - n'_\rho v'_\rho}{e'_\rho}$. So,

$$e'_{\rho-1}v_\rho + (\pi_{\rho+1} - 1)e'_\rho e_\rho < e'_{\rho-1}v_\rho + \pi_{\rho+1}e'_\rho e_\rho =$$

$$e'_{\rho-1}v_\rho + \frac{v'_{\rho+1} - n'_\rho v'_\rho}{e'_\rho}e'_\rho e_\rho = e_\rho v'_{\rho+1} = \mathrm{I}_\infty.$$

$\square$

8.3   MAXIMAL CONTACT

Let (f) be a fixed branch with semigroup of values G of genus g. If (h) is any branch, we will denote the genus of $S(h)$ by $\mathrm{gen}(h)$. We want to introduce a notion of maximal contact among (f) and all branches (h), under the restriction that $\mathrm{gen}(h)$ is less or equal to a given integer γ, where $\gamma < g$.

There is no hope to define this maximal contact by taking

$$\sup\{\mathrm{I}(f,h); \ \ \mathrm{gen}(h) \leq \gamma\},$$

since this last set is not bounded.

Indeed, let $G = \langle v_0, v_1, v_2, \ldots \rangle$ of genus at least 2. If we take $H_n = \langle n, n+1 \rangle$, we have $\rho(G, H_n) = 0$, and since $v_0(n+1) < nv_1$, for sufficiently large n, then the element I_∞ of $\Gamma(G, H_n)$ is $v_0(n+1)$, for sufficiently large n. This shows that

$$\sup\{\mathrm{I}(f,h); \ \ \mathrm{gen}(h) \leq 1\} \geq \sup\{v_0(n+1); \ \ n \in \mathrm{I\!N}\} = \infty.$$

In contrast we have the following result.

PROPOSITION 8.7 *Let (f) be a given branch with semigroup of values $G = \langle v_0, \ldots, v_g \rangle$. Let γ be an integer less than g. Then*

$$M_\gamma(f) := \max\left\{\frac{\mathrm{I}(f,h)}{\mathrm{mult}(f)\mathrm{mult}(h)}; \ \ \mathrm{gen}(h) \leq \gamma\right\} = \frac{e_\gamma v_{\gamma+1}}{v_0^2}.$$

Proof: We clearly have

$$M_\gamma(f) = \sup_{i,H} \left\{ \frac{\max \Gamma(G,H)}{\mathrm{mult}(G)\mathrm{mult}(H)};\ \ i \le \gamma,\ \ \mathrm{gen}(H) = i \right\}.$$

Suppose that $H = \langle v_0', \ldots, v_i' \rangle$ with $i \le \gamma$. From the Theorem (8.5), for $\Gamma(G,H)$, we have

$$\mathrm{I}_\infty = \begin{cases} e_\rho v_{\rho+1}' & \text{if } v_0 v_{\rho+1}' < v_0' v_{\rho+1} \\ e_\rho' v_{\rho+1}, & \text{if } v_0 v_{\rho+1}' > v_0' v_{\rho+1}. \end{cases}$$

Since in any case we have

$$\frac{\mathrm{I}_\infty}{v_0 v_0'} \le \frac{e_\rho v_{\rho+1}}{v_0^2},$$

with equality holding if $v_0 v_{\rho+1}' > v_0' v_{\rho+1}$, and since from (8.7) the sequence $(e_{i-1} v_i)$ is increasing, it follows that

$$M_\gamma(f) = \frac{e_\gamma v_{\gamma+1}}{v_0^2}.$$

$\square$

We will say that a branch (h) such that $\mathrm{gen}(h) \le \gamma$ has *maximal contact* of order γ with (f) if we have

$$\frac{\mathrm{I}(f,h)}{\mathrm{mult}(f)\mathrm{mult}(h)} = M_\gamma(f).$$

Define

$$G_\gamma = \langle \frac{v_0}{e_\gamma}, \ldots, \frac{v_\gamma}{e_\gamma} \rangle,$$

which is strongly increasing semigroup.

Since $\rho(G, G_\gamma) = \gamma$ and $v_{\gamma+1}' = \infty$, it follows that in $\Gamma(G, G_\gamma)$ we have

$$\mathrm{I}_\infty = e_\gamma' v_{\gamma+1} = v_{\gamma+1}.$$

Hence the maximal contact of order γ with (f) occurs for a branch (h) such that $S(h) = G_\gamma$.

BIBLIOGRAPHY

[1] G. Angermüller. Die Wertehalbgruppe einer ebenen irreduziblen algebroiden Kurve. Math. Zeitschr. 153:267–282, 1977.

[2] R. Apéry. Sur les Branches Superlinéaires des Courbes Algébriques. CRAS, 222:1198–1200, 1946.

[3] A. Azevedo. The Jacobian Ideal of a Plane Algebroid Curve. PhD. Thesis, Purdue, 1967.

[4] V. Bayer. Semigroup of Two irreducible Algebroid Plane Curves. Manuscripta Math. 39: 207–241, 1985.

[5] E. Brieskorn, H. Knorrer. Plane Algebraic Curves. Birkhäuser Verlag, 1986.

[6] A. Campillo. Algebroid Curves in Positive Characteristic. Springer Lecture Notes in Mathematics 813, 1980.

[7] A. Dickenstein, C. Sessa. An Integral Criterion for the Equivalence of Plane Curves. Manuscripta Math. 37:1–9, 1982.

[8] A. Hefez, M.E. Hernandes. Computational Methods in the Local Theory of Curves. Publicações Matemáticas, IMPA, Rio de Janeiro, 2001.

[9] M. Merle. Invariants Polaires des Courbes Planes. Inventiones Math. 41:103–111 1977.

[10] J. Milnor. Singular Points of Complex Hypersurfaces. Annals of Mathematics Studies, 61, Princeton Univ. Press, 1968.

[11] M. Puiseux. Recherches sur les Fonctions Algébriques. Journal de Math. 1. 15:365–480, 1850.

[12] J.J. Risler. Sur L'idéal Jacobien D'une Courbe Plane. Bull. Soc. Math. France, 99:305–311, 1971.

[13] R. Walker. Algebraic Curves. Springer-Verlag, 1978.

[14] O. Zariski. General Theory of Saturation and of Saturated Local Rings II. Am. J. Math. 93:872–964, 1971.

[15] O. Zariski. Le Problème des Modules pour les Branches Planes. Hermann, Paris, 1986.

[16] O. Zariski, P. Samuel. Commutative Algebra, Volume 2, Van Nostrand, 1960.

Openness and Multitransversality

C.T.C. WALL Department of Mathematical Sciences The University of Liverpool, Liverpool L69 7ZL, United Kingdom. E-mail: ctcw@liv.ac.uk

The transversality theorems of Thom and Mather have been used in a wide variety of situations to show that the family of maps satisfying certain families of conditions is dense (even, a dense G_δ-set) in the space of all maps. However although the original result of Thom showed that the set of maps was also open, this is not the case for later extensions of the result, and in particular theorems showing that a set of maps defined by multitransversality conditions is open are very few.

We begin this paper by a fuller discussion of the problem, and then a review of previous work in this area, concluding with a simple criterion due to du Plessis. In the rest of the paper we demonstrate the effectiveness of this criterion by applying it to a number of problems showing the openness of families of curves in a plane defined by collections of multitransversality conditions. Similar applications to curves in 3-space will be discussed elsewhere.

1 THE PROBLEM OF OPENNESS

The main thrust of Thom's original transversality theorem, and of its numerous later variants and extensions, is to approximate a map $f : X \to Y$ by one transverse to one or more submanifolds of Y; in other words, to show that the set of maps satisfying the prescribed transversality condition is dense in the space $C^\infty(X, Y)$ of all maps, provided with some appropriate topology. In the case of a single submanifold W closed in Y it is immediate that the set of maps transverse to W is also open in $C^\infty(X, Y)$, with any of the usual topologies.

The case of a collection of submanifolds is less trivial. It is natural to suppose that these stratify a subset W of Y, and then Thom suggested a condition for the set of transverse maps to be open. The final result appears in a paper [18] of Trotman which includes also a rather full list of references to previous papers on the topic, most of which contain errors; thus illustrating the rather delicate nature of the question.

THEOREM 1.1 *Let S be a locally finite stratification of a closed subset W of a C^1-manifold Y. Then the following conditions are equivalent:*

  S is A-regular in the sense of Whitney;
  For every C^1-manifold X, $\{z \in J^1(X,Y) : z \pitchfork S\}$ is open in $J^1(X,Y)$;
  For every C^1-manifold X, $\{f \in C^1(X,Y) : f \pitchfork S\}$ is open in $C^1(X,Y)$ with the strong C^1-topology.

For the non-specialist, the result can be summarised (inaccurately) by the slogan: 'Transversality to a stratification is an open condition if and only if the stratification is A-regular'.

The common error to which Trotman draws attention is as follows. Say that S satisfies condition (t) if every C^1 map of a manifold to Y which is transverse to S at a point $x \in X$ is also transverse to S in a neighbourhood of x. It was thought that (t) was sufficient to imply that the set of maps transverse to S was open. The situation was clarified in [18] where it was shown that although condition (A) implies (t), the converse is false in general; whereas, by Theorem 1.1, (A) is the condition ensuring openness of transversality.

This is the more annoying as attempts to prove openness lead naturally to looking at condition (t). The two conditions are indeed equivalent for subanalytic stratifications S by [17], and this is sufficient for many purposes.

For the multitransversality theorem of Mather [13] one first defines $X^{(r)}$ to be the subset of X^r consisting of r-tuples of distinct points, and $_rJ^k(X,Y)$ to be the preimage of $X^{(r)}$ under the natural projection $(J^k(X,Y))^r \to X^r$. Then a C^k map $f : X \to Y$ induces by restriction of $(j^k f)^r$ a section $_r j^k f$ to the projection $_r J^k(X,Y) \to X^{(r)}$. If W is a submanifold of $_r J^k(X,Y)$, then the theorem states that the set of maps f such that $_r j^k f$ is transverse to W is dense in $C^k(X,Y)$.

To establish openness of the set of transverse maps in this case is much less straightforward. What one can see easily is that if W is compact (we allow W to have boundary, so it may be a submanifold of a larger manifold $\tilde{W}$ with boundary) then multitransversality to W is indeed an open condition. However, the submanifolds of $_r J^k(X,Y)$ of interest are never compact: their closures in $(J^k(X,Y)^r$ always contain points which project to the fat diagonal in X^r. We seek to develop methods of proving that certain sets defined by collections of multitransversality conditions are open.

In the next section we discuss general methods. This excludes special devices, such as the ingenious use in [2] of an auxiliary map to deduce openness of the set of maps there considered from openness of stable maps.

We suppose throughout that the source manifold X is compact. Then the ordinary and strong C^∞ topologies on $C^\infty(X,Y)$ coincide, and we may use this topology from now on without further mention. Most positive results may be extended without difficulty to proper maps if X is not compact, using the strong C^∞ topology; we refer the reader to [15, Chapter 4] for a discussion of some of the issues that arise in further situations.

2 REVIEW OF KNOWN METHODS

The idea of using condition (t) may be rescued to some extent as follows. We deal with the case $r = 2$ of 'bigerms'; essentially the same argument will deal with jets of any multiplicity.

LEMMA 2.1 *Suppose X is compact and we are given A-regular stratifications S of a closed subset A of $J^k(X,Y)$ and T of a closed subset B of $_2J^k(X,Y)$.*
(i) Suppose that for any $f : X \to Y$ transverse to S there exist a neighbourhood U of f and a neighbourhood V of the diagonal in $X \times X$ such that for any $g \in U$, $j^k g$ is transverse to S and the image of $_2j^k g$ is disjoint from $B \cap \pi^{-1}V$. Then the set of f with $j^k f \pitchfork S$ and $_2j^k f \pitchfork T$ is open in $C^\infty(X,Y)$.
(ii) The same follows if we suppose merely that for any $f : X \to Y$ transverse to S and any $\mathbf{x} \in X$ there exist a neighbourhood U_x of f and a neighbourhood V_x of $\mathbf{x}$ in X such that for any $g \in U_x$, $j^k g|_{V_x}$ is transverse to S and the image of $_2j^k g|_{V_x^{(2)}}$ is disjoint from B.

Proof: (i) Let V_1 be a compact neighbourhood of the diagonal in $X \times X$ with $\overline{V_1} \subset V$, and V_2 a compact subset of $X \times X$, disjoint from the diagonal, with $V_1 \cup V_2 = X \times X$. Then the set of g with $_2j^k g|_{V_2}$ transverse to T is open, and its intersection with U is a neighbourhood of f consisting of maps transverse everywhere to T.

(ii) We may suppose the V_x open. Since X is compact, it is covered by finitely many of them. The union of the corresponding $V_x \times V_x$ is then a neighbourhood V of the diagonal in $X \times X$, and the intersection of the U_x is an open neighbourhood U of f. These satisfy the hypotheses of (i), so the conclusion follows from the first part. □

It is not enough to suppose that the image of $_2j^k f$ is disjoint from $B \cap \pi^{-1}V$: it seems certain that examples similar to Trotman's, pointing out the error noted above, can be constructed.

An approach using stratifications of function spaces is given in Looijenga's thesis [12]. We recall some of his definitions.

A subspace T of a Fréchet space $\mathcal{F}$ is *a weak codimension k submanifold* if, for any $f \in T$, there exist a neighbourhood U_f of f in $\mathcal{F}$ and a C^∞-submersion $p_f : U_f \to \mathbb{R}^k$ such that $p_f^{-1}(0) = T \cap U_f$. There is an obvious notion of transversality to such a submanifold, and for any C^∞-manifold U, the T-transversal maps are open and dense in $C^\infty(U,F)$.

If U is an open subset of $\mathcal{F}$, and S a locally finite partition of U into weak submanifolds of $\mathcal{F}$ of finite codimension, then we call (U,S) a *weak Whitney stratification* if, for any smooth mapping G of a (finite dimensional) manifold U into U which is transverse to the members of S, $G^{-1}(U,S)$ defines a Whitney stratification of U. (In fact, A-regularity is enough for our purposes.) The following can be proved.

PROPOSITION 2.2 *If (U,S) is a weak Whitney stratification, and M is a compact manifold, then the set of maps transverse to S is open and dense in $C^\infty(M,U)$.*

Looijenga establishes a weak Whitney stratification for the space of maps from a manifold into $\mathbb{R}$ by combining basic results on $\mathcal{A}$-unfoldings of map-germs into $\mathbb{R}$ with some technical theorems of Sergeraert [16] which yield local product structures in function space. This approach has two disadvantages: it makes heavy use of the apparatus of Fréchet spaces, which does not appear very relevant to the problem, and it uses several results specific to the case when the target has dimension 1. However, our function spaces are indeed Fréchet, and not Banach spaces: the results of [16] are applicable here, while those of Abraham [1] are not.

To make further progress using this method, it would be necessary to give direct constructions of local product structures in function spaces: a likely method would be to develop the theory of $\mathcal{A}$-unfoldings far enough to give continuous maps of function spaces. This appears to be an interesting question.

The major known result concerning openness of multitransversality concerns the specific, but very important, example of stability. A map is said to be C^r-stable if it has a neighbourhood such that all maps in that neighbourhood are equivalent to it under C^r-diffeomorphisms of source and target. It thus follows from the definition that the set of C^r-stable maps is open. The following, however, is far from trivial. If we write MT-stable to denote maps which are C^0-stable in the sense of Thom and Mather – we recall that this class of maps is defined by a family of multitransversality conditions – then the result states that MT-stable maps form an open set in $C^\infty(X,Y)$.

There are two proofs of this result. The argument of Mather [14] can be summarised as follows. Write Ω for the set of MT-stable maps. The multitransversality conditions imposed are strong enough to imply finite singularity type, so any $f \in \Omega$ has a proper stable unfolding $\{f' : X' \to Y'; i,j\}$, where 'unfolding' means that

$$
\begin{array}{ccc}
X & \xrightarrow{f} & Y \\
i \downarrow & & j \downarrow \\
X' & \xrightarrow{f'} & Y'
\end{array}
$$

is a pullback diagram. It is shown in the theory that any proper stable map has a canonical stratification, and this gives in particular a regular stratification $\mathcal{S}_2$ of Y'. Now by the theory of versal unfoldings there exist a neighbourhood $\mathcal{U}$ of f and a continuous map

$$(\Phi,\Psi) : \mathcal{U} \longrightarrow C^\infty(X,X') \times C^\infty(Y,Y')$$

such that for all $g \in \mathcal{U}$, $(F, \Phi(g), \Psi(g))$ is an unfolding of g. Now transversality to $\mathcal{S}_2$ is an open condition, so there is an open neighbourhood $\mathcal{W}$ of j in $C^\infty(Y,Y')$ such that any $\psi \in \mathcal{W}$ is transverse to $\mathcal{S}_2$. Then $\Psi^{-1}(\mathcal{W})$ is a neighbourhood of f contained in Ω.

This proof has the merit of reducing multitransversality of f to ordinary transversality of j where we can apply the previous result. However to do this it requires the existence of a common unfolding F of all maps near f, and a regular stratification of F. Moreover while for this particular purpose it suffices to use $\mathcal{K}$-versal unfoldings, for other problems one tends to need $\mathcal{A}$-versal unfoldings and level-preserving families, so this leads back to the questions mentioned above.

The second proof is due in principle to Looijenga [9]. First, if Ω is not a neighbourhood of f, then there exists a sequence $\{f_j\}$ of maps not in Ω and converging to f (this is justified in [15, 3.4.11] despite the fact that the topology does not have countable neighbourhood bases). Passing to a subsequence, we may embed these in a 1-parameter family forming a C^∞ map $F : X \times I \to Y \times I$ of the form $F(x,t) = (f_t(x),t)$ with $f_0 = f$ and $f_{1/n}$ the sequence. A contradiction is now obtained by showing that all f_t for t small enough satisfy the multitransversality condition.

The above mentioned justification uses the following criterion, which is due to du Plessis.

PROPOSITION 2.3 *[15, Corollary 3.4.12] Let N and P be C^∞-manifolds, with N compact. Then a subset $\mathcal{U}$ is open in $C^\infty(N,P)$ if and only if for all 1-parameter families $\{f_t : N \to P \mid t \in U\}$ (for U a neighbourhood of 0 in $\mathbb{R}$) defining a smooth map $F \mid N \times U \to P \times U$, the set $\{t \in U \mid f_t \in \mathcal{U}\}$ is open in $\mathbb{R}$.*

In the remainder of this paper we will show how to use this criterion to obtain specific results. However, the question of obtaining a full analogue of Theorem 1.1 remains both open and challenging.

To apply this result to a particular problem, we take $\mathcal{U}$ as the set of maps satisfying a given set of multitransversality conditions. To show that $\mathcal{U}$ is open we consider a 1-parameter family $\{f_t\}$ of maps with f_0 satisfying the conditions, and need to deduce that they hold for f_t for all sufficiently small t. This does not require us to go further into function space technicalities, and is close in spirit to the inadequate but simple argument using condition (t). We shall spend the remainder of this paper in illustrating this method for proving openness. We confine ourselves here to problems concerning plane curves: applications to families of space curves will be considered elsewhere.

3 CONTACT WITH LINES

We consider plane curves as given by parametrisations $f : S^1 \to M$, where M is some model of the plane, with $f(S^1) = \Gamma$. The stable maps form the subset consisting of immersions whose only self-intersections are normal crossings (thus with only 2 branches at each). The openness of this set thus follows. However we now show this directly – i.e. without using the characterisation of stable maps.

LEMMA 3.1 *Let $\mathcal{C}_0$ denote the set of immersions $f : S^1 \to M$ whose only singularities are normal crossings. Then $\mathcal{C}_0$ is open in $C^\infty(S^1, M)$.*

Proof: We will use Proposition 2.3, so consider a smooth 1-parameter family $F = \{f_t : S^1 \to M \mid t \in U\}$, with $f_0 \in \mathcal{C}_0$, and suppose there exist arbitrarily small values of t for which $f_t \notin \mathcal{C}_0$: we seek a contradiction. We may choose a sequence of values $t_i \to 0$ with $f_{t_i} \notin \mathcal{C}_0$.

First suppose, for infinitely many values of i, f_{t_i} not an immersion: then it has a critical point P_i. Since S^1 is compact we may suppose (passing to a subsequence of the t_i if necessary) that P_i converges to a limit P. Since F is smooth, it follows by continuity of first derivatives that P is a critical point of f_0: a contradiction.

Next suppose there exist infinitely many values of i for which f_{t_i} has a non-transverse double point given, say, by $f_{t_i}(P_i) = f_{t_i}(Q_i)$. By compactness, we may again suppose that P_t converges to a limit P_0 and Q_t to a limit Q_0. If $P_0 \neq Q_0$ then $f_0(S^1)$ has a non-transverse double point, contrary to hypothesis. But if $P_0 = Q_0$

then since $F : S^1 \times U \to P^2 \times U$ is an immersion, it embeds some neighbourhood of $(P_0, 0)$, and this also is a contradiction.

Similarly if there are infinitely many values of i giving a triple point $f_{t_i}(P_i) = f_{t_i}(Q_i) = f_{t_i}(R_i)$, then we may suppose that all three sequences converge: $P_i \to P_0$ etc. If the limits P_0, Q_0 and R_0 are all distinct, then f_0 has a triple point: a contradiction. But if they are not, e.g. $P_0 = Q_0$, we obtain the same contradiction as in the previous case. □

There are numerous extensions to this result: here is the simplest, where we give a different style of argument which clearly extends to many other cases.

LEMMA 3.2 *Let $\mathcal{C}_0'$ denote the set of maps $f : S^1 \to M$ whose only singularities are normal crossings and simple cusps. Then $\mathcal{C}_0'$ is open in $C^\infty(S^1, M)$.*

Proof: Again consider a smooth 1-parameter family $F = \{f_t : S^1 \to M \mid t \in U\}$, with $f_0 \in \mathcal{C}_0'$, and suppose there exist arbitrarily small values of t for which $f_t \notin \mathcal{C}_0'$, and seek a contradiction. If there is a sequence $t_i \to 0$ such that $f_{t_i}(S^1)$ has a singular point $f_{t_i}(P_i) = X_i$ which is not a node or cusp, we may suppose as before that P_i converges to a point P, and hence X_i converges, to X, say.

Now if Γ_0 has a node at X, then as in the previous proof we see that no singularities worse than nodes can appear nearby. If we have a cusp, then the family F in a neighbourhood of P is induced from the versal unfolding (for right-left equivalence) of a cusp singularity, and since no nearby curve in the versal unfolding has a singularity other than an ordinary node or cusp, the same must hold here. □

We next take M to be the projective plane $P^2 = P^2(\mathbb{R})$, and impose a genericity condition on the family of intersections of Γ with lines. Let us first suppose that Γ is already stable. Each of the following conditions defines a residual, hence dense, set by Mather's multitransversality theorem:

there are no hyperflexes,

there are no flecnodes,

an inflexional tangent cannot be a bitangent,

an inflexional tangent may not pass through a node,

there are no tritangents,

a bitangent may not pass through a node,

a nodal tangent may not touch the curve elsewhere,

a nodal tangent may not pass through another node,

the line joining two nodes may not touch the curve,

no line passes through three nodes.

We may give a unified statement of these conditions as follows. If the local intersection number of curves Γ and C at a point P is k, define the *contact number* of C with Γ at P to be $k-1$, and define the *total contact number* $\kappa(C, \Gamma)$ of C with Γ to be the sum of the contact numbers at all their common points, or equivalently, the total intersection number $C.\Gamma$ diminished by the number $\#(C \cap \Gamma)$ of points of intersection. Then the above conditions on Γ are equivalent to requiring that the total contact number with any line is at most 2.

Here we use 'intersection number' in the real sense: if C is given by an equation $f(x, y) = 0$ and Γ by a parametrisation $(x, y) = (a(t), b(t))$ then the intersection number at $t = 0$ is the multiplicity of 0 as root of $f(a(t), b(t)) = 0$, i.e. the order of $\phi(t) = f(a(t), b(t))$ at $t = 0$. If $\phi(t)$ has order k at 0, and we have a 1-parameter family, leading to $\phi_u(t)$, then the sum of orders of vanishing of ϕ_u at values of t near 0 is $\leq k$, as we can reduce to the case of a polynomial $\phi_u(t) = t^k + \sum_1^k c_i(u)t^{k-i}$. We will use this semicontinuity property below.

PROPOSITION 3.3 *Let $\mathcal{C}_1$ be the family of smooth curves $f : S^1 \to P^2$ such that for each line L in the plane, the total contact number of L with Γ is at most 2. Then $\mathcal{C}_1$ is an open set.*

Proof: Let $F = \{f_t : S^1 \to P^2 \,|\, t \in U\}$ be a smooth 1-parameter family, with $f_0 \in \mathcal{C}_1$. Suppose there exist arbitrarily small values of t for which there is a line L_t whose total contact number with Γ_t is at least 3. Since we are working in projective space, the set of all lines forms the dual projective space $P^{2\vee}$ and is, in particular, compact. Thus we may pick a sequence of values of t converging to 0 for which L_t converges to a line L_0.

For each point $P \in L_0 \cap \Gamma_0$, with intersection number $a + 1$, say, choose a neighbourhood U of P not containing any other point of $L_0 \cap \Gamma_0$. Then the sum of the intersection numbers of Γ_t with L_t at points of U is upper semicontinuous. For small t either the number of points of $L_t \cap U \cap \Gamma_t$ is 0, so the intersection number and contact number are also 0, or it is ≥ 1, so as the total intersection number is $\leq a + 1$, the contact number is $\leq a$. So the total contact number also is upper semicontinuous. Hence the total contact number of L_0 with Γ_0 is ≥ 3, a contradiction. $\quad\square$

In this proof we did not make any assumptions on the singularities of Γ: assuming in addition that f is stable, or that it is an embedding, defines the intersection of two open subsets of function space, which is then also open. However, the condition on contact numbers itself imposes a sharp condition on the possible singularities.

We can contemplate various notions of 'singularity' of a plane curve, but the classifications most familiar to the author refer to singularities of a defining equation of the curve. We thus indicate briefly (adapting [6, p. 56]) how to obtain a defining equation from a parametrisation.

The *multiplicity* of Γ at a point is the minimum of its local intersection numbers with a line through the point. Suppose this finite: explicitly, suppose Γ has a single branch having intersection number $k < \infty$ at the origin with the x-axis. In the local parametrisation $x = a(t)$ and $y = b(t)$ with b of order k at 0. We can thus write $b(t) = u^k$ and choose u as a new local parameter: say $x = A(u), y = u^k$. Now by the Malgrange preparation theorem we may find C^∞ functions $A_i(y)$ such that $A(u) = \sum_0^{k-1} u^i A_i(y)$. For $0 \leq j \leq k - 1$ we may multiply by u^j to obtain the system

$$\sum_{i=k-j}^{k-1} u^{i+j-k} y A_i(y) + u^j (A_0(y) - x) + \sum_{i=1}^{k-j-1} u^{i+j} A_i(y) = 0,$$

which we regard as simultaneous linear equations in $\{u^i \,|\, 0 \leq i \leq k - 1\}$. Equating the determinant to 0 gives a monic equation of degree k in x whose coefficients are C^∞ functions of y.

Provided the local intersection number of Γ with $y = 0$ at the origin is finite, there are only a finite number of branches. We may use this procedure for each branch, and then multiply the equations together. We thus have an equation, and may use the standard (Arnol'd) notations of singularity theory to describe the cases that will arise. For example, we have

LEMMA 3.4 *If $\Gamma \in \mathcal{C}_1$ then any singularity of Γ has type A_1 or A_2. Thus $\mathcal{C}_1 \subset \mathcal{C}_0'$.*

Proof: Suppose $\Gamma \in \mathcal{C}_1$. If Γ had a singular point P of order ≥ 3 then choosing L to be tangent to some branch at P would give $(L.\Gamma)_P \geq 4$, so $\kappa(L,\Gamma)_P \geq 3$, a contradiction. If P has multiplicity 2, and does not have type A_1, take a local equation $f(x,y) = 0$. We may suppose f has 2-jet y^2. Now if the coefficient c of x^3 in f is non-zero, we have an A_2 singularity. But if $c = 0$, the local intersection number of Γ with $y = 0$ is at least 4, and again we have a contradiction.   □

We have introduced the use of the total contact number as a convenient device for formulating a collection of multitransversality conditions. Essentially the same argument could be given in more detail, and would then show that several subsets of the above collection of conditions also define open sets in the space of curves. The argument yields also the following generalisation, which is useful in certain situations.

PROPOSITION 3.5 *Suppose Δ a (perhaps reducible) curve such that for all lines L we have $\kappa(L,\Delta) \leq 2$. Then the set of curves Γ such that for all lines L we have $\kappa(L, \Gamma \cup \Delta) \leq 2$ is open.*

4   REMARKS ON DUALITY

We have just shown that any curve $\Gamma \in \mathcal{C}_1$ has only node and cusp singularities. We next show that the same holds for the dual. Recall that a curve Γ is said to have *ordinary singularities* in the sense of Plücker if neither Γ nor its dual curve has singularities other than ordinary nodes and cusps. Here, of course, the dual curve $\Gamma^\vee$ is the family of tangents to Γ, considered as a curve in the dual projective plane $P^\vee$.

LEMMA 4.1 *If $\Gamma \in \mathcal{C}_1$, then $\Gamma^\vee$ has only node and cusp singularities, so Γ has ordinary singularities.*

Proof: We recall some results from [20]. If G denotes a curve-germ at P with well-defined tangent L, $G^\vee$ its dual (with centre L and unique tangent P) write $m = m(G)$ for the multiplicity of G and $i = i(G)$ for the local intersection number of G with L; similarly for $m^\vee$ and $i^\vee$. Then, according to Zeuthen [21], $i^\vee = i = m + m^\vee$.

Let L be a (singular) point of $\Gamma^\vee$, let the tangents there be P_i and the corresponding germs $\Gamma_i^\vee$. Thus the line L is tangent to Γ at the points P_i, and the corresponding germs are Γ_i. Applying Zeuthen's formula to each of these germs (with an obvious notation) and summing, we obtain

$$2 \geq \kappa(L,\Gamma) = L.\Gamma - \#(L \cap \Gamma)$$

which is no less than

$$L.\Gamma - \sum_{Q \in L \cap \Gamma} m_Q(\Gamma) = \sum_i (i_i - m_i) = \sum_i m_i^\vee = \mathrm{mult}_L(\Gamma^\vee).$$

Thus $\Gamma^\vee$ has no points of multiplicity above 2. Now further results of [20] show that if $\Gamma^\vee$ has an A_k singularity with $k \geq 3$, then Γ either has the same, or has a singular point of higher multiplicity. But neither of these is the case.   □

By a result of Bruce [4], the local intersection number of two touching curves is equal to that of their duals at the corresponding point, provided that at least one of the curves has non-zero curvature. It is thus natural to seek a self-dual theory. Define $\mathcal{C}_1^\vee$ to consist of curves Γ such that $\Gamma^\vee \in \mathcal{C}_1$. Were the duality map continuous on (e.g.) the set of maps with ordinary singularities, we could deduce the openness of $\mathcal{C}_1^\vee$ from Proposition 3.3. However, the duality map is far from continuous, as we see by considering, for example, the family $f_u(t) = (u + t^2, ut + t^3)$ of cubic curves Γ_u. For $u \neq 0$ we have a nodal cubic, whose dual has degree 4; for $u = 0$ a cuspidal cubic, with dual of degree 3. We thus have to treat the dual cases separately.

We review which further situations are excluded by the $\mathcal{C}_1$ hypothesis. A line with $\kappa(L, \Gamma) = 3$ would have intersections with Γ of multiplicities 4, 3 and 2, or 2 and 2 and 2. In the first case we either have a hyperflex tangent (giving an E_6 on the dual) or a flecnode, with the tangent as a flexional tangent to one branch.

For the remaining cases introduce the notations A, B, C for the 3 types of local intersection number 3: inflexional tangent, nodal tangent, cuspidal tangent respectively; and U, V, W for the types of local intersection number 2: ordinary tangent, nodal chord, cuspidal chord. Eight of the 19 cases correspond to similar configurations on the dual:

Figure on Γ:	AU	AV	AW	BU	CU	UUU	UUV	UUW
Figure on $\Gamma^\vee$:	D_5	UUW	AW	BU	*flecnode*	D_4	UUV	AV

Of the remaining 11 cases, 4 configurations on $\Gamma^\vee$ correspond to a bitangent (BV, CV) or inflexional tangent (BW, CW) of Γ passing through a point of contact of a bitangent (BV, BW) or inflexional tangent (CV, CW). The next 3 correspond to 2 bitangents (UVV), a bitangent and a flex tangent (UVW) or 2 flex tangents (UWW) meeting on Γ, and the final 4 cases (VVV, VVW, VWW, WWW) to the concurrence of 3 lines, each a bitangent or flex tangent.

We do not prove openness of $\mathcal{C}_1^\vee$, but establish

LEMMA 4.2 *The set* $\mathcal{C}_1^* := \mathcal{C}_1 \cap \mathcal{C}_1^\vee$ *is open.*

Proof: By Proposition 3.3, $\mathcal{C}_1$ is open. The additional explicit conditions defining $\mathcal{C}_1^*$ are the exclusions of the 11 situations just listed. We thus consider a 1-parameter family Γ_u in $\mathcal{C}_1$, with $\Gamma_0 \in \mathcal{C}_1^*$. Suppose, if possible, that for arbitrarily small u, Γ_u presents one of these 11 situations. Each of these refers to the concurrence of 3 curves, each a line or Γ. We may thus also suppose that each of the lines, and the point of concurrence, also converge, while of course the curves Γ_u converge to $\Gamma = \Gamma_0$.

A sequence of inflexional tangents converges to the tangent at a flex or hyperflex; and a sequence of bitangents converges to a bitangent or hyperflex tangent. Since $\Gamma_0 \in \mathcal{C}_1^*$, it does not present any of the configurations in question. It follows that two (at least) of the limiting lines must coincide. But also Γ has no hyperflex. As the number of flex tangents and bitangents can only vary when the family of curves passes through a hyperflex, we have our contradiction.   □

5   CONTACT WITH CIRCLES

Our next example considers the conditions studied in a series of papers by Bruce, Giblin and Gibson starting with [5]; see also [3]. The *symmetry set* of the plane curve Γ is defined to be the locus of centres of circles C whose total contact number $\kappa(C,\Gamma) \geq 2$. For several results they impose the condition that this number is at most 3 for any circle. Certainly this defines a dense set of curves: we now seek to prove it open.

Take our model M of the plane to be the inversive plane S^2, which may be identified with $P^1(\mathbb{C})$. The group of the geometry is $PGL_2(\mathbb{C})$, which preserves the family of circles but does not respect distances. Coordinates may be obtained by stereographic projection of S^2 to a plane: this maps circles to circles and straight lines. We will study the contact of Γ with the family of all circles in the plane. We show

THEOREM 5.1 *Let $\mathcal{C}_2$ be the family of smooth curves $f : S^1 \to S^2$ such that for each circle C in S^2, the total contact number of C with Γ is at most 3. Then $\mathcal{C}_2$ is an open set.*

All the arguments used in the proof of Proposition 3.3 apply here with one exception: we need to know that the family of all circles is compact, so that a sequence of circles contains a convergent subsequence.

If we identify S^2 with the unit sphere in $\mathbb{R}^3$, then the circles are the intersections of S^2 with the planes $ax + by + cz = d$ where we may normalise $a^2 + b^2 + c^2 = 1$ and then require $|d| < 1$ for the intersection to be a circle. We compactify the space of circles by allowing $|d| \leq 1$, thus including 'point' circles. It is then necessary to consider the total contact number of Γ with a sequence of circles shrinking to a point. Thus we need to study the structure of Γ near a point.

We recall from elementary geometry that if P is a smooth point of a plane curve Δ then among the circles touching Δ at P there is a unique one with local intersection number ≥ 3: the 'circle of curvature' at P.

LEMMA 5.2 *If Γ is a curve in $\mathcal{C}_2$, then Γ has no singularities other than (ordinary) nodes and cusps: we have $\mathcal{C}_2 \subset \mathcal{C}_0'$.*

Proof: We argue by contradiction. If P is a singular point of Γ of multiplicity ≥ 4, we take a circle C touching one of the branches of Γ at P: then the local intersection number $(C.\Gamma)_P \geq 5$, so $\kappa(C,\Gamma) \geq 4$: a contradiction. If P is a singular point of multiplicity 3 such that one of the branches of Γ at P is smooth, then the circle of curvature C of that branch again has $(C.\Gamma)_P \geq 5$, giving a contradiction. If Γ has multiplicity 3 and just one branch at P, we can make a linear change of coordinates to put its equation in the form $y^3 + ax^4 + \text{higher terms} = 0$. If $a \neq 0$ here, we have a singularity of type E_6: we defer consideration of this case. If $a = 0$ take C as the circle $y = 0$, which again gives a contradiction since $(C.\Gamma)_P \geq 5$.

Now suppose P is a singular point of multiplicity 2. If there are 2 branches B_1, B_2 at P, then both are smooth and if the singularity is not a simple node, the two are tangent. Take C as the circle of curvature of B_1: this has local intersection number ≥ 3 with B_1 and ≥ 2 with B_2, giving at least 5 in all: a contradiction. If there is a single branch at P, with a singular point of type A_{2k}, the equation may be written in the form $(y - \phi(x))^2 = x^{2k+1}$, and if $k \geq 2$ then the circle of curvature of $y = \phi(x)$ has local intersection number at least 5 with Γ at P. Thus only singularities of types A_1 and A_2 remain.

We must now return to the case when Γ has a singular point P of type E_6. For any circle C having the same tangent as Γ at P, the local intersection number is 4, which does not of itself yield a contradiction. Consider the family of all such curves C, which are the intersections of S^2 with a variable plane through a fixed line tangent to the sphere: begin with the plane tangent to S^2 (so having intersection a point circle) and rotate continuously through an angle π till we return to the original plane. We may suppose the local equation of Γ to be $y^3 = x^4$ added to higher order terms. So a small circle below the x-axis meets Γ only at P. A small circle $x^2 + y^2 = 2ay$ just above the axis (i.e. with $a > 0$ small) meets it where (substituting for x^2 and ignoring the higher order terms) $y^3 = (y^2 - 2ay)^2$. Apart from the root $y = 0$ this has one root near $y = 1$ and one with $y = 4a^2 + \ldots$. Such circles thus cross Γ at least twice. During the rotation of the plane there must be a first point where the circle C of intersection with S^2 meets — and hence is tangent to — Γ. For this circle we have an additional contribution to κ and now have the desired contradiction. $\qquad\square$

This final argument is an essential point in our proof: it can be shown by direct calculation that for any curve Γ_0 with a singular point of type E_6 there exist 1-parameter families Γ_t of curves and C_t of circles shrinking to a point circle C_0 at the singularity such that $\kappa(C_t, \Gamma_t) = 4$.

The proof of Theorem 5.1 is completed by the following lemma.

LEMMA 5.3 *Let $F = \{f_u : S^1 \to S^2\}$ be a smooth 1-parameter family of maps with $f = f_0$; write $\Gamma := f(S^1)$ and let $P \in S^1$, $X = f(P)$.*
(i) Let X be a smooth point of Γ. Then there exist neighbourhoods V of X in S^2 and U of 0 in $\mathbb{R}$ such that if C is a circle contained in V and $u \in U$, then $\kappa(C, f_u(S^1)) \leq 1$.
(ii) Let X be a simple node of Γ. Then there exist neighbourhoods V of X in S^2 and U of 0 in $\mathbb{R}$ such that if C is a circle contained in V and $u \in U$, then $\kappa(C, f_u(S^1)) \leq 2$.
(iii) Let X be a simple cusp of Γ. Then there exist neighbourhoods V of X in S^2 and U of 0 in $\mathbb{R}$ such that if C is a circle contained in V and $u \in U$, then $\kappa(C, f_u(S^1)) \leq 3$.

Proof: (i) We take local coordinates with respect to which f takes the form $f(t) = (t + \phi(t), \psi(t))$, where ϕ and ψ denote functions of higher order (i.e. divisible by t^2). The deformed curve f_u may be written in the same form, but now the error terms $\phi(t, u), \psi(t, u)$ belong to the ideal generated by t^2 and u. We consider intersections with the circle $(x - a(u))^2 + (y - b(u))^2 = r(u)^2$ where all of $a(u), b(u)$ and $r(u)$ must tend to 0 with u. Intersection points are given by

$$(t + \phi(t, u) - a(u))^2 + (\psi(t, u) - b(u))^2 - r(u)^2 = 0.$$

When u becomes zero, the left hand side of the equation reduces to t^2 added to higher order terms. The local intersection number is thus at most 2 in the limit. It follows that for all small enough u, $\kappa \leq 1$.

(ii) We have just seen that for a smooth branch, small circles cannot be either circles of curvature or bitangent. Thus when we have 2 smooth branches, the only possibilities for contributions to κ are for circles simply touching one or other branch, or passing through the point of intersection. Since a circle cannot do all three (as the branches are not tangent at the point of intersection), we must have $\kappa \leq 2$.

(iii) We follow the same approach as in (i). We have a family of curves given by $(x, y) = (t^2 + \phi(t, u), t^3 + \psi(t, u))$, with $\phi \in \langle t^3, u \rangle$, $\psi \in \langle t^4, u \rangle$, and consider the contact with the same family of circles as before. This leads to the equation

$$(t^2 + \phi(t, u) - a(u))^2 + (t^3 + \psi(t, u) - b(u))^2 - r(u)^2 = 0.$$

When u vanishes, the left hand side of this equation has order 4 in t. There are thus (at most) 4 intersection points in the limit, and hence for small values of u, and so $\kappa \leq 3$, as asserted.   □

6   CONTACT WITH CONICS

We next consider the analogous question for contact with conics. This is in part motivated by work of Izumiya and Sano [10], [11] and Giblin and Sapiro [7], [8] on affine differential geometry. We define $\mathcal{C}_3$ to be the set of curves whose total contact number with any *smooth* conic is at most 5. We aim to prove openness of a set (to be defined) closely related to $\mathcal{C}_3$.

It is clear from the work on circles that to prove openness it will be necessary to treat with some care sequences of conics converging in some sense to a singular conic. If we take conics to be given by equations, then a singular conic is either a line-pair or a repeated line. To deal with the case of a line-pair we will take the total contact numbers with the two lines separately, with a correction if Γ passes through the point of intersection. We will thus restrict to $\mathcal{C}_1 \cap \mathcal{C}_3$. Thus by Lemma 3.4, our curves will have at most ordinary nodes and cusps. (In fact we can show that for all $\Gamma \in \mathcal{C}_3$, Γ has no singularities of multiplicity ≥ 4, and none of multiplicity 3 of type other than D_4 or E_6.)

If we treat a repeated line naively, then since any intersection with Γ has multiplicity ≥ 2, each would contribute to the contact number, which would thus be unbounded. We will thus need to study the behaviour of $\kappa(C_u, \Gamma_u)$ as C_u converges to a repeated line L^2. Recall that another compactification of the space of smooth conics is that of Schubert: we consider together with the sequence $\{C_u\}$ of conics the sequence $\{C_u^\vee\}$ of their duals, and say that $\{C_u\}$ converges only if both the underlying loci, of points and of lines, converge. There are three types of degenerate conic: a line-pair (with the point of intersection a repeated point), a point-pair (with the join a repeated line), and a flag consisting of a line and a point on it (both repeated). In our situation we may assume (by the usual compactness argument) that $\{C_u^\vee\}$ converges, to a pair (A, B) of points (possibly coincident) on L.

We might hope that for $P \in L$ distinct from A and B, the nearby curves Γ_u would have contact with only one branch of C_u near P, so that the local κ would be semicontinuous, while near $P = A$ the local intersection number $C_u.\Gamma_u$ is at most double $L.\Gamma$, thus providing a bound in this case. While the latter hope can be justified, we are disabused of the former by the example where Γ_u is given locally by $(x, y) = (t^2, t^3 - ut)$ and C_u by $x(x - u) = 0$. The line L given by $x = 0$ has $\kappa(L, \Gamma_0) = 1$ but if $u > 0$ the curve Γ_u has contact with both $x = 0$, which passes through the node, and $x = u$ which is a tangent.

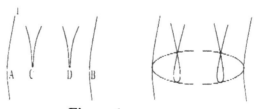

Figure 1

Building on this, Figure 1 illustrates that if the line L is transverse to Γ at two cusps C, D and two further points A, B we can find a smooth family Γ_u of curves through Γ and a smooth family C_u of conics converging to L^2 (with $C_u^\vee$ converging to AB) such that $\kappa(C_u, \Gamma_u) = 6$.

We thus see that in fact $\mathcal{C}_1 \cap \mathcal{C}_3$ is *not* open. However we now show

THEOREM 6.1 *The set $\mathcal{C}_0 \cap \mathcal{C}_1 \cap \mathcal{C}_3$ is open.*

Proof: Consider a 1-parameter family Γ_u with $\Gamma_0 \in \mathcal{C}_0 \cap \mathcal{C}_1 \cap \mathcal{C}_3$. Since $\mathcal{C}_0 \cap \mathcal{C}_1$ is open, we may suppose that all Γ_u belong to it. Suppose, if possible, that for arbitrarily small u, $\Gamma_u \notin \mathcal{C}_3$: then there is a smooth conic C_u with $\kappa(C_u, \Gamma_u) \geq 6$. Choose a sequence $u_n \to 0$ of such values of u; write C_n for C_{u_n} and Γ_n for Γ_{u_n}. We may suppose that both C_n and its dual converge to conics C_0, $C_0^\vee$.

First suppose C_0 non-singular. Then, as in the proof of Proposition 3.3, it follows that $\kappa(C_0, \Gamma) \geq 6$: a contradiction.

Next suppose C_0 consists of two distinct lines L, M. Then, as in the preceding case, $\kappa(L \cup M, \Gamma) \geq 6$. By hypothesis, $\kappa(L, \Gamma) \leq 2$ and $\kappa(M, \Gamma) \leq 2$. But we see directly from the definition that $\kappa(L \cup M, \Gamma)$ is equal to $\kappa(L, \Gamma) + \kappa(M, \Gamma)$ or to $\kappa(L, \Gamma) + \kappa(M, \Gamma) + 1$ according as Γ does not or does pass through $L \cap M$. We thus have a contradiction.

It remains to consider the case $C_0 = L^2$, where $C_0^\vee = AB$, with the points A, B (which may coincide) on the line L.

We require a precise analysis of possibilities. At a point P of L, the curve Γ may be smooth or have a node; the contact number $\kappa(L, \Gamma)_P$ may be 0, 1 or 2. We thus have 5 cases, which we label $(S0), (S1), (S2), (N1)$ and $(N2)$ where S refers to simple, N to nodal, and the integer is the contact number. We also need to distinguish whether or not P is one of the points A, B.

LEMMA 6.2 *Assume $\Gamma \in \mathcal{C}_0 \cap \mathcal{C}_1 \cap \mathcal{C}_3$. Suppose the sequences $C_n \to L^2$, $C_n^\vee \to AB$, $\Gamma_n \to \Gamma$ as above. Let $P \in L \cap \Gamma$. Then there exists a neighbourhood U of P in the plane such that, for all large enough n, $\kappa(C_n \cap U, \Gamma_n) \leq \beta$, where β is given by the following table.*

Type of P	S0	S1	S2	N1	N2
$P \neq A, B$	0	1	2	1	2
$P = A$	1	2	4	2	4

Proof: First suppose $P \neq A, B$. For large enough n, $C_n \cap U$ approximately consists of two lines both parallel to L. Take L as the x-axis. If Γ is smooth at P and $(L.\Gamma)_P = k$, we may parametrise Γ as $(x, y) = (a(t), t^k)$. Thus for Γ_u we have $y = b(t, u)$ an unfolding of t^k. The order of contact with any line $y = c$ is the multiplicity of c as root of $\partial b(t, u)/\partial t = 0$. Letting U tend to 0, we see that in the limit the sum of all these orders does not exceed $k - 1$. This gives the first 3 entries in the upper line of the table.

If Γ has a simple node at P, this is stable: each nearby Γ_n has a simple node (and no other singularity) nearby. For the $(N1)$ case, $U \cap \Gamma$ is the union of two smooth curves, each of type $(S0)$, so having 0 contact in the limit with the C_n. Thus the only contact occurs when C_n passes (necessarily transversely) through the node. Similarly for the $(N2)$ case, the branch of Γ tangent to L contributes at most 1 to the contact number, and a further 1 is provided if C_n passes through the node: the total cannot exceed 2.

We now suppose $P = A$ (B, if distinct, plays the same role as A). If $(L.\Gamma)_P = k$, semicontinuity shows that for large enough n, $(C_n \cap U).\Gamma_n \le 2k$, so that $\kappa(C_n \cap U, \Gamma_n) \le 2k - 1$. This proves the result for the $(S0)$ case; in the others we need to improve this estimate by 1. We thus suppose that equality holds, and seek a contradiction. Equality here implies that we have intersection number $2k$ concentrated at a single point.

For $(N1)$ this means that C_n is tangent to each branch of Γ_n at the node, which clearly cannot occur.

For the case of $(S1)$, observe that near P the conic C_n has a very sharp turn: see figure.

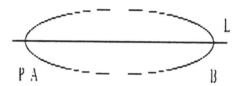

Figure 2

Since Γ has intersection number 2 with L at P (so resembles $y = x^2$), it cannot lie on the inside of C_n. As there is only one point of intersection, it must be outside. But then the curvatures of C_n and Γ_n at the point of intersection are in opposite senses, so in fact the intersection number ≤ 2.

The argument for $(N2)$ is now like that for $(N1)$: the case of equality could only occur if we had intersection number 2 with one branch and 4 with the other at the node, which is clearly impossible. (We cannot improve the estimate to $\kappa < 4$ since it remains possible to touch the 'vertical' branch at the node and have intersection number 3 with the 'horizontal' branch of Γ_n at a further point in U.)

It remains to consider case $(S2)$. First note that Γ has an ordinary inflexion at P, and ordinary inflexions are stable. We make this precise as follows. We have a smooth parametrisation $(x, y) = (\phi(t, u), \psi(t, u))$ of Γ_u such that $\phi(t, 0)$ has a simple zero and $\psi(t, 0)$ a triple one at $t = 0$. The condition for a flex is that
$$\Delta(t, u) := \begin{vmatrix} \partial\phi/\partial t & \partial\psi/\partial t \\ \partial^2\phi/\partial t^2 & \partial^2\psi/\partial t^2 \end{vmatrix} = 0,$$ and it follows from the preceding sentence
that $\partial\Delta(t, u)/\partial t$ is non-zero at $t = 0$. By the implicit function theorem there is a smooth function-germ $t_0(u)$ such that $\Delta(t_0(u), u) = 0$. Change the parametrisations by $t' := t - t_0(u)$; then Γ_u has a flex at $t = 0$. Shifting coordinates (by smooth functions of u), we may suppose that this flex is at $(0, 0)$; rotating coordinates, we may suppose the flex tangent to be $y = 0$ in each case. We now change parametrisation again, taking $t'' := x$ as new parameter on each curve. Since $y = 0$ is an inflexional tangent at 0, the parametrisation takes the form $y = t^3\chi(t, u)$; and since Γ had an ordinary flex, $\chi(0, 0) \ne 0$.

Substituting in the equation $ax^2 + 2hxy + by^2 + 2gx + 2fy + c = 0$ of a generic conic, where we may normalise $b = 1$ since the conic is converging to y^2, gives

$$t^6\chi^2(t,u) + 2ht^4\chi(t,u) + 2ft^3\chi(t,u) + at^2 + 2gt + c = 0.$$

For this to have a root of multiplicity 6 at $t = \tau$ the left hand side, together with its first 5 derivatives with respect to t, must vanish at that value. We solve the first 5 of these equations for c, g, a, f and h in turn, obtaining formulae which imply successively

$$c \in \langle g\tau, a\tau^2, f\tau^3, h\tau^4, \tau^6 \rangle, \quad g \in \langle a\tau, f\tau^2, h\tau^3, \tau^5 \rangle,$$

$$a \in \langle f\tau, h\tau^2, \tau^4 \rangle, \quad f \in \langle h\tau, \tau^3 \rangle, \quad h \in \langle \tau^2 \rangle.$$

Substituting in the final equation gives $\tau = 0$. But then the conic becomes $y^2 = 0$, contradicting the hypothesis that it is smooth. □

We return to the proof of Theorem 6.1. Write $\kappa(L, \Gamma) = a + b + c$, where a, b are the respective contributions of A, B and c is the sum of the others. It follows from the lemma that, for all large enough n, $\kappa_n := \kappa(C_n, \Gamma_n) \leq \max(2a, 1) + \max(2b, 1) + c$. Now by hypothesis, $a + b + c \leq 2$. If $a = b = 0$, $\kappa_n \leq 2 + c \leq 4$. If $a > 0, b > 0$, then $\kappa_n \leq 2a + 2b + c \leq 2(a + b + c) \leq 4$. Otherwise we may suppose $a > 0 = b$. If $c > 0$ then $\kappa_n \leq 2a + c + 1 \leq 2a + 2c \leq 4$. In the final case $c = 0$, if $a = 2$ the upper bound is $4 + 1 + 0 = 5$ (in all other cases it was strictly smaller). But this already contradicts our hypothesis $\kappa_n \geq 6$, and thus proves the theorem. □

As in the case of Theorem 5.1, Theorem 6.1 has an extension in the spirit of Proposition 3.5, which is proved by the same method.

REFERENCES

[1] R. Abraham. Transversality in manifolds of mappings. Bull. Amer. Math. Soc. 69:470-474, 1963.

[2] T. Banchoff, T. Gaffney, C. McCrory. Counting tritangent planes of space curves. Topology 24:15-24, 1985.

[3] T. Banchoff, P.J. Giblin. Global theorems for symmetry sets of smooth curves and polygons in the plane. Proc. Roy. Soc. Edinburgh 106A:221–231, 1987.

[4] J.W. Bruce. On contact of hypersurfaces. Bull. London Math. Soc. 13:51–54, 1981.

[5] J.W. Bruce, P.J. Giblin, C.G. Gibson. Symmetry sets. Proc. Roy. Soc. Edinburgh 101A:163–186, 1985.

[6] D. Eisenbud, W. Neumann. Three dimensional link theory and invariants of plane curve singularities. Princeton University Press, 1985.

[7] P.J Giblin, G. Sapiro. Affine invariant distances. envelopes and symmetry sets, Technical Report HPL-96-93, Hewlett-Packard, June 1996; Geom. Dedicata. 71:237–261, 1998.

[8] P.J. Giblin, G. Sapiro. Affine versions of the symmetry set. In: Real and Complex Singularities, eds. J.W.Bruce and F.Tari, Chapman and Hall/CRC Research Notes in Mathematics 412. pp 173–187, 1999.

[9] C.G. Gibson, K. Wirthmüller, A.A. du Plessis, E.J.N. Looijenga. Topological stability of smooth mappings. Springer lecture notes 552, 1976.

[10] S. Izumiya, T. Sano. Generic affine differential geometry of plane curves. Proc. Edinburgh Math. Soc. 41:315–324, 1998.

[11] S. Izumiya, T. Sano. Generic affine differential geometry of space curves. Proc. Roy. Soc. Edinburgh. 128A:301–314, 1998.

[12] E.J.N. Looijenga. Structural stability of smooth families of C^∞-functions. Doctoral thesis, Universiteit van Amsterdam, 1974.

[13] J.N. Mather. Transversality, Advances in Math. 4:301-335, 1970.

[14] J.N. Mather. Stratifications and mappings. In: Proceedings of the conference on dynamical systems, ed. M.M. Peixoto, Academic Press. pp 195-232, 1973.

[15] A.A. du Plessis, C.T.C. Wall. The geometry of topological stability. OUP, 1995.

[16] F. Sergeraert. Un théorème de fonctions implicites pour les espaces de Fréchet. Ann. Sci. Ec. Norm. Sup. 5:599-660, 1972.

[17] D.J.A. Trotman. A transversality property weaker than Whitney (A)-regularity. Bull. London Math. Soc. 8:225-228, 1976.

[18] D.J.A. Trotman. Stability of transversality to a stratification implies Whitney A-regularity. Invent. Math. 50:273-277, 1979.

[19] C.T.C. Wall. Geometric properties of generic differentiable manifolds. In: Geometry and Topology, III Latin American school of mathematics, ed. J. Palis, M.P. do Carmo, Springer lecture notes in Mathematics 597. pp 707-774, 1977.

[20] C.T.C. Wall. Duality of singular plane curves. Jour. London Math. Soc. 50:265-275, 1994.

[21] H.G. Zeuthen. Note sur les singularités des courbes planes. Math. Ann. 10:210-220, 1876.

The Distribution $\int_A f^s \square$ and the Real Asymptotic Spectrum

DANIEL BARLET   Université Henri Poincaré (Nancy 1) and Institut Universitaire de France, Institut Elie Cartan UMR 7502 CNRS-INRIA-UHP, BP 239-F-54506, Vandœuvre-lès-Nancy Cedex, France E-mail: barlet@iecn.u-nancy.fr

AHMED JEDDI   Université Henri Poincaré (Nancy 1) and Institut Universitaire de Formation des Maîtres de Lorraine, Institut Elie Cartan UMR 7502 CNRS-INRIA-UHP, BP 239-F-54506, Vandœuvre-lès-Nancy Cedex , France
E-mail: jeddi@iecn.u-nancy.fr

Abstract

We give here a survey of our recent results on the asymptotics of fiber-integrals associated to the singularity of a real analytic germ $f : (\mathbb{R}^{n+1}, 0) \to (\mathbb{R}, 0)$. By Mellin's transform they are described by poles of the meromorphic continuation of the distribution $\int_A f^s \square$ on $\mathbb{R}^{n+1}$ and by Fourier's transform they give the asymptotics when $\tau \longrightarrow \pm\infty$ for the oscillating integral $\int_A \exp(i\tau f) \square$ where $A = \sum a_q A_q$ is a linear combination of the connected components of the local complement of $\{f = 0\}$ in $\mathbb{R}^{n+1}$.

1   INTRODUCTION

Let $n \geq 1$ and $f : (\mathbb{R}^{n+1}, 0) \to (\mathbb{R}, 0)$ be a non constant germ of real analytic function with a critical point at 0. For $X_{\mathbb{R}}$ a small open semi-analytic set around 0 we consider $A := \sum_q a_q A_q$ a finite linear combination of connected components A_q

137

of the open set $X_{\mathbb{R}} \setminus \{f = 0\}$. Then we shall study the corresponding distribution $\int_A f^s \, \square$ and the oscillating integral $\int_A \exp(i\tau f) \, \square$ which are respectively, for $\varphi \in \mathcal{C}_c^\infty(X_{\mathbb{R}})$ a test function, the Mellin and Fourier transform of the fiber integral $t \mapsto \int_{A \cap \{f=t\}} \varphi(x) dx/df$, $t \neq 0$ and small enough. Using Bernstein identity for f^s (see [9, 10]) one obtains a meromorphic extension to the complex plane for $\int_A f^s \, \square$ with possible poles located inside a set $-B_f = \{-p/r(f), \ p \in \mathbb{N}^*\}$, where $r = r(f)$ is a positive integer independent of φ. Each pole of order k at $-\alpha$ corresponds, on the oscillating integral side, to a non zero term $(\pm\tau)^{-\alpha}(\log \pm\tau)^k$ in the asymptotic expansion of $\int_A \exp(i\tau f) \, \square$ when $\tau \longrightarrow \pm\infty$. Following Palamodov ([21] for $A = \sum_q A_q = X_{\mathbb{R}} \setminus \{f = 0\}$) the *real asymptotic spectrum* for f at 0 corresponds to the leading terms in the oscillating integral expansion (see Section 4), taking in account the power of $\log \pm\tau$.

Our approach is topological : using Gauss-Manin theory for the complexification $f_{\mathbb{C}}$ of f, the position of some *real cycles* ($V(A)$ in Section 2 and $\delta(A)$ in Section 3) associated to A in the cohomology of the Milnor fiber of $f_{\mathbb{C}}$ at 0 and the saddle point method, we give informations (modulo $\mathbb{Z}$) on this *real asymptotic spectrum*. This approach is coordinate free and so it is disjoint from Newton polyhedra classical approach (using non degeneracy condition).

The general problem is far to be solved; in particular the interaction between strata are not taken in account here (see [4]). The aim of this survey is to state and discuss some recent results which give a new significative step in this classical subject.

2   DISTRIBUTION $\int_A f^s \, \square$

2.1   Fiber-integral and Mellin's transform

Let $f_{\mathbb{C}} : (\mathbb{C}^{n+1}, 0) \to (\mathbb{C}, 0)$ be the complexification of a non constant analytic germ $f : (\mathbb{R}^{n+1}, 0) \to (\mathbb{R}, 0)$. For $0 < \eta << \epsilon << 1$, $X_{\mathbb{C}} = B_{\mathbb{C}}(0, \epsilon) \cap f_{\mathbb{C}}^{-1}(D)$, $D = D(0, \eta)$, we consider the map $f_{\mathbb{C}} : X_{\mathbb{C}} \to D$, usually called Milnor representative of this complexification. Then $f : X_{\mathbb{R}} \to]-\eta, \eta[$ will denote the restriction of $f_{\mathbb{C}}$ to $X_{\mathbb{R}} := X_{\mathbb{C}} \cap \mathbb{R}^{n+1}$. Let $A := \sum_q a_q A_q$ in $H^0(X_{\mathbb{R}} \setminus \{f = 0\}, \mathbb{C})$ and define, for a test function $\varphi \in \mathcal{C}_c^\infty(X_{\mathbb{R}})$, the fiber-integral :

$$t \in]-\eta, \eta[\setminus\{0\} \longmapsto \int_{A \cap \{f=t\}} \varphi(x) dx/df,$$

in the following way. First,

$$\int_{A \cap \{f=t\}} \varphi(x) dx/df := \sum_q a_q \int_{A_q \cap \{f=t\}} \varphi(x) dx/df,$$

where the sum is finite because $X_{\mathbb{R}} \setminus \{f = 0\}$ is a semi-analytic open set, so has finitely many connected components A_q. Now we have to precise the orientation chosen for the (smooth) hypersurface $A_q \cap \{f = t\}$ and also the meaning of dx/df. Our convention is fixed by the fact that we want that

$$\int_{-\eta}^{+\eta} \left(\int_{A_q \cap \{f=t\}} \varphi(x) dx/df \right) dt = \int_{A_q} \varphi(x) dx,$$

where A_q is oriented in canonical way in $\mathbb{R}^{n+1}$ (so dx is a positive volume form for this orientation).

We shall use the following definition for the Mellin transform.

DEFINITION 2.1 *For $g \in C^\infty(\mathbb{R}^*)$ bounded and compactly supported in $\mathbb{R}$, the Mellin transform M_g of g is the holomorphic function defined for $\Re s > 0$ by*

$$i\pi M_g(s) = \int_0^{+\infty} t^{s-1} g(t)dt - e^{-i\pi s} . \int_0^{+\infty} t^{s-1} g(-t)dt.$$

Using this definition, we define $\int_A f^{s-1}\varphi(x)dx$ as the Mellin transform of the fiber-integral $\int_{A\cap\{f=t\}} \varphi(x)dx/df$ for any test function $\varphi \in C_c^\infty(X_\mathbb{R})$. This leads to the following formula

$$\int_A f^s(x)\varphi(x)dx =$$

$$= \sum_{A_q\subset\{f>0\}} a_q \int_{A_q} f^s(x)\varphi(x)dx + e^{-i\pi s} . \sum_{A_q\subset\{f<0\}} a_q \int_{A_q} (-f)^s(x)\varphi(x)dx.$$

Then, to see the meromorphic extension of this distribution which holomorphically depends on s for $\Re s > 0$, it is enough to explain the case of the function $\int_{A_q} f^s(x)\varphi(x)dx$ where $A_q \subset \{f > 0\}$. We recall (i) there exists a polynomial $b(s)$ (Bernstein-Sato polynomial of f at 0) and a finite order holomorphic operator $\mathcal{P} \in \mathcal{D}_{X_\mathbb{C}}[s]$ such that the Bernstein-Sato identity $\mathcal{P}f^{s+1} = b(s)f^s$ holds (see [9, 10]), (ii) the classical Stokes' formula works also for integration on semi-analytic sets (see [12]). We conclude easily by these two results that the function above admits a meromorphic extension to $\mathbb{C}$. Let us denote by $\int_A f^s \square$ the distribution obtained in such a way for any A. Moreover we obtain that, because roots of b are negative rational numbers (see [19]), the set of possible poles of this meromorphic extension is contained in $-B_f = \{-p/r(f), \ p \in \mathbb{N}^*\}$ for some positive integer $r = r(f)$ which does not depend on φ. This could also be deduced from the asymptotics of the fiber-integrals $\int_{A_q\cap\{f=t\}} \varphi(x)dx/df$ when $t \to 0^\pm$ (see [13]).

2.2 A criterion using the homology of the real Milnor fiber

To simplify we shall only discuss the case of a component A of the open set $\{f > 0\}$ in $X_\mathbb{R}$. Choosing $t_0 \in]0, \eta[$ as base point in $D(0, \eta)$, we can identify the complex Milnor fiber F with $f_\mathbb{C}^{-1}(t_0)$. So the real Milnor fiber for A, which is $F_A := \mathbb{R}^{n+1} \cap f_\mathbb{C}^{-1}(t_0)$ has a natural embedding $j_A : F_A \hookrightarrow F$. Let us denote by $V(A)$ the image of the application $j_A : H_*(F_A) \longrightarrow H_*(F)$ where H_* means homology in positive degrees. The following theorem is proven in [5].

THEOREM 2.2 *let $\lambda = \exp(-2i\pi u)$, with $u \in [0, 1[\cap\mathbb{Q}_+$, be an eigenvalue of the monodromy T acting on $H_*(F, \mathbb{C})$. Denote by $V(A)_\lambda$ the intersection of $V(A)$ with $H_*(F, \mathbb{C})_\lambda$ the λ-spectral subspace of T.*

Then, if k is the nilpotency order of $T - \lambda Id$ acting on $V(A)_\lambda$ (the largest integer h such that $(T - \lambda Id)^{h-1}(V(A)_\lambda) \neq \{0\}$ if $V(A)_\lambda \neq 0$ and 0 if $V(A)_\lambda = 0$), the meromorphic extension of $\int_A f^s \square$ has a pole of order $\geq k$ at $-u - \nu$, for $\nu \in \mathbb{N}$ and large enough.

The proof is based on the fact that a Jordan block for T acting on the cohomology can be expressed by holomorphic forms satisfying some differential relations (see [7]). For instance if $e \in H^p(F, \mathbb{C})_\lambda$ is an eigenvector, there exists a p-holomorphic form ω on $X_\mathbb{C}$, an integer m such that (i) $d\omega = (m + u)df_\mathbb{C}/f_\mathbb{C}) \wedge \omega$ on $X_\mathbb{C}$, (ii) $e = [\omega/f_\mathbb{C}^{m+u}]_{|F}$ in $H^p(F, \mathbb{C})$. Then one can show that the lack of poles for the current $\int_A f^s \omega \wedge \square$ (defined for $\Re s > 0$ and $\psi \in \varphi \in \mathcal{C}_c^\infty(X_\mathbb{R})$ of degree n, by the formula $\langle \int_A f^s \omega \wedge \square, \psi \rangle =< \int_A f^s \square, \omega \wedge \psi >$ at $-\mathbb{N} - u$ contradicts the fact that there exists $\gamma \in H_p(F_A, \mathbb{C})$ with $< \gamma, e > \neq 0$. Note that in this setting, the eigenvalue 1 does not produce, in general, order ≥ 2 poles at negative (large enough) integers as in the complex case (see [6]).

This criterion is not always useful, nor necessary. For instance, in the isolated singularity case, the component A (for f changing sign) can be contractible and then $V(A) = 0$ for positive degrees.

The following result shows that for an isolated singularity the only interesting case for applying it is when the real Milnor fiber is compact [16].

THEOREM 2.3 *Assume that f has an isolated singularity at 0 in $X_\mathbb{R}$. Then for any component A of $X_\mathbb{R} \setminus \{f = 0\}$, the vector space $V(A)$ is in the invariant part of the monodromy in degrees $p \in [1, n - 1]$.*

The proof uses essentially two facts : (i) the open set $X_\mathbb{R} \setminus \{0\}$ is an homological sphere of dimension n, (ii) apart from residues at negative integers, the polar parts of $\int_A f^s \square$ are distributions supported by $\{0\}$.

In fact, theorem 1.1 in [16] says that the asymptotics of the fiber-integrals $\int_{A \cap \{f=t\}} \psi$, for $t \longrightarrow 0^\pm$, where $\psi \in \mathcal{C}_c^\infty(X_\mathbb{R})$ is a test form of degree n, are generated by a finite family of n-forms on $X_\mathbb{C}$ which generates the Gauss-Manin bundle. This proves that poles $\int_A f^s \square$ only depend on the cohomology in degree n of the Milnor fiber. This also explains the relationship between the theorem above and the criterion.

3   ALMOST ISOLATED SINGULARITY

Let us first recall the definition of this notion introduced in [3, 4].

DEFINITION 3.1 *Let λ be an eigenvalue of the monodromy acting on the cohomology of the Milnor fiber at 0 of a non constant germ $f_\mathbb{C} : (\mathbb{C}^{n+1}, 0) \longrightarrow (\mathbb{C}, 0)$.*

We say that the singularity of f at 0 is "almost isolated for λ" if, and only if, for each $z \neq 0$ near enough 0, λ is not an eigenvalue of the monodromy acting on the reduced cohomology of the Milnor fiber of $f_\mathbb{C}$ at z.

Note that when $\lambda = 1$, to take the reduced cohomology is essential. It is clear that an isolated singularity is almost isolated for any λ. But, of course, the singular set of $\{f_\mathbb{C} = 0\}$ can be positive dimensional for an almost isolated singularity for a given λ.

As in the case of an isolated singularity, the poles at $-u - \nu$ for $\nu \in \mathbb{N}$ and $u \in [0, 1[\cap\mathbb{Q}$ such that $\lambda = \exp(-2i\pi u)$, come only from the cohomology in degree n of the Milnor fiber of $f_\mathbb{C}$ at the origin. This is not so surprising because the assumption means that $(R\Psi_{f_\mathbb{C}})_\lambda$, the λ-spectral part of the vanishing cycle complex $R\Psi_{f_\mathbb{C}}$ of $f_\mathbb{C}$ has cohomology concentrated at $\{0\}$. By perversity this implies that it has only degree n cohomology sheaf.

To state the results let us first define, for any $A = \sum_q a_q A_q$ in $H^0(X_{\mathbb{R}} \backslash \{f = 0\}, \mathbb{C})$ a closed n-cycle $\delta(A)$ in $H^n(F, \mathbb{C})$ where F is the Milnor fiber of $f_{\mathbb{C}}$ at 0.

DEFINITION 3.2 *Assume that $F = f_{\mathbb{C}}^{-1}(t_0)$ where the base point t_0 in $D(0, \eta) \backslash \{0\}$ is chosen in $]0, \eta[$.*
For $A_q \subset \{f > 0\}$ let $\delta(A_q) = A_q \cap F$ (with the orientation chosen before).
For $A_q \subset \{f < 0\}$ let us consider the closed C^∞ embedding induced by following a C^∞-trivialisation of Milnor's fibration along the half circle $t_0 e^{i\theta}$, $\theta \in [\pi, 2\pi]$ (for more details see [2]). Then define $\delta(A_q) := -\theta_*(A_q \cap f_{\mathbb{C}}^{-1}(-t_0))$ and $\delta(A) := \sum_q a_q \delta(A_q)$.

We remark that the definition of $\delta(A)$ is not the convention of [8, 17] because the choice in 2.1 is not the same that in [2].

First we consider the case where $\lambda \neq 1$.

THEOREM 3.3 *Assume that the complexified germ $f_{\mathbb{C}} : (\mathbb{C}^{n+1}, 0) \longrightarrow (\mathbb{C}, 0)$ of f has an almost isolated singularity at 0 for the eigenvalue $\lambda = \exp(-2i\pi u)$ where $u \in]0, 1[\cap \mathbb{Q}$.*
Then we have equivalence between:
(i) the distribution $\int_A f^s \square$ has a pole of order $\geq k_0$ at $-u - \nu$ for $\nu \in \mathbb{N}$, ν large enough,
(ii) $(T - \lambda Id)^{k_0 - 1}(\delta(A)_\lambda) \neq 0$ in $H^n(F, \mathbb{C})_\lambda$, where $\delta(A)_\lambda$ is the λ-component of $\delta(A)$ in the spectral decomposition of the monodromy T acting on $H^n(F, \mathbb{C})$.

Of course in condition (ii) we mean that $\delta(A)_\lambda = 0 \iff k_0 = 0$.
The proof of (i) $\implies$ (ii) is given in [8] and its converse in [17]. Basically, five key points are used .
(1) Each vector $e \in H_c^n(F, \mathbb{C})$ is the class of a semi-meromorphic n-form ω with poles along $\{f_{\mathbb{C}} = 0\}$, $f_{\mathbb{C}}$-proper support and relatively closed ($d\omega \wedge df_{\mathbb{C}} \equiv 0$).
(2) Each Jordan block of size k for the monodromy T acting on $H_c^n(F, \mathbb{C})_\lambda$ can be induced by semi-meromorphic n-forms, with f-proper supports, $\omega_1, ..., \omega_k$ as in (1) satisfying the relation :
$$\forall j \in [1, k], \quad d\omega_j = (m + u)(df_{\mathbb{C}}/f_{\mathbb{C}}) \wedge \omega_j + (df_{\mathbb{C}}/f_{\mathbb{C}}) \wedge \omega_{j-1}, \omega_0 \equiv 0,$$
for some $m \in \mathbb{N}$.
(3) By integration of such representative of a Jordan block on $A \cap \{f = t\}$, we obtain fiber-integral of the form $|t|^{m+u} P(\log |t|)$ where P is a polynomial of degree $\leq k-1$. Choosing now a Jordan block which is not orthogonal to $(T - \lambda Id)^{k_0 - 1}(\delta(A))_\lambda$ for the Poincaré duality on F, allows to produce a polynomial of degree $\geq k_0 - 1$. We easily conclude by Mellin's transform.
   The converse needs also the following key points.
(4) Each distribution associated to a polar part at $-u - \nu$, $\nu \in \mathbb{N}$, for $\int_A f^s \square$ is supported by $\{0\}$.
(5) The canonical map $Can : H_c^n(F, \mathbb{C})_\lambda \longrightarrow H^n(F, \mathbb{C})_\lambda$ is an isomorphism.
As a consequence of this strategy it is not very hard to obtain, using that all our homological constructions are defined over $\mathbb{R}$ (see [15]):

COROLLARY 3.4 *In the same setting of the previous theorem, for $\nu \in \mathbb{N}$ large enough, the order of the pole for $\int_A f^s \square$ at $-u - \nu$ and $u - \nu$ are the same.*

Remark that by theorem 3.1 they do not depend on ν large enough; for some examples see [16].
   Now we have a result analogous to theorem 3.1 for the case $\lambda = 1$ (see [2]).

THEOREM 3.5 *Assume that the singularity of the complexified germ of f has an almost isolated singularity relatively to the eigenvalue 1 of the monodromy, then we have equivalence between :*

(i) $\delta(A)$ has a non zero component $\delta(A)_1$ on $H^n(F, \mathbb{C})_1$,

(ii) the meromorphic extension to the complex plane of the holomorphic distribution s, $\Re s > 0 \longmapsto \int_A f^s \, \square$ admits a pole of order ≥ 2 at $s = -(n+1)$.

Moreover, the order of the pole at $\nu \in \mathbb{N}$, $\nu \geq n+1$, of

$$s \mapsto \frac{1}{\Gamma(s)} \int_A f^s \, \square$$

is exactly the nilpotency order of $T - 1$ acting on $\delta(A)_1$.

We just give here the key point in the proof. The main difficulty of the $\lambda = 1$ case compared to the previous case ($\lambda \neq 1$) is that, because the canonical map $Can : H^n_c(F, \mathbb{C})_1 \longrightarrow H^n(F, \mathbb{C})_1$ is no longer an isomorphism, the intersection form coming from the Poincaré duality $H^n_c(F, \mathbb{C}) \times H^n(F, \mathbb{C}) \longrightarrow \mathbb{C}$ has to be replaced by the canonical (non degenerated) hermitian form (see [3])

$$h : H^n(F, \mathbb{C}) \times H^n(F, \mathbb{C}) \longrightarrow \mathbb{C}.$$

This hermitian form is related to the *variation* map $Var : H^n(F, \mathbb{C})_1 \longrightarrow H^n_c(F, \mathbb{C})_1$ which has been defined and computed in term of differential forms in [3] under the present assumption.

4   OSCILLATING INTEGRALS

Now apply Fubini's theorem to the Fourier transform of the fiber-integral, we obtain for $\varphi \in \mathcal{C}^\infty_c(X_\mathbb{R})$:

$$\int_{-\eta}^{\eta} \exp(i\tau t)(\int_{A \cap \{f=t\}} \varphi(x)dx/df)dt = \int_A \exp(i\tau f)\varphi(x)dx.$$

Using the asymptotic expansion of the fiber-integral when $t \to 0^\pm$ leads to an asymptotic expansion for the oscillating integral

$$\int_A \exp(i\tau f)\varphi(x)dx = \sum_q a_q \int_{A_q} \exp(i\tau f)\varphi(x)dx,$$

when $\tau \to \pm\infty$ in the scale $(\pm\tau)^{-\alpha}(\log \pm\tau)^k$ where $\alpha \in B_f$ and $k \in [0, n]$.

In [21] Palamodov introduced the *real asymptotic spectrum* of the previous oscillating integral (in the case $A = \sum_q A_q = X_\mathbb{R} \setminus \{f = 0\}$). This is the subset $S_\mathbb{R}(f, 0)$ in B_f of β satisfying : (a) there exists a non zero term of the type $(\pm\tau)^{-\beta}(\log \pm\tau)^k$, $k \in [0, n]$, in the asymptotics when $\tau \to \pm\infty$ for some $\varphi \in \mathcal{C}^\infty_c(X_\mathbb{R})$, (b) each *spectral element* is endowed with a order k_β which is the largest k such that $(\pm\tau)^{-\beta}(\log \pm\tau)^{k-1}$ appears, (c) the $\mathbb{C}[[1/\pm\tau]]$-module of asymptotics is generated by the finite family (of leading terms) $(\pm\tau)^{-\beta}(\log \pm\tau)^q$, $\beta \in S_\mathbb{R}(f, 0)$, $q < k_\beta$. Of course $\beta \in S_\mathbb{R}(f, 0)$ has order $k_\beta \geq 1$ if, and only if, $\int_A f^s \, \square$ has a pole of order $k_\beta \geq 1$ at $-\beta - m$ for some integer $m \in \mathbb{N}$. Palamodov has conjectured that for any germ $f : (\mathbb{R}^{n+1}, 0) \longrightarrow (\mathbb{R}, 0)$

singular at 0 (i.e $df(0) = 0$) the real asymptotic spectrum of f at 0 is non empty. It means that there exists some $\varphi \in C_c^\infty(X_\mathbb{R})$ such that $\int_{X_\mathbb{R}} \exp(i\tau f)\varphi(x)dx$ is not, as a function of τ, in the Schwartz space $\mathcal{S}(\mathbb{R})$. The second author proved this conjecture in [14] and we shall give an outline the proof of it. Remark that, by complex conjugacy, we can restrict ourselves to the asymptotics for $\tau \longrightarrow +\infty$.

4.1 Real isolated singularity

By a real isolated singularity we mean that for $x \in X_\mathbb{R}$, $df(x) = 0 \iff x = 0$. We remark that if the singularity is isolated in the complex, the answer to the conjecture would be positive if the cycle $\delta(A) \neq 0$ for $A = \sum_q a_q A_q = X_\mathbb{R} \setminus \{f = 0\}$. But even in this restrictive case, we don't know how to check whether that cycle is trivial or not. In fact, the proof is based on Malgrange idea [20] which consists to relate the oscillating integral to the saddle point method.

For $0 << 2\eta' << \eta << 1$, we denote here by $f_\mathbb{C} : X_\mathbb{C} \longrightarrow D$, $D = D(0, \eta)$, a Milnor representative of the complexified germ of f. For a $(n+1)$-chain $\Gamma \subset X_\mathbb{C}$ such that $\partial\Gamma \subset X_\mathbb{C}^i$, $X_\mathbb{C}^i := \{z \in X_\mathbb{C}, \Im f_\mathbb{C}(z) > 0\}$, and a $(n+1)$-holomorphic form Ω, the integral $\int_\Gamma \exp(i\tau f)\Omega$ admits, for $\tau \longrightarrow +\infty$, the same type of asymptotic expansion as before. This expansion depends only, for fixed Ω, on $[\Gamma]$ the class of Γ in the relative homology $H_{n+1}(X_\mathbb{C}, X_\mathbb{C}^i, \mathbb{C})$, because for every compact $K \subset X_\mathbb{C}^i$ the function $\tau \longmapsto \int_K \exp(i\tau f)\Omega$ is flat at $+\infty$.

The following result precises and extends to real isolated singularity case the proposition 7 ([20]).

THEOREM 4.1 *Assume that the germ has a real isolated singularity at 0 and let ρ be any function in $C_c^\infty(X_\mathbb{R})$ such that $\rho \equiv 1$ near 0.*
Then there exists a $(n+1)$-chain $\delta_\mathbb{R}$ (called real) satisfying :
(i) $\partial\delta_\mathbb{R} \subset X_\mathbb{C}^i$, $\delta_\mathbb{R} \cap \mathbb{R}^{n+1}$ is a real neighbourhood of 0,
(ii) the class of $\delta_\mathbb{R}$ in $H_{n+1}(X_\mathbb{C}, X_\mathbb{C}^i, \mathbb{C})$ is not trivial, and
(iii) for any $(n+1)$-holomorphic form Ω the two oscillating integrals

$$\int_{X_\mathbb{R}} \exp(i\tau f)\rho\Omega, \quad \int_{\delta_\mathbb{R}} \exp(i\tau f)\Omega$$

have the same asymptotic expansion for $\tau \longrightarrow +\infty$.

We first construct a $\tilde{\delta}_\mathbb{R}$ satisfying the condition (i), by pushing a real small sphere (around 0) toward $X_\mathbb{C}^i$ with the help of the gradient of the imaginary part of $f_\mathbb{C}$. This is possible because $\Im f_\mathbb{C}$ and $f_\mathbb{C}$ have the same critical set which can supposed to be in the complex singular fiber $\{f_\mathbb{C} = 0\}$. As each coefficient of $\tau^{-\alpha}(\log\tau)^k$ is a distribution supported by the origin, the condition (iii) is automatically fulfilled for $\tilde{\delta}_\mathbb{R}$. But it is more difficult to see whether the condition (ii) is fulfilled.

Secondly, we construct a smooth vector field S^∞ (resp. T^∞) on an open complex neighbourhood of $\{f = 0\} \setminus \{0\}$ in such a way that its scalar product in $\mathbb{C}^{n+1}$ with the gradient of the real part (resp. gradient of the imaginary part) is equal 1 and $[S^\infty, T^\infty] = 0$ outside a small complex ball. By using the integral curves of such fields, we build a smooth manifold $\Delta(t_0) \subset X_\mathbb{C}^i$ ($0 < t_0 << 1$) of real dimension $n+2$, and with quasi-regular boundary $\cup_{j=1}^4 \Delta^j(t_0)$ such that : (a) $\Delta^1 = \mathbb{R}^{n+1} \cap \overline{\Delta(t_0)}$ is a fixed closed neighbourhood of 0 in $\mathbb{R}^{n+1}$, independent of t_0, with $f(\Delta^1) = [-\eta', \eta']$,

(b) the volume of $\Delta^j(t_0)$, $j = 2;3$, tends to 0 for $t_0 \to 0^+$, (c) $\Delta^4(t_0) \subset \{z, -2\eta' \le \Re f_{\mathbb{C}} \le 2\eta', \Im f_{\mathbb{C}} = t_0\}$.

Furthemore, by using the fibration given by $\Re f_{\mathbb{C}}$ and Fubini's formula on $\Delta^4(t_0)$, we can see that for all $v \in [-2\eta', 2\eta']$, $\gamma_{v+it_0} := \Delta^4(t_0) \cap \{\Re f_{\mathbb{C}} = v\}$ is a n-cycle in the fiber over $v + it_0$. And particularly, there is an homotopy in $\overline{\Delta(t_0)}$ between the cycle γ_{it_0} and $\partial\Delta^1$ which is a real n- sphere.

Let now Ω be a $(n+1)$-holomorphic form such that $\int_{\Delta^1} \Omega \ne 0$ (take for instance $\Omega \not\equiv 0$ and positive on $\mathbb{R}^{n+1}$). Then using the property (b), Stokes' formula on $\Delta(t_0)$ and Gauss-Manin theory we get : (1) $\int_{\gamma_{v+it_0}} \Omega/df_{\mathbb{C}} \ne 0$ for all $v \in [-2\eta', 2\eta']$.

Let $h(w, .)$, $w \in [0, 1]$ be the previous homotopy. We set by definition $\delta_{\mathbb{R}} := \Delta^1(t_0) \cup \{h(w, x), (w, x) \in [0, 1] \times \partial\Delta^1(t_0)\}$. The intersection $\delta_{\mathbb{R}} \cap \tilde{\delta}_{\mathbb{R}}$ is a neighbourhood of 0 in $\mathbb{R}^{n+1}$; then, by the saddle point method, one can easily conclude that the two chains have the same class in the relative cohomology. Recall that, as $X_{\mathbb{C}}$ is Stein and contractible, by Milnor's fibration we have an isomorphism between $H_{n+1}(X_{\mathbb{C}}, X_{\mathbb{C}}^i, \mathbb{C})$ and the homology in degree n of the fiber of $f_{\mathbb{C}}$ at it_0, and it is clear that the cycle γ_{it_0} is the image of the class $[\delta_{\mathbb{R}}]$ by this isomorphism. The condition (b) implies that this class is not trivial and, for Ω satisfying (1), the integral $\int_{\delta_{\mathbb{R}}} \exp(i\tau f)\Omega$ is not flat at $+\infty$. Finally by (iii), we get an oscilating integral $\tau \longmapsto \int_{X_{\mathbb{R}}} \exp(i\tau f)\rho\Omega$ which is not in $\mathcal{S}(\mathbb{R})$.

4.2 General case

The procedure is to reduce the question to the isolated singularity case, by considering the restriction of the distribution T_α^k, wich are the coefficient of $\tau^{-\alpha}(\log \tau)^k$, $(\alpha, k) \in B_f \times [0, n]$. By the Fourier and Mellin inverse transforms, these distributions are in the vector space spanned by the distributions $U_{\beta,\pm}^q, (\beta, q) \in B_f \times [1, n+1]$ corresponding to the poles of $f_\pm^s$, $< f_\pm^s, \varphi >:= \int_{\pm f>0} (\pm f)^s(x)\varphi(x)dx$ for $\Re s > 0$ and $\varphi \in \mathcal{C}_c^\infty(X_{\mathbb{R}})$. For a generic point $a \in (df)^{-1}(0) \setminus \{0\}$ and a generic real hyperplan $H \subset \mathbb{R}^{n+1}$ through a, we denote by $T_{\alpha,H}^k$ and $U_{\beta,\pm,H}^q$ the similar distributions associated to the singularity of the germ $f_{|H} : (H, a) \longrightarrow (\mathbb{R}, 0)$ given by the restriction of f to H.

THEOREM 4.2 *Assume that the real singularity is non isolated. Then*

(i) there exists a generic real hyperplan H, through a generic point of $(df)^{-1}(0) \setminus \{0\}$, which is non characteristic for all distributions $f_\pm^s$ and $U_{\beta,\pm}^k$,

(ii) each restriction on H: $(f_\pm^s)_{|H}$ $(U_{\beta,\pm}^k)_{|H}$ is well defined and we have $(f_\pm^s)_{|H} = (f_{|H})_\pm^s$, $(U_{\beta,\pm}^k)_{|H} = U_{\beta,\pm,H}^k$, $B_{f_{|H}} \subset B_f$.

The two following crucial facts are used in the proof.

(a) For a fixed complex number λ, the characteristic variety of the holonomic $\mathcal{D}_{X_{\mathbb{C}}}$-module $\mathcal{D}_{X_{\mathbb{C}}} f_{\mathbb{C}}^\lambda = \{\mathcal{D}_{X_{\mathbb{C}}}/P \in \mathcal{D}_{X_{\mathbb{C}}}, Pf_{\mathbb{C}}^\lambda = 0$, at a generic point $\}$ is contained in the variety $\mathcal{V}_{f_{\mathbb{C}}}^* \cap f_{\mathbb{C}}^{-1}(0) \times (\mathbb{C}^{n+1} \setminus 0)$ where $\mathcal{V}_{f_{\mathbb{C}}}^* := \overline{\cup_{s \ne 0} T_{f_{\mathbb{C}}^{-1}(s)}^* X_{\mathbb{C}}}$ is the characteristic variety of the $\mathcal{D}_{X_{\mathbb{C}}}$-module $\mathcal{D}_{X_{\mathbb{C}}}[s]f_{\mathbb{C}}^s$, $T_{f_{\mathbb{C}}^{-1}(s)}^* X_{\mathbb{C}}$ being the conormal of the Milnor fiber over s ([19], proposition 6.1). It is interesting to note that if for all $j \in \mathbb{N}^*$, $\lambda - j$ is not a root of the b-function of $f_{\mathbb{C}}$ then we have

$$\mathcal{D}_{X_{\mathbb{C}}} f_{\mathbb{C}}^\lambda \xrightarrow{\sim} \mathcal{D}_{X_{\mathbb{C}}}[s]f_{\mathbb{C}}^s/(s - \lambda)\mathcal{D}_{X_{\mathbb{C}}}[s]f_{\mathbb{C}}^s \text{ [loc.cit]};$$

we also refer to [10] (theorem 7.6.1) for further results on the regular holonomic distributions.

(b) Locally at 0, the complex singular fiber $f_{\mathbb{C}}^{-1}(0)$ can be endowed with a stratification $\sum = (V_j)_j$, so called *good stratification* which particularly satisfyies the condition (see [11]): if ξ is non zero cotangent vector to the stratum V_j at z, then

$$(z, \xi) \notin \overline{\cup_{s \neq 0} T^*_{f_{\mathbb{C}}^{-1}(s)} X_{\mathbb{C}}}.$$

(1) Firstly, by using the definition of the distribution $f_\pm^s$ one can show that all wave front sets $WF_\infty(f_\pm^s)$, and then $WF_\infty(U_{\beta,\pm}^k)$ (resp. $WF_\infty(T_\alpha^q)$), are contained in the intersection $\mathcal{V}_{f_{\mathbb{C}}}^* \cap f^{-1}(0) \times (\mathbb{R}^{n+1} \setminus 0)$ (resp. $\mathcal{V}_{f_{\mathbb{C}}}^* \cap (df)^{-1}(0) \times (\mathbb{R}^{n+1} \setminus 0)$).

(2) Secondly, as $\mathcal{D}_{X_{\mathbb{C}}} f_{\mathbb{C}}^\lambda$ is holonomic, there exists a generic non characteristic complex hyperplan $H_{\mathbb{C}}$ through a generic point $a \in (df_{\mathbb{C}})^{-1}(0) \setminus \{0\}$ and by (b) we can choose it in such a way that $a \in (df)^{-1}(0) \setminus \{0\}$ and $H = H_{\mathbb{C}} \cap \mathbb{R}^{n+1}$ is a real hyperplan. Then H is non characteristic for all the distributions above.

(3) Accordingly, the restriction of each distribution to H is well defined by the limit (in the space of distributions on H) of the restriction of its regularizing sequence (see [18]). The fact that for $f_\pm^s$ this convergence is uniform on any compact subset of $\mathbb{C} \setminus B_f$ and the Bernstein identity imply the remainder of the statement.

COROLLARY 4.3 *The real asymptotic spectrum is not empty.*

We remark that the case of $dim_{\mathbb{R}}((df)^{-1}(0)) = n$ reduces to the one variable case, for which the proof is trivial. So we can assume $dim_{\mathbb{R}}((df)^{-1}(0)) < n$. In this later case, by induction on the dimension of the singular locus, the non characteristic real hyperplan given by the previous theorem at a generic point of the biggest stratum gives the reduction to the real isolated singularity case.

5 EXAMPLES

In the previous sections, we exactly deal with the three approachs (a) embedding criterion, (b) almost isolated singularity and (c) saddle point method. This allows in principle to determine (as far as the computation of the complex Gauss-Manin system is known) poles of the concerned distributions or singular terms of the oscillating integral corresponding to a fixed eigenvalue of the monodromy. The method (b) complements the method (a). The last method don't need any hypothesis on the singularity but we still can't (for the moment) use it for a connected component. All these methods determine *spectral elements* up to shift by an integer. For this raison, in the following examples, we will essentially compute $\exp(-2i\pi S_{\mathbb{R}}(f, 0))$.

In this section we illustrate each method by an example. Let $\hat{dz_j}$, $0 \leq j \leq n$ denotes the form of degree n: $(dz_0 \wedge ... \wedge dz_n)/dz_j$ where $z_0, ..., z_n$ are the coordinates in $\mathbb{C}^{n+1}$. The first main difficulty which arises is to compute Milnor numbers and Jordan bases of the monodromy on the vector spaces $H^*(F, \mathbb{C})$, which is not easy in general. Nevertheless, by using the Leray exact sequence for the pair $(X_{\mathbb{C}}, F)$ and some properties of the constructible sheaf given by the cohomology of degree 1, we succeed in some quasi-homogeneous cases (see [16]).

EXAMPLE 5.1 Compact fiber

On $\mathbb{C}^4$, the monodromy of the quasi-homogeneous polynomial $f_\mathbb{C}$:

$$f_\mathbb{C}(z_0,\, z_1,\, z_2,\, z_3) = \left(z_0^2 + z_1^2\right)^2 + z_2^2 + z_3^2$$

is semi-simple with eigenvalues $-1;\, 1$. The restriction f of $f_\mathbb{C}$ to $\mathbb{R}^4$ has a compact fiber and an isolated singularity at 0. Using the same arguments as in the example 4.1 of [16], we have $dim_\mathbb{C} H^3(F,\, \mathbb{C}) = 2$, F being the Milnor fiber. Consider now the holomorphic 3-form $\omega = z_0 \stackrel{\wedge}{dz_0} - z_1 \stackrel{\wedge}{dz_1} + 2z_2 \stackrel{\wedge}{dz_2} - 2z_3 \stackrel{\wedge}{dz_3}$. Then it satisfies on $\mathbb{C}^4$ the relations :

$$d\omega = \frac{3}{2}\frac{df_\mathbb{C}}{f_\mathbb{C}} \wedge \omega, \quad d[(z_0^2 + z_1^2)\omega] = 2\frac{df_\mathbb{C}}{f_\mathbb{C}} \wedge (z_0^2 + z_1^2)\omega.$$

Denote respectively by x, y, z and w the real part of z_0, z_1, z_2 and z_3. Then by a change of variables and Stokes' formula, we obtain for $t > 0$:

$$\int_{\{f=t\}} \omega = \left[6\int_{\{f\leq 1\}} dx \wedge dy \wedge dz \wedge dw\right].t^{3/2}, \text{ and}$$

$$\int_{\{f=t\}} (x^2 + y^2)\omega = \left[6\int_{\{f\leq 1\}} (x^2 + y^2)dx \wedge dy \wedge dz \wedge dw\right].t^2.$$

As the integrals on $\{f \leq 1\}$ are positive, we conclude :

(i) the vectors $e_1 = [\omega/f_\mathbb{C}^{\frac{3}{2}}]_{|F}$ and $e_2 = [(z_0^2 + z_1^2)\omega/f_\mathbb{C}^2]_{|F}$ are respectively the eigenvectors corresponding respectively to -1 and 1 and provide a base of the cohomology in degree 3 of Milnor's fiber,

(ii) the $\mathbb{C}[[t]]$-module of asymptotic expansions of $\int_{f=t}\varphi$, for t tending to 0^+, φ test form of degree 3 in $\mathcal{C}_c^\infty(\mathbb{R}^4)$, is generated by t^2 and $t^{\frac{3}{2}}$.

Finally, we get $\exp(-2i\pi S_\mathbb{R}(f,0)) = \{1;\, -1\}$ and each spectral element has order one.

EXAMPLE 5.2 Almost isolated singularity

On $\mathbb{C}^3$, the monodromy of the homogeneous polynomial $f_\mathbb{C}$:

$$f_\mathbb{C}(z_0,\, z_1,\, z_2) = z_0^2 + z_1^2 - z_2^2$$

is semi-simple with one eigenvalues -1. This is a complex isolated singularity (then almost isolated); it is clear that the cohomology in degree 2 of Milnor's fiber is generated by the vector $[\omega/f_\mathbb{C}^{\frac{3}{2}}]_{|F}$ where $\omega = z_0 \stackrel{\wedge}{dz_0} - z_1 \stackrel{\wedge}{dz_1} + z_2 \stackrel{\wedge}{dz_2}$. Set $f(x,\, y,\, z) = x^2 + y^2 - z^2$ the restriction of $f_\mathbb{C}$ on $\mathbb{R}^3$. Then the semi-analytic set $\mathbb{R}^3 \setminus \{f = 0\}$ admits three connected components $A_1 = \{f > 0\}$, $A_2 = \{f < 0,\, z > 0\}$ and $A_3 = \{f < 0,\, z < 0\}$. We have $\delta(A_1) = 0$ and $\delta(A_2) = \delta(A_3) \neq 0$. Computation done in [8] shows that for $A = a_1 A_1 + a_2 A_2 + a_3 A_3$, the distribution $\int_A f^s \square$ admits a simple pole located in $-\mathbb{N} - \frac{1}{2}$ if, and only if, $a_2 + a_3 \neq 0$.

EXAMPLE 5.3 Oscillating integral

On $\mathbb{C}^3$, the monodromy of the quasi-homogeneous polynomial $f_{\mathbb{C}}$:

$$f_{\mathbb{C}}(z_0, z_1, z_2) = z_0^2 z_1^2 - z_2^3$$

is semi-simple. Consider now the holomorphic 2-forms $\omega_1 = 3z_0 \, \overset{\wedge}{dz_0} - 3z_1 \, \overset{\wedge}{dz_1}$ $+4z_2 \, \overset{\wedge}{dz_2}$, $\omega_2 = z_2\omega_1$, $\omega_3 = z_0 z_1 \omega_1$ and $\omega_4 = z_0 z_1 z_2 \omega_1$. Then they satisfy the relations :

$$d\omega_j = \alpha_j \frac{df_{\mathbb{C}}}{f_{\mathbb{C}}} \wedge \omega_j$$

where $\alpha_1 = \frac{5}{6}$, $\alpha_2 = \frac{7}{6}$, $\alpha_3 = \frac{4}{3}$ and $\alpha_4 = \frac{5}{3}$. On the other hand, by the same arguments as in example 4.1 in [16], one can show that $dim_{\mathbb{C}} H^2(F, \mathbb{C}) = 4$. For $1 \le j \le 4$, we set $e_j = [\omega_j / f_{\mathbb{C}}^{\alpha_j}]_{|F}$. The Milnor fiber of $f_{\mathbb{C}}$ is biholomorphic to the Milnor fiber of the function $g_{\mathbb{C}}$ defined by $g_{\mathbb{C}}(z_0, z_1, z_2) = (z_0^2 + z_1^2)^2 - z_2^3$. Thus, by considering on $\mathbb{R}^3$ (see example 4.3 in [16]) the distributions $\int_A g^s \,\square$, $g(x, y, z) = (x^2 + y^2)^2 - z^3$, we can see that the cohomology in degree 2 of Milnor's fiber of $f_{\mathbb{C}}$ is generated by the eigenvectors e_1, e_2, e_3 and e_4 corresponding respectively to the eigenvalue $\lambda_1 = \exp(-5i\pi/3)$, $\lambda_2 = \overline{\lambda}_1$, $\lambda_3 = \exp(-2i\pi/3)$ and $\lambda_4 = \overline{\lambda}_3$.

The singularity of $f_{\mathbb{C}}$ is almost isolated at 0 relatively to λ_3 and λ_4; the transverse singularity along the critical locus is a cusp for which eigenvalues of the monodromy are precisely λ_1 and λ_2. A direct computation on the oscillating integral shows that $\exp(-2i\pi S_{\mathbb{R}}(f, 0)) = \{\lambda_1, \lambda_2\}$, each element having order 2. This corresponds to the phenomenon described in [4] in the complex setting. Remark that as in the proof of corollary 1, we can show that for any singularity, the set of complex numbers $\exp(-2i\pi S_{\mathbb{R}}(f, 0))$ is self-conjugate.

EXAMPLE 5.4 Oscillating integral

In [22] Scherk considered the non quasi-homogeneous polynomial $f_{\mathbb{C}}$:

$$f_{\mathbb{C}}(z_0, z_1) = z_0^5 + z_1^5 + z_0^2 z_1^2.$$

He computed Milnor number ($\mu = 11$) and gave a base of the local algebra. This is an example where the monodromy is not semi-simple and non compact real fiber.
We remark that in [1] the first author used the theory of (a, b)-modules to show the existence of a $(2, 2)$-Jordan block of the monodromy for the family $z_0^8 + z_1^8 + w z_0^2 z_1^2$, $w \in \mathbb{C}^*$.
By considering the chain $\delta_{\mathbb{R}}$ of the theorem 4.1 and the form $dz_0 \wedge dz_1$ one can see that $\frac{1}{2} \in S_{\mathbb{R}}(f, 0)$, but our method here doesn't give any information about its multiplicity. It is to be expected that Brieskorn's formula (D) (in [22], page 30) permits to get more *spectral elements*.

The authors would like to thank the referee for his remarks which permit to improve this paper.

REFERENCES

[1] D. Barlet. Un critère d'existence de bloc de Jordan pour un (a, b)−module. Prépublications de l'Institut Elie Cartan no 10 (2000).

[2] D. Barlet. Multiple Poles at Negative Integers for $\int_A f^\lambda \square$ in the Case of an Almost isolated Singularity. Publ. RIMS, Kyoto Univ. 35:571-584, 1999.

[3] D. Barlet. La variation pour une hypersurface ayant une singularité presque isolée relativement à la valeur propre 1. Revue de l'Inst. Cartan 15:5–29, 1997.

[4] D. Barlet. Emmêlements de strates consécutives pour les cycles évanescent. Ann. Scient. École Norm. Sup. 4ième série, t. 24:401–506, 1991.

[5] D. Barlet. Contribution effective dans le cas réel. Compositio Mathematica 56:351-359, 1985.

[6] D. Barlet. Contribution du cup-produit de la fibre de Milnor aux pôles de $\mid f \mid^{2\lambda}$. Annales de l'Institut Fourier, t. 34, Fasc. 4:75–107, 1984.

[7] D. Barlet. Contribution effective de la monodromie aux développements asymptotiques. Ann. Scient. École Norm. Sup. 17:293–315, 1984.

[8] D. Barlet, A. Mardhy. Un critère topologique d'existence des pôles pour le prolongement méromorphe de $\int_A f^\lambda \square$. Ann. Inst. Fourier (Grenoble) 43 3:743–750, 1993 (et son Erratum, Ann. Inst. Fourier(Grenoble), 44 2:629–630, 1994.)

[9] I.N. Bernstein. The analytic continuation of generalized functions with respect to a parameter. FAP 6 4:26–40, 1972.

[10] J.E. Björk. Analytic $\mathcal{D}$-module and Applications. Kluwer Academic Publishers. Netherlands, 1993.

[11] H.A. Hamm, L.D. Tráng. Un théorème de Zariski du type Lefschetz. Ann. scient. Éc. Norm. Sup., 4ième série, t. 6:317-366, 1973.

[12] M.E. Herrera. Integration on semi-analytic set, Bull. Soc. Math. France 94:141-180, 1966.

[13] P. Jeanquartier. Développement asymptotique de la distribution de Dirac attachée à une fonction analytique. C. R. Acad. Sc. Paris 271:1159–1161, 1970.

[14] A. Jeddi. Preuve d'une conjecture de V. P. Palamodov, to appear: Topology 41 2.

[15] A. Jeddi. Pôles de $\int_A f^s \bullet$ et conjugaison complexe. C.R. Acad. Sci. Paris, t. 320:165-168, Série I, 1995.

[16] A. Jeddi. Singularité réelle isolée. Ann. Ins. Fourier, Grenoble 41:1:87-116, 1991.

[17] A. Jeddi, A Mardhy. Pôles de $\int_A f^s \bullet$ pour une singularité presque isolée. Manuscripta Math. 97:435–452, 1998.

[18] L. Hörmander. The Analysis of Linear Partial Differential Operators. I. Distribution Theory and Fourier Analysis. Springer-Verlag, Berlin-Heidelberg-New York, 1983.

[19] M. Kashiwara. *B*-functions and holonomic systems, Invent. Math. 38:33–53, 1977.

[20] B. Malgrange. Intégrales asymtotiques et monodromie. Ann. Scient. Éc. Norm. Sup. 4ième série, t. 7:405–430, 1974.

[21] V.P. Palamodov. Asymptotic expansions of integrals in complex and real regions. Math USSR, vol. 55:1:207-236, 1986.

[22] J. Scherk. On the Gauss-Manin Connection of an Isolated Hypersurface Singularity. Math. Ann. 238:23-32, 1978.

Deformations of Boundary Singularities and Non-Crystallographic Coxeter Groups

INNA SCHERBAK   School of Mathematical Sciences, Tel Aviv University, Ramat Aviv 69978, Israel   E-mail: scherbak@tau.ac.il

Abstract

Singularities connected with Coxeter groups $I_2(k)$, H_3 and G_2 are studied. At first, such singularities were found by O. Lyashko in the classification of critical points of non-singular functions on a singular hypersurface. We establish a link between these critical points and boundary singularities. We describe a class of deformations of boundary singularities which provides miniversal deformations of critical points of non-singular functions on a singular hypersurface. In particular, Coxeter groups $I_2(k)$ and H_3 turn out to be connected with unimodal boundary singularities B_{k-1}^3 and F_4^4, respectively, and group G_2 is connected with simple boundary singularity F_4.

INTRODUCTION

Since 1972 it has been known that singularities of holomorphic functions are closely related to the geometry of Coxeter groups. To describe this relation, we recall briefly some data about singularities and about Coxeter groups.

BIFURCATION DIAGRAMS OF SINGULARITIES   Let f be a holomorphic function germ at a critical point. Its *deformation* is a holomorphic family germ $F(\cdot, \lambda)$ such that $F(\cdot, O) = f$, $\lambda \in \mathbb{C}^\mu$. For an equivalence relation on the set of the holomorphic germs, deformation F is called *versal* with respect to the relation, if F contains (for values of parameter λ close to O) representatives of *all* equivalence classes close enough to f. A versal deformation with the minimal possible dimension of the parameter space, $\mathbb{C}^\mu$, is called *miniversal*. The values of λ's such that the corresponding germ in a miniversal deformation has zero as its critical value form a hypersurface in $\mathbb{C}^\mu$ called a *bifurcation diagram*. A classical example is provided by an ordinary

151

singularity, which is a class of *stable equivalency*. It is well known that for a singularity of finite multiplicity, miniversal deformations exist, the bifurcation diagram is unique up to a diffeomorphism, and its structure contains a lot of information about the singularity ([5]).

MANIFOLDS OF NON-REGULAR ORBITS OF COXETER GROUPS A finite group generated by reflections in μ-dimensional euclidean space, i.e. a *Coxeter group*, has a basis of invariants, that is, the manifold of the orbits of the complexified action of the group is furnished with a natural structure of a smooth algebraic variety $\mathbb{C}^\mu$. The number of points in the orbit of a typical point is equal to the number of elements in this group. However, some orbits are smaller. These *non-regular* orbits form an algebraic hypersurface in the orbit space called a *manifold of non-regular orbits*. The list of the crystallographic Coxeter groups (or of the Weyl groups of the simple Lie groups) contains groups A_μ, D_μ ($\mu \geq 4$), E_6, E_7, E_8 having root systems with only roots of equal length, and groups B_μ ($\mu \geq 2$), C_μ ($\mu \geq 3$), F_4 and G_2 with inhomogeneous root systems. Besides the Weyl groups, the list of Coxeter groups contains non-crystallographic groups $I_2(p)$ ($p \geq 5$) - the symmetry groups of regular p-gons, H_3 - the symmetry group of an icosahedron, and H_4 - the symmetry group of a hyper-icosahedron ([6]).

In [1] it was established that the bifurcation diagrams of the simple ordinary singularities A_μ, D_μ ($\mu \geq 4$), E_6, E_7, E_8 are diffeomorphic to the manifolds of the non-regular orbits of the corresponding reflection groups acting on the complex space.

The extension of this connection to include other reflection groups has been a problem stimulating a deep research in the singularity theory.

Deformations which are miniversal with respect to other equivalence relations appear naturally by considering singularities with additional structures as boundaries, obstacles, symmetries etc. Among the corresponding bifurcation diagrams one can recognize the manifolds of the non-regular orbits of other reflection groups. After papers [3], [12], [7] this is a "standard" way of finding reflection groups in the singularity theory.

Namely, miniversal deformations of the simple singularities on manifolds with boundary are to be identified with deformations of the simple singularities $A_{2\mu-1}$, $D_{\mu+1}$, E_6 which are miniversal in the class of $\mathbb{Z}_2$-invariant functions ([3], [5]). The corresponding bifurcation diagrams are diffeomorphic to the manifolds of the non-regular orbits of reflection groups B_μ, C_μ, F_4.

A similar example is given by singularities of the distance function in the problem of avoiding an obstacle [12]. The manifolds of the non-regular orbits of Coxeter groups $I_2(5)$, H_3, H_4 are realized as the bifurcation diagrams of deformations of the simple singularities A_4, D_6, E_8, respectively, which are miniversal in the class of the singularities of even multiplicity.

Recently, the link between unitary reflection groups and singularities of functions with cyclic simmetry was established by Goryunov [7].

Unlike the previous cases, singularities connected with Coxeter groups $I_2(p)$ ($p \geq 5$), H_3, G_2, have appeared in [8] in a different way. These are singularities of critical points of functions on a singular hypersurface, and the corresponding critical points are not simple but unimodal.

In the present paper we propose a construction that allows us to include groups $I_2(p)$, H_3 and G_2 into the generic framework of the singularity theory described above.

We establish a link between critical points of non-singular functions on a singular hypersurface and boundary singularities. It turns out that there is one-to-one correspondence between simple boundary singularities and simple critical points on a singular surface. We prove that the bifurcation diagram of a critical point of a non-singular function on a singular surface is the bifurcation diagram of a certain deformation of the restriction of the function to the surface which is miniversal with respect to the stable equivalence of functions on a singular surface. This deformation is related to the corresponding boundary singularity.

In particular, unimodal singularities $I_2(p)$, $p \geq 5$, and H_3 are stable equivalence classes of critical points of non-singular functions on hypersurfaces of types A_{p-1} and A_3, respectively. These critical points correspond to the unimodal boundary singularities B_{p-1}^3 and F_4^4, respectively. The bifurcation diagrams of these critical points are the bifurcation diagrams of certain deformations of simple singularities A_3 and A_4 related to the boundary singularities B_{p-1}^3 and F_4^4, respectively. The simple critical point G_2 on a hypersurface of type A_2 corresponds to the simple boundary singularity $F_4 \equiv F_2^2$, and the manifold of the non-regular orbits G_2 appears as the bifurcation diagram of a deformation of A_2 related to F_4.

Main results of this paper are announced in [11]. The structure of the paper is as follows. In Sec. 1 we recall the theory of critical points of functions on a singular hypersurface ([8], [4]). Sec. 2 is devoted to the case of non-singular functions on a singular hypersurface. In Sec. 3 we recall some facts from the boundary singularity theory ([5], [13]). In Sec. 4 a connection between non-singular functions on a singular hypersurface and boundary singularities is described. In Sec. 5, the case when the hypersurface has a singularity of type A_k is considered. We prove that the origin is a non-critical point for a function germ on a singular hypersurface if and only if the corresponding boundary germ is of type B_k. In Sec. 6, for a wide class of critical points on a singular hypersurface, we prove that the local ring is isomorphic to the local ring of the ordinary singularity given by the restriction of the corresponding boundary germ to the boundary. In Sec. 7 we describe miniversal deformations of a critical point on a singular hypersurface in terms of certain deformations of the restriction of the corresponding boundary germ to the boundary. As a corollary, we get (Sec. 8) that the manifolds of the non-regular orbits of reflection groups $I_2(p)$ and H_3 are diffeomorphic to generic hyperplane sections of the corresponding bifurcation diagrams. In Sec. 9 we consider the critical point of type G_2.

1   PRELIMINARIES

1.1   Critical points of functions on a singular hypersurface

In this section we recall the theory of critical points of functions on a manifold with singular boundary of [8]. We use slightly different terminology and call these critical points "critical points of functions on a singular hypersurface" in order not to mess them up with critical points on a manifold with boundary participating in our considerations.

A function germ f on a singular hypersurface V is a triple $(f, V, \mathbb{C}^n)$ where
- $f : (\mathbb{C}^n, O) \to (\mathbb{C}, 0)$ is a germ of a holomorphic function;
- $V = \{z \in \mathbb{C}^n \mid h(z) = 0\}$ is a germ of a hypersurface with an isolated singular point at O.

Germs of diffeomorphisms $(\mathbb{C}^n, O) \to (\mathbb{C}^n, O)$ act on the set of the triples, and two triples are *equivalent* if they lie in the same orbit of this action.

For $m > n$, denote by π the natural projection $\pi : \mathbb{C}^m = \mathbb{C}^n \times \mathbb{C}^{m-n} \to \mathbb{C}^n$,

$$\pi(z_1, \ldots, z_n, z_{n+1}, \ldots, z_m) = (z_1, \ldots, z_n).$$

In $\mathbb{C}^m$, we define hypersurface $\tilde{V} = \{h(z_1, \ldots, z_n) + z_{n+1}^2 + \cdots + z_m^2 = 0\}$ and function germ $\tilde{f} = \pi^* f$. The triple $(\tilde{f}, \tilde{V}, \mathbb{C}^m)$ is called *the stabilization* of $(f, V, \mathbb{C}^n)$.

Functions on singular hypersurfaces are *stable equivalent* if they have equivalent stabilizations. Stable equivalent triples have the same singularity of hypersurfaces.

We say that O is a *non-critical* point of $(f, V, \mathbb{C}^n)$ if $V_0 = f^{-1}(O)$ is a germ of a smooth hypersurface (i.e. f is *non-singular* at O) transversal to V at O. In the opposite case we say that O is a *critical point of the triple* $(f, V, \mathbb{C}^n)$.

The transversality at a singular point of V means the following. Consider $PT^*\mathbb{C}^n$, the projectivization of the cotangent bundle of $\mathbb{C}^n$, and the canonical projection

$$\rho : PT^*\mathbb{C}^n \to \mathbb{C}^n, \quad \rho(z_1, \ldots, z_n, p_1 : \cdots : p_n) = (z_1, \ldots, z_n).$$

For any subvariety $M = \{\phi(z) = 0\}$ in $\mathbb{C}^n$, denote by PM the image of M under the embedding in $PT^*\mathbb{C}^n$:

$$(z \in M) \mapsto (z, \text{tangent plane to } M \text{ at } z).$$

In other words, PM is the following subvariety of $PT^*\mathbb{C}^n$:

$$PM = \{\phi(z) = 0, \quad p_i \frac{\partial \phi}{\partial z_j} - p_j \frac{\partial \phi}{\partial z_i} = 0, \ 1 \le i < j \le n\}.$$

In particular, if V has an isolated critical point at O, then PV is reducible and consists of $PT^*\{O\}$ and $V_1 = \overline{PV \setminus PT^*\{O\}}$, the closure of $PV \setminus PT^*\{O\}$. We say that $V_0 = f^{-1}(O)$ is *transversal* to V at O, if V_0 is smooth and $V_1 \cap PV_0 \cap \rho^{-1}(U) = \emptyset$, where U is a small neighborhood of the origin in $\mathbb{C}^n$. This means that the tangent plane to V_0 at the origin, considered as a point in $PT^*\mathbb{C}^n$, does not belong to V_1.

In the case when hypersurface V has an isolated simple singularity, the classification of critical points on V reduces to the description of the orbits of the Lie group of diffeomorphisms preserving V. The Lie algebra of this group, T_V, consists of the vector fields preserving V. Vector field $\bar{v}$ *preserves hypersurface* V, $\bar{v} \in T_V$, if the directional derivative $L_{\bar{v}} h$ belongs to the principal ideal (h), i.e. $< \bar{v}, \text{grad} h > = gh$ for some smooth germ g. We define the ideal

$$I_{f|V} = \{L_{\bar{v}} f \mid \bar{v} \in T_V\}$$

and the local algebra

$$Q_{f|V} = \mathcal{O}_n / I_{f|V},$$

where $\mathcal{O}_n$ is the ring of the holomorphic germs $(\mathbb{C}^n, O) \to (\mathbb{C}, 0)$. The *multiplicity* $\mu(f, V)$ of the critical point O of $(f, V, \mathbb{C}^n)$ is defined as

$$\mu(f, V) = \dim_{\mathbb{C}} Q_{f|V} - 1.$$

For any stabilization $(\tilde{f}, \tilde{V}, \mathbb{C}^{n+k})$ of $(f, V, \mathbb{C}^n)$, the local algebras $Q_{f|V}$ and $Q_{\tilde{f}|\tilde{V}}$ are isomorphic, and thus the multiplicity $\mu(f, V, \mathbb{C}^n)) = \mu(\tilde{f}, \tilde{V}, \mathbb{C}^{n+k}) = \mu(f, V)$ does not depend on dimension.

By the usual way, the notions of modality, versal deformation, bifurcation diagram can be defined for this situation. In particular, the *modality* of $(f, V, \mathbb{C}^n)$ is the minimal number m such that a small neighborhood of the orbit of this germ (under the action of diffeomorphisms preserving V) is covered by a finite number of m-parameter families of orbits. When m is equal to 0 or 1, a critical point is called *simple* or *unimodal*, respectively.

In the case $\mu < \infty$, one can take a versal deformation of $(f, V, \mathbb{C}^n)$ in the form $(F, V, \mathbb{C}^n)$, where F is the family of functions

$$F(z, \lambda) = f(z) + \lambda_0 e_0 + \cdots + \lambda_\mu e_\mu,$$

here $e_0, \ldots, e_\mu$ are representatives of a basis of the local algebra $Q_{f|V}$ over $\mathbb{C}$, and $\lambda = (\lambda_0, \ldots, \lambda_\mu) \in \mathbb{C}^{\mu+1}$ is the parameter of versal deformation.

The bifurcation diagram $\Sigma(f, V)$ of the critical point $(f, V, \mathbb{C}^n)$ is a hypersurface in the base of versal deformation, $\mathbb{C}^{\mu+1}$, formed by the parameter values $\lambda \in \mathbb{C}^{\mu+1}$ such that 0 is a critical value of $(F(\cdot, \lambda), V, \mathbb{C}^n)$.

It turns out that simple and unimodal critical points appear only on a hypersurface with a simple singularity of type A_k. In [8], it was proven that for a critical point of modality 1, the bifurcation diagram is analytically trivial along the strata $\mu = \text{const}$, and the classification of the simple and unimodal critical points on a hypersurface of type A_k was obtained.

A part of the classification is connected with reflection groups H_3 and $I_2(p)$. Namely, a critical point of type $I_2(p)$, $p \geq 4$, is given by the germ $x + \epsilon y + z^2$, $\epsilon \neq 0$, on the hypersurface $xy = z^p$ of type A_{p-1}, and a critical point of type H_3 is given by the germ $x + y + \epsilon z^3$ on the hypersurface $xy = z^5$ of type A_4. These critical points are unimodal and ϵ is a parameter along the strata $\mu = \text{const}$. The main result of [8] is the following theorem.

THEOREM 1.1 *The intersection of the bifurcation diagram of a critical point of type* $I_2(p)$, $p \geq 4$, *or* H_3, *with a hyperplane in the base of versal deformation which is transversal to the stratum* $\mu = \text{const}$, *is biholomorphic equivalent to the manifold of the non-regular orbits of the corresponding group generated by reflections, acting on the complexification of the Euclidean space* $\mathbb{C}^\mu$.

1.2 Non-singular functions on a singular hypersurface

As it was pointed out in [8], if the number of variables is greater then two, then simple and unimodal critical points on a singular hypersurface can appear only for function germs with non-zero 1-jet. Moreover, simple critical points appear only on a hypersurface of type A_k. On a boundary of type D_k or E_k, there are only unimodal non-critical points.

For that reason, in the paper, we study triples $(f, V, \mathbb{C}^n)$ where $f : (\mathbb{C}^n, O) \to (\mathbb{C}, 0)$ is a non-singular germ, i.e. $V_0 = f^{-1}(0)$ is a germ of smooth hypersurface.

Let $z_1, \ldots, z_n$ be coordinates in $\mathbb{C}^n$ such that $\partial f / \partial z_1$ does not vanishes at the origin. The change of variables $x = f$, $y_1 = z_2$, $\ldots$, $y_{n-1} = z_n$, gives an equivalent triple

$$(x, V = \{g(x, y_1, \ldots, y_{n-1}) = 0\}, \mathbb{C}^n).$$

We denote $y = (y_1, \ldots, y_{n-1})$, $g_0(y) = g(0, y_1, \ldots, y_{n-1})$. In this case $V_0 = \{x = 0\}$ and thus

$$PV_0 = \{x = 0, \ p_1 = \cdots = p_{n-1} = 0\}.$$

The tangent plane to V_0 at the origin is point $(0, \ldots, 0; 1 : 0 : \cdots : 0)$ in $PT^*\mathbb{C}^n$ with coordinates $(x, y_1, \ldots, y_{n-1}; p_0 : p_1 : \cdots : p_{n-1})$. Further, PV is given by

$$PV = \{g(x, y) = 0, \ p_0 \frac{\partial g}{\partial y_i} = p_i \frac{\partial g}{\partial x}, \ p_j \frac{\partial g}{\partial y_i} = p_i \frac{\partial g}{\partial y_j}, \ 1 \le i < j \le n\}.$$

Therefore the intersection $PV_0 \cap PV$ is

$$PV_0 \cap PV = \{g(x, y_1, \ldots, y_{n-1}) = 0, \ x = 0, \ p_0 \frac{\partial g}{\partial y_i} = 0, \ p_i = 0, \ i = 1, \ldots, n-1\}$$

and thus

$$PV_0 \cap V_1 = \{g(x, y_1, \ldots, y_{n-1}) = 0, \ x = 0, \ \frac{\partial g}{\partial y_i} = 0, \ p_i = 0, \ i = 1, \ldots, n-1\}.$$

EXAMPLE 1.2 Let $V = \{x^{k+1} = y^2\} \subset \mathbb{C}^2$, $k \ge 2$. Then for the triple $(x, V, \mathbb{C}^2)$, the origin is a non-critical point, whereas for the triple $(y, V, \mathbb{C}^2)$, the origin is a critical point. Indeed,

$$PV = \{x^{k+1} = y^2, \ (k+1)x^k p_1 = 2y p_0\} = \{y = x^{\frac{k+1}{2}}, \ (k+1)x^k p_1 = 2x^{\frac{k+1}{2}} p_0\}.$$

If $x \ne 0$, we get

$$p_0 = \frac{k+1}{2} p_1 x^{\frac{k-1}{2}}.$$

For $f = x$, we have $\underline{PV_x = \{x = p_1 = 0\}}$ and the corresponding point in $PT^*\mathbb{C}^2$ is $(0, 0; 1 : 0) \notin V_1 = \overline{PV \setminus PT^*\{O\}}$, whereas for $f = y$, $PV_y = \{y = p_0 = 0\}$ and the corresponding point in $PT^*\mathbb{C}^2$ is $(0, 0; 0 : 1) \in V_1$.

1.3   Boundary singularities

Here we recall some basic facts of boundary singularities theory [13].

A *boundary germ* is a triple $(g, Y, \mathbb{C}^n)$, where Y is a germ at O of a smooth hypersurface, called *a boundary*, and $g : (\mathbb{C}^n, O) \to (\mathbb{C}, 0)$ is a holomorphic germ such that both g and $g_0 = g|_Y : (Y, O) \to (\mathbb{C}, 0)$ have isolated critical points. In appropriate local coordinates $(x, y_1, \ldots, y_{n-1})$ of $\mathbb{C}^n$, the boundary is $Y = \{x = 0\}$, and $g_0(y) = g(0, y_1, \ldots, y_{n-1})$.

A *stabilization* of $(g, Y, \mathbb{C}^n)$ is a boundary germ $(\tilde{g}, \tilde{Y}, \mathbb{C}^m)$, $m > n$, where $\tilde{g}(x, y_1, \ldots, y_{m-1}) = g(x, y_1, \ldots, y_{n-1}) + y_n^2 + \cdots + y_{m-1}^2$, $\tilde{Y} = \{x = 0\}$.

A *boundary singularity* is a boundary germ considered up to germs of diffeomorphisms preserving the boundary and up to stabilizations.

The *multiplicity* $\mu(g, Y)$ of the critical point of the boundary germ $(g, Y, \mathbb{C}^n)$ is the dimension over $\mathbb{C}$ of the *local ring* $Q(g, Y) = \mathcal{O}_n / I(g, Y)$, where $I(g, Y)$ is the ideal generated by $x \partial g/\partial x, \partial g/\partial y_1, \ldots, \partial g/\partial y_{n-1}$:

$$\mu(g, Y) = \dim_{\mathbb{C}} Q(g, Y).$$

The multiplicity of a boundary germ and of any its stabilization is the same.

In a natural sense, boundary singularity $(g, Y, \mathbb{C}^n)$ is an extension of two ordinary singularities given by germs g and g_0. These two ordinary singularities are called the *decomposition* of $(g, Y, \mathbb{C}^n)$.

Recall that for an ordinary singularity given by a holomorphic germ $f(z_1, \ldots, z_n)$ at the critical point O, the multiplicity is the dimension over $\mathbb{C}$ of the local ring $Q(f) = \mathcal{O}_n / I(f)$, where ideal $I(f)$ is generated by the partial derivatives $\partial f / \partial z_1, \ldots, \partial f / \partial z_n$:

$$\mu(f) = \dim_{\mathbb{C}} Q(f).$$

For boundary singularity $(g, Y, \mathbb{C}^n)$ with decomposition (g, g_0):

$$\mu(g, Y) = \mu(g) + \mu(g_0).$$

A versal deformation of boundary singularity $(g, Y, \mathbb{C}^n)$ one can take in the form

$$G(x, y, \lambda) = g(x, y) + \Sigma \lambda_i' e_i' + x \Sigma \lambda_j'' e_j'',$$

where $\{e_i', 1 \leq i \leq \mu(g_0)\}$ represent a basis of $Q(g_0)$, and $\{e_j'', 1 \leq j \leq \mu(g)\}$ represent a basis of $Q(g)$. The bifurcation diagram of the boundary singularity $(g, Y, \mathbb{C}^n)$ has two irreducible components, which are bifurcation diagrams of g and g_0 respectively multiplying by complex spaces of appropriate dimensions.

In generic case, the decomposition does not define a boundary singularity (see example 4), but boundary singularities with decomposition of type (A_k, A_l) are well-defined by their decomposition ([10]). In particular, if g_0 is a Morse function (i.e. of type A_1) and g is of type A_k, then $(g, Y, \mathbb{C}^n)$ is of type B_{k+1}. It can be given by the function germ $x^{k+1} + Q(y)$, where $Q(y) = Q(y_1, \ldots, y_{n-1})$ is a Morse function, the boundary is $x = 0$.

2 LINK BETWEEN NON-SINGULAR FUNCTIONS NON SINGULAR HYPERSUPERFACES AND BOUNDARY SINGULARITIES

If $(f(z), V = \{g(z) = 0\}, \mathbb{C}^n)$ is a germ of non-singular function f on a singular hypersurface V, such that $\mu(f, V) < \infty$, then triple $(g(z), V_0 = \{f(z) = 0\}, \mathbb{C}^n)$ is a boundary germ.

Conversely, any boundary germ $(g(x, y), Y = \{x = 0\}, \mathbb{C}^n)$ defines a germ $(x, V = \{g(x, y) = 0\}, \mathbb{C}^n)$ of a non-singular function on a singular hypersurface.

PROPOSITION 2.1 *Equivalent germs of non-singular functions on a singular hypersurface define the same boundary singularity*

Proof: If $(f_i, V_i, \mathbb{C}^n)$, $i = 1, 2$, are equivalent triples and f_1, f_2 are germs of non-singular functions, then there exist germs of diffeomorphisms $\phi, \phi_1, \phi_2 : (\mathbb{C}^n, O) \to (\mathbb{C}^n, O)$, such that ϕ sends $(f_1, V_1, \mathbb{C}^n)$ to $(f_2, V_2, \mathbb{C}^n)$ and ϕ_i sends $(f_i, V_i, \mathbb{C}^n)$ to $(x, \{g_i(x, y) = 0\}, \mathbb{C}^n)$, $i = 1, 2$.

The corresponding boundary germs are $(g_i(x, y), Y, \mathbb{C}^n)$ $(i = 1, 2)$ with the boundary $Y = \{x = 0\}$, and one goes to another by the diffeomorphism $\phi_2 \circ \phi \circ \phi_1^{-1}$ which obviously preserves the boundary. $\square$

2.1 Simple critical points

Comparing the lists of the simple critical points on singular surfaces [8] and of the simple boundary singularities [3], we get the following

PROPOSITION 2.2 *The simple critical points on a singular hypersurface correspond to the simple boundary singularities C_k, $k \geq 2$, F_4. Non-critical points on a singular hypersurface correspond to the simple boundary singularities B_k, $k \geq 2$.*

EXAMPLE 2.3

- Simple critical point G_2 appears in [8] for non-singular germ $x + y$ on a singular surface $xy = z^3$. It is easy to check that the corresponding (simple) boundary singularity is F_4. It can be given by germ $x^2 + y^3$ and boundary $\{x = 0\}$.

- Critical points $I_2(p), p \geq 4$, appear in the Lyashko classification on a hypersurface of type A_{p-1}. These critical points are unimodal. The normal form of the function germ is $f = x + \epsilon y + z^2$, where $\epsilon \neq 0$ is a parameter along the strata $\mu = \text{const}$. The hypersurface V is given by $V = \{xy = z^p + Q\}$, where Q is a Morse function in additional variables.

  The corresponding boundary singularity has decomposition (A_{p-1}, A_3). Indeed, A_{p-1} is the type of the hypersurface. Consider $f^{-1}(O) = V_0 = \{x + \epsilon y + z^2 = 0\}$. The intersection with the boundary, $V \cap V_0$, is given by

  $$V \cap V_0 = \{z^p + (\epsilon y + z^2)y = 0\}.$$

  We have: $z^p + (\epsilon y + z^2)y = z^p - \alpha z^4 + (\sqrt{\alpha}z^2 + \sqrt{\epsilon}y)^2$, where $\alpha = 1/4\epsilon$, therefore this is a singularity of type A_3.

  Thus the boundary singularity corresponding to the critical point of type $I_2(p)$ is the unimodal boundary singularity of type B_{p-1}^3 (in notations of [10]).

- H_3-singularity.
  In the Lyashko list, a critical point of type H_3 is given by germ $x + y + \epsilon z^3$ on a hypersurface $xy = z^5$ of type A_4 (ϵ is a parameter along the strata $\mu = \text{const}$). After the change of variables

  $$(x, \ y, \ z) \ \rightarrow \ (X = x + y + \epsilon z^3, \ y, \ z),$$

  we get germ $z^5 - (X - y - \epsilon z^3)y$ on the boundary $X = 0$. It is easy to see that this boundary singularity has decomposition (A_4, A_4). This is unimodal boundary singularity F_4^4 (in notations of [10]).

2.2 Singular hypersurface of type A_k

Here we consider triples $(f, V, \mathbb{C}^n)$ such that f is a non-singular germ, and hypersurface V has a simple singularity of type A_k at the origin. We call such triples *L-germs*. Note that the critical points on a singular hypersurface, which are connected with reflection groups G_2, $I_2(p), p \geq 4$, and H_3, are L-germs.

For a L-germ, we can choose coordinates such that the triple is of the form $(f(x, y), \{x^{k+1} = Q(y)\}, \ \mathbb{C}^n)$, where $Q(y) = Q(y_1, \dots, y_{n-1})$ is a Morse function.

THEOREM 2.4 *The origin is a non-critical point of the triple* $(f(x,y),\ V = \{x^{n+1} = Q(y)\},\ \mathbb{C}^n)$, *if and only if* $\partial f/\partial x|_0 \neq 0$.

Proof: As it follows from [2], [9], if $\partial f/\partial x$ does not vanish at the origin, then f can be reduced to x by a diffeomorphism preserving $\{x^{n+1} = Q(y)\}$, and we get a triple which is stable equivalent to the triple $(x, V, \mathbb{C}^2)$ of the example 1.

If $\partial f/\partial x|_0 = 0$, then $\partial f/\partial y_i|_0 \neq 0$ for some $1 \leq i \leq n-1$ (recall that f is non-singular at the origin). We can assume

$$\frac{\partial f}{\partial y_1}|_0 \neq 0,\ \frac{\partial f}{\partial x}|_0 = \frac{\partial f}{\partial y_i}|_0 = 0,\ i = 2,\ldots,n-1.$$

Indeed, if $f = a_1 y_1 + \cdots + a_{n-1} y_{n-1} +$ (terms of order ≥ 2), and $a_1 \neq 0$, then the required change of variables is

$$(x, y_1, \ldots, y_{n-1}) \mapsto (x, Y_1 = a_1 y_1 + \cdots + a_{n-1} y_{n-1}, y_2, \ldots, y_{n-1}).$$

The tangent plane to $\{f = 0\}$ at the origin corresponds to point $(0,\ldots,0;0:1:0:\cdots:0)$. This point is obviously in V_1 (the intersection with plane $\{y_2 = \cdots = y_{n-1} = 0\}$ reduce this case to the case of two variables considered in example 1).   □

THEOREM 2.5 *The origin is a non-critical point for a L-germ, if and only if the corresponding boundary germ is of type* B_{k+1}.

Proof: Consider L-germ $(f(x,y),\ \{x^{n+1} = Q(y)\},\ \mathbb{C}^n)$, where Q is a Morse function. The condition that the origin is a non-critical point for the L-germ means that $\partial f/\partial x|_0 \neq 0$, i.e. the equation $f = 0$ can be solved with respect to x: $x = \phi(y)$, where $\phi(y)$ is a holomorphic function germ. Then the restriction of the function $x^{k+1} - Q(y)$ to the boundary, $(\phi(y))^{k+1} - Q(y)$, has a non-degenerate quadratic part and therefore defines an ordinary singularity of type A_1. Therefore boundary singularity $(x^{k+1} - Q(y), f = 0, \mathbb{C}^n)$ has decomposition (A_k, A_1), i.e. this is a boundary singularity of type B_{k+1}.   □

If the hypersurface, V, has another simple singularity, i.e. a singularity of type D_k $(k \geq 4)$ or E_k $(k = 6,7,8)$, then one can prove that a germ $(x, V = \{g(x,y) = 0\}, \mathbb{C}^{n+1})$ does not have the origin as a critical point if the germ $g_0(y) = g(0,y)$ defines a singularity of type A_2.

EXAMPLE 2.6   The following germs on the boundary $x = 0$ define unimodal non-critical points $(x, V = \{g(x,y) = 0\}, \mathbb{C}^2)$ on a hypersurface of type D_k, $k \geq 4$, and E_k, $k = 6,7,8$, respectively:

$$
\begin{aligned}
(D_k, A_2): \quad & g(x,y) = xy^2 + y^3 + ax^{k-1},\ a \neq 0, \\
(E_6, A_2): \quad & g(x,y) = x^4 + y^3 + ax^3 y, \\
(E_7, A_2): \quad & g(x,y) = x^3 y + y^3 + ax^5, \\
(E_8, A_2): \quad & g(x,y) = x^5 + y^3 + ax^4.
\end{aligned}
$$

Consider hypersurface $V = \{x^a + y^b = 0\}$, $a > b$. Direct calculations show that O is a non-critical point for triple $(x, V, \mathbb{C}^2)$ and O is a critical point for triple $(y, V, \mathbb{C}^2)$. More generically, we have the following proposition.

PROPOSITION 2.7 *Let hypersurface* $V = \{g(x_1, \ldots, x_n) = 0\}$ *be given by a quasi-homogeneous function* g *with* $\deg x_i = \alpha_i$, $1 \leq i \leq n-1$, $\alpha_1 > \alpha_2 \geq \cdots \geq \alpha_n$. *Then the origin is a critical point of* $(x_1, V, \mathbb{C}^n)$.

Proof: Note first of all, that if g is a quasi-homogeneous function with $\deg g = d$, then $\partial h / \partial x_i$ is a quasi-homogeneous function as well and $\deg \partial h / \partial x_i = d - \alpha_i, i = 1, \ldots, n$. We have $d - \alpha_1 < d - \alpha_2 \leq \cdots \leq d - \alpha_n$, and therefore for the curve

$$x_i = a_i t^{\alpha_i}, \; g(a_1, \ldots, a_n) = 0,$$

lying on V, the tangent planes correspond to the points

$$(x; t^{d-\alpha_1} : t^{d-\alpha_2} : \cdots : t^{d-\alpha_n}) = (x; 1 : t^{\alpha_1-\alpha_2} : \cdots : t^{\alpha_1-\alpha_n}) \to (O; 1 : 0 \cdots : 0)$$

as $t \to 0$. Point $(O; 1 : 0 : \cdots : 0)$ corresponds to the function x_1.   □

2.3   Local rings

It appears that in many cases, the local ring $Q_{f|V} = \mathcal{O}_n / I_{f|V}$ of the triple $(f, V = \{h(z) = 0\}, \mathbb{C}^n)$ is isomorphic to the local ring $Q(h_0)$ of the ordinary singularity given by $h_0 = h|_{f=0}$. First we establish this result in the quasi-homogeneous case.

THEOREM 2.8 *Let* $(g(x,y), Y = \{x = 0\}, \mathbb{C}^n)$ *be a boundary germ given by a quasi-homogeneous function* g, *and* $V = \{g(x,y) = 0\}$. *Then the local rings* $Q_{x|V}$ *and* $Q(g_0)$ *are isomorphic.*

Proof: Consider germ $(x, V, \mathbb{C}^n)$ on a singular hypersurface corresponding to the given boundary germ. Let the weights of the variables be $\deg x = \alpha, \deg y_i = \beta_i, i = 1, \ldots, n-1$. Then, as it is proven in [8], the module of tangent vector fields, T_V, is generated by the Euler vector field

$$v_0 \doteq \alpha x \frac{\partial}{\partial x} + \beta_1 y_1 \frac{\partial}{\partial y_1} + \cdots + \beta_{n-1} y_{n-1} \frac{\partial}{\partial y_{n-1}},$$

and by the Hamiltonian vector fields

$$v_k = \frac{\partial g}{\partial x} \frac{\partial}{\partial y_k} - \frac{\partial g}{\partial y_k} \frac{\partial}{\partial x}, \;\; 1 \leq k \leq n-1,$$

$$v_{ij} = \frac{\partial g}{\partial y_i} \frac{\partial}{\partial y_j} - \frac{\partial g}{\partial y_j} \frac{\partial}{\partial y_i}, \;\; 1 \leq i, j \leq n-1.$$

Applying these vector fields to the function x, we get the generators of $I_{x|V}$ which are $x, \partial g / \partial y_k, k = 1, \ldots, n-1$. Therefore

$$Q_{x|V} = \mathcal{O}_n / < x, \frac{\partial g}{\partial y_1}, \ldots, \frac{\partial g}{\partial y_{n-1}} > \cong \mathcal{O}_{n-1} / I(g_0) = Q(g_0).$$ □

COROLLARY 2.9 *If* $(g(x,y), Y = \{x = 0\}, \mathbb{C}^n)$ *is a boundary germ given by a quasi-homogeneous function* g, *then* $\mu(x, V) = \mu(g_0) - 1$

The similar statement is hold for L-germs of modality 0 or 1.

THEOREM 2.10 *If* $(f, V = \{h(z) = 0\}, \mathbb{C}^n)$ *is a L-germ of modality 0 or 1, then:*

(i) $Q_{f|V} \cong Q(h|_{f=0})$;

(ii) $\mu(f, V) = \mu(h|_{f=0}) - 1$.

The proof of the theorem can be obtained by direct calculations using the list of normal forms of simple and unimodal critical poins and their versal deformations given in [8].

We say that $(f, V = \{h(z) = 0\}, \mathbb{C}^n)$ is *a good triple* if it satisfies condition *(i)* (and, hence, *(ii)*). In particular, triples of types H_3 and $I_2(p)$ are good.

Using the parametrized Morse lemma, one can prove the similar result for triples $(x, V = \{g(x, y) = 0\}, \mathbb{C}^n)$ such that the corresponding boundary germs $(g(x, y), Y = \{y = 0\}, \mathbb{C}^n)$ have decomposition (A_k, A_l).

Unfortunately we do not know a proof working for all these cases. It would be interesting to get such a proof and to understand general conditions for triple to be good.

Note that for L-triples, the equivalence class of a critical point $(f, V = \{h(z) = 0\}, \mathbb{C}^n)$ does not in general defined by the equivalence class of the germ $h|_{f=0}$, as the following example shows.

EXAMPLE 2.11 Consider two germs on $\mathbb{C}^3$ with the boundary $\{x = 0\}$:

$$h_1 = xz - z^4 - zy^2 - y^4, \quad h_2 = xz - z^4 - y^4 - zy^3.$$

As it follows from [10], they define non-equivalent boundary singularities with the same decomposition (A_3, D_5). This means that the critical points $(x, V_i = \{h_i = 0\}, \mathbb{C}^3)$ $(i = 1, 2)$ on a singular hypersurface of type A_3 have the same singularity, namely D_5, of the restrictions to the bundary, $h_i|_{x=0}$, but they are non-equivalent L-germs.

For holomorphic functions, an isolated critical point always has a finite multiplicity. Next example provides a non-singular function having a critical point of infinite multiplicity at an isolated singular point of a hypersurface.

EXAMPLE 2.12 Consider the following critical point on a singular hypersurface:

$$(x, \ V = \{x^3 + xz^2 + y^2z = 0\}, \ \mathbb{C}^n).$$

It is clear that hypersurface V has an isolated singularity at O, but the multiplicity of the critical point $\mu(x, V) = \infty$. Indeed, V is given by a homogeneous function, therefore theorem 3 gives $Q_{x|V} \cong \mathcal{O}_{n-1}/I(g_0)$, where $g_0 = y^2z$. Function g_0 has a non-isolated critical point at 0 (in fact, the line $y = 0$ is the line of critical points), that means that $\dim_{\mathbb{C}} Q_{x|V} = \infty$.

2.4 Versal deformations and bifurcation diagrams of good triples

Let $L = (x, V = \{g(x, y) = 0\}, \mathbb{C}^n)$ be a good triple. A versal deformation of L one can take in the form

$$L_\lambda = (x + \lambda_0 e_0 + \cdots + \lambda_\mu e_\mu, V, \mathbb{C}^n),$$

where $\{e_i = e_i(y), \ 0 \le i \le \mu\}$ represent a basis of the local ring $Q(g_0)$.

We define a deformation $G_L(x, y, \lambda)$ of boundary germ $(g(x, y), \{x = 0\}, \mathbb{C}^n)$ which corresponds to the versal deformation of the germ L as the family

$$G_L(x, y, \lambda) = g(x + \lambda_0 e_0 + \cdots + \lambda_\mu e_\mu, y).$$

The restriction of this family to the boundary,

$$G_L^0(y, \lambda) = G_L(0, y, \lambda) = g(\lambda_0 e_0 + \cdots + \lambda_\mu e_\mu, y),$$

is a deformation of $g_0 = g(0, y)$ related to the boundary germ $(g(x, y), \{x = 0\}, \mathbb{C}^n)$. We call this deformation *boundary deformation of* g_0 and the corresponding bifurcation diagram *the boundary diagram*. One can see that the boundary diagrams of equivalent boundary germs are diffeomorphic. It turns out that the bifurcation diagram of this deformation is the bifurcation diagram of L.

THEOREM 2.13 *The bifurcation diagram of L is*

$$\Sigma(L) = \{\lambda \in \mathbb{C}^{\mu+1} \mid 0 \text{ is a critical value of } G_L^0(\cdot, \lambda)\}$$

Proof: Change of variables

$$(x, y) \mapsto (X = x - \lambda_0 e_0 - \cdots - \lambda_\mu e_\mu, y)$$

gives an equivalent family

$$\tilde{L}_\lambda = (X, V_\lambda = \{g(X + \lambda_0 e_0 + \cdots + \lambda_\mu e_\mu, y) = 0\}, \mathbb{C}^n).$$

The bifurcation diagram of this family is given by

$$\Sigma(L) = \{\lambda \in \mathbb{C}^{\mu+1} \mid 0 \text{ is a critical value of } \tilde{L}_\lambda\}.$$

Consider $PT^*\mathbb{C}^n$ with coordinates $(x, y_1, \ldots, y_{n-1}; p_0 : p_1 : \cdots : p_{n-1})$. Variety PV_λ of tangent planes to V_λ is given by

$$G_L(X, y, \lambda) = 0, \ p_0 \frac{\partial G_L}{\partial y_i} = p_i \frac{\partial G_L}{\partial X}, \ p_i \frac{\partial G_L}{\partial y_j} = p_j \frac{\partial G_L}{\partial y_i}, \ i, j = 1, \ldots, n - 1.$$

The intersection with $PV_0 = \{X = p_1 = \cdots = p_{n-1} = 0\}$ gives

$$PV_\lambda \cap PV_0 = \{X = p_1 = \cdots = p_{n-1} = 0, \ G_L(X, y, \lambda) = 0, \ p_0 \partial G_L / \partial y_i = 0\}.$$

Thus the intersection $V_1 \cap PV_0$ is given by

$$V_1 \cap PV_0 = \{G_L(X, y, \lambda) = 0, \ X = 0, \ \partial G_L / \partial y_i = 0, \ p_i = 0, \ i = 1, \ldots, n - 1\} =$$

$$= \{G_L^0(y, \lambda) = 0, \ \partial G_L^0 / \partial y_i = 0, \ p_i = 0, \ i = 1, \ldots, n - 1\}.$$

It is non-empty if and only if 0 is the critical value of $G_L^0(\cdot, \lambda)$. $\square$

2.5 Critical points G_2, H_3 and $I_2(p)$

The simple critical point G_2 appears in the Lyashko classification for non-singular germ $x + y$ on a singular surface $xy = z^3$. The corresponding simple boundary singularity F_4 is given by germ $x^2 + y^3$ and boundary $\{x = 0\}$. The boundary deformation is given by $G_G^0(y, \lambda) = (\lambda_1 + \lambda_2 y)^2 + y^3$. Simple calculations show

THEOREM 2.14 *The bifurcation diagram of the deformation G_G^0 is diffeomorphic to the manifold of the non-regular orbits of the group G_2*

Consider function $f = x$ having a critical point of type H_3 at the origin on a hypersurface $V = \{x^2 + y^5 = 0\}$. Its versal deformation is $x + \lambda_3 y^3 + \lambda_2 y^2 + \lambda y + \lambda_0$. Thus the corresponding deformation of the unimodal boundary singularity F_4^4 is

$$G_H(x, y, \lambda) = y^5 + (x + \lambda_3 y^3 + \lambda_2 y^2 + \lambda y + \lambda_0)^2.$$

As it is proven in[8], the bifurcation diagram of this critical point is analytically trivial along the stratum $\mu = $ const and λ_3 is a parameter along this stratum. The bifurcation diagram of this critical point is the bifurcation diagram of the deformation

$$G_H^0(y, \lambda) = y^5 + (\lambda_3 y^3 + \lambda_2 y^2 + \lambda y + \lambda_0)^2$$

of simple singularity A_4.

A critical point of type $I_2(p)$, $p \geq 5$, can be given by germ $f = x$ on a hypersurface $V = \{(x + y^2)^2 + y^p = 0\}$. Its versal deformation is $x + y^2 + \lambda_2 y^2 + \lambda_1 y + \lambda_0$. Again, as it follows from [8], the bifurcation digram is analytically trivial along the stratum $\mu = $ const and λ_2 is a parameter along this stratum. The corresponding deformation of the unimodal boundary singularity B_{p-1}^3 is

$$G_I(x, y, \lambda) = y^p + (x + y^2 + \lambda_2 y^2 + \lambda_1 y + \lambda_0)^2.$$

The bifurcation diagram of this critical point is the bifurcation diagram of the deformation

$$G_I^0(y, \lambda) = y^p + (y^2 + \lambda_2 y^2 + \lambda_1 y + \lambda_0)^2$$

of simple singularity A_3.

Triples H_3 and $I_2(p)$ are good, therefore λ_3 (resp. λ_2) is a parameter along the stratum $\mu = $ const for the deformation G_H (resp. G_I) of the unimodal boundary singularity F_4^4 (resp. B_{p-1}^3) as well. The bifurcation diagram of critical point H_3 (resp. $I_2(p)$) is the component of the bifurcation diagram of G_H (resp. G_I) which corresponds to the restriction to the boundary. Thus get the following result.

THEOREM 2.15 *The bifurcation diagram of the deformation G_I^0 (G_H^0 resp.) of a simple singularity A_3 (A_4 resp.) is diffeomorphic to the manifold of the non-regular orbits of the group $I_2(p)$ (H_3 resp.) multiplying by a complex line*

Acknowledgment: A part of this paper was written at the Isaac Newton Institute for Mathematical Sciences, Cambridge. Author thanks the Institute and the organizers of the Programme on Singularity Theory for their hospitality and support.

REFERENCES

[1] V.I. Arnol'd. Normal forms of functions near degenerate critical points, the Weyl groups A_k, D_k, E_k and Lagrangian singularities. Functional Anal. Appl. 6:254–272, 1972.

[2] V.I. Arnol'd. Wave front evolution and equivariant Morse lemma. Comm. Pure Appl. Math. 29:557–582, 1976.

[3] V.I. Arnol'd. Critical points of function on a manifold with a boundary, the simple Lie groups B_k, C_k, F_4, and singularities of evolutes. Russian Math. Surveys 33:91–105, 1978.

[4] V.I. Arnol'd, V.V. Goryunov, O.V. Lyashko, V.A. Vassiliev. Singularities II. Encycl. Math. Sci. 39, Springer Verlag, 1993.

[5] V.I. Arnol'd, S.M. Gusein-Zade, A.N. Varchenko. Singularities of Differentiable Maps, Vol. I. Birkhäuser, Basel, 1985.

[6] N. Bourbaki. Groupes et algèbres de Lie, Chapters IV, V and VI. Hermann, Paris, 1968.

[7] V. Goryunov. Unitary reflection groups associated with singularities of functions with cyclic symmetry. Russian Math. Surveys 54 5:873–893, 1999.

[8] O. Lyashko. Classification of critical points of functions on a manifold with singular boundary. Functional Anal. Appl. 17 3:187–193, 1983.

[9] O. Lyashko. Geometry of bifurcation diagrams. Journal of Soviet Math. 27:2536–2759, 1984.

[10] I. Scherbak. Boundary singularities with simple decomposition. Journal of Soviet Math. 60:1681–1693, 1992.

[11] I. Scherbak. Noncristallographic Coxeter Groups and Boundary Singularities. Functional Anal. Appl. 35:2, 2001.

[12] O. Scherbak. Wave fronts and reflection groups. Russian Math. Surveys 43 3:149–194, 1988.

[13] I. Scherbak, A. Szpirglas. Boundary singularities: topology and duality. Advances in Soviet Math. 21:213–223, 1994.

Transversal Whitney Topology and Singularities of Haefliger Foliations

SHYUICHI IZUMIYA   Department of Mathematics, Hokkaido University, Sapporo 060-0810, Japan   E-mail: izumiya@math.sci.hokudai.ac.jp

KUNIHIDE MARUYAMA*   Department of Mathematics, Hokkaido University, Sapporo 060-0810, Japan   E-mail: k-maruyama@mug.biglobe.ne.jp

Dedicated to the memory of Professor Luiz A. Favaro

Abstract

In order to study singularities of Haefliger foliation, we define the notion of transversally Whitney C^∞-topology modulo a regular foliation on the set of C^∞-mappings into a foliated manifold. We prove a kind of transversality theorem with respect to this new topology. All arguments we use here are analogous to those of the theory for the ordinary Whitney C^∞-topology. However, this is the first attempt for the study of generic properties of Haefliger foliations

1   INTRODUCTION

In his paper [2] Prof. Luiz A. Favaro studied *$\mathcal{F}$-stability* of C^∞-mappings into a foliated manifold. The definition of $\mathcal{F}$-stability is given as follows: Let M, N be C^∞-manifolds and N is regularly foliated by $\mathcal{F}$. We consider the space of C^∞-mappings $C^\infty(M, N)$ with the Whitney C^∞ topology. We say that $f \in C^\infty(M, N)$ is *$\mathcal{F}$-stable*

* Currrent address: NEC nogawaryou 211, miyamae-ku nogawa 3139, kawasaki 216-0001, Japan.

if there exists a neighbourhood V_f of f in $C^\infty(M,N)$ such that for any $g \in V_f$, there exist a diffeomorphism $h : M \longrightarrow M$ and a diffeomorphism $k : N \longrightarrow N$ taking leaves of $\mathcal{F}$ to leaves of $\mathcal{F}$ such that $g = k \circ f \circ h^{-1}$.

The infinitesimal version of the above stability is defined as follows: $f \in C^\infty(M,N)$ is *infinitesimally $\mathcal{F}$-stable* if for any $w \in \Gamma^\infty(f^*TN)$, there exist $u \in \Gamma^\infty(TM)$, $v \in \Gamma^\infty(TN)$ with $\pi(v)$ is locally constant along the leaves and $w = df \circ u + v \circ f$, where $\pi : TN \longrightarrow TN/T\mathcal{F}$ is the canonical projection.

One of the purpose in his paper is to consider that the above two definitions are equivalent or not. He has shown that it is true in the local sense. Of course his assertion is rather clear in the present time, because there are a lot of tools and results in singularity theory now (c.f., Damon [1]). Global $\mathcal{F}$-stability is, however, still an open question now. Unfortunately, we can not contribute to the global $\mathcal{F}$-stability in this paper. Let us consider the meaning of the above stability of mappings. We say that a C^∞-mapping $F : M \longrightarrow N$ is $\mathcal{F}^\perp$-nonsingular at $x \in M$ if f is transversal to $\mathcal{F}$ at x. Otherwise, we say that f is $\mathcal{F}^\perp$-singular at x. The above notion of stability corresponds to the stability of $\mathcal{F}^\perp$-singularities. One of the motivations to study $\mathcal{F}^\perp$-singularities is given in the study of singular Haefliger foliations. It is clear that if f is $\mathcal{F}^\perp$-nonsingular at any point, $f^{-1}\mathcal{F}$ is a regular foliation. If f has $\mathcal{F}^\perp$-singularities, then $f^{-1}\mathcal{F}$ is a singular Haefliger foliation. We now review the definition of Haefliger foliations. Suppose that there exist an open covering $M = \cup_{i \in I} V_i$ and a local C^∞-mapping $\phi_i : V_i \longrightarrow \mathbb{R}^q$ on each V_i with the following properties:

(1) If $V_i \cap V_j \neq \emptyset$, there exist open neighbourhoods U_i of $\phi_i(V_i \cap V_j)$, U_j of $\phi_j(V_i \cap V_j)$ and a diffeomorphism $\psi_{ij} : U_j \longrightarrow U_i$ such that $\psi_{ij} \circ \phi_i(x) = \phi_j(x)$ for $x \in V_i \cap V_j$.

(2) If $V_i \cap V_j \cap V_k \neq \emptyset$, $\psi_{ij}, \psi_{jk}, \psi_{ik}$ satisfy the cocycle condition: $\psi_{ik}(x) = \psi_{ij} \circ \psi_{jk}(x)$ for any $x \in V_i \cap V_j \cap V_k$.

In this case, $\mathcal{H} = \{(\phi_i, V_i) \mid i \in I\}$ is called *a Haefliger foliation* and q is called *the codimension of $\mathcal{H}$*. We can extend each level set $\phi_i^{-1}(c) \subset V_i$ to V_j if $\phi_i^{-1}(c) \cap V_j \neq \emptyset$. The maximal connected component consisting of such level sets is called *a leaf of $\mathcal{H}$*. We say that $x \in M$ is *a singular point* of $\mathcal{H}$ if x is a singular point of ϕ_i for $x \in V_i$. It has been known (and easy to prove) the following proposition.

PROPOSITION 1.1 *Let $\mathcal{H}$ be a Haefliger foliation on M. Then there exist a smooth manifold N, a regular foliation $\mathcal{F}$ and a C^∞-mapping $f : M \longrightarrow N$ such that $f^{-1}\mathcal{F} = \mathcal{H}$.*

This proposition gives a strong motivation for the study of $\mathcal{F}^\perp$-singularities of C^∞-mapping $f : M \longrightarrow N$. In the definition of $\mathcal{F}$-stability and infinitesimal $\mathcal{F}$-stability, the directions for perturbations of $f \in C^\infty(M,N)$ contains too many information for the study of $\mathcal{F}^\perp$-singularities. It can be perturbed into the tangent direction of leaves of $\mathcal{F}$. In this paper, we study the stability of singularities of the Haefliger foliation $f^{-1}\mathcal{F}$. For the purpose, we do not need the perturbation corresponding to the tangent direction of leaves. Our main purpose is summarised that we introduce a certain topology on $C^\infty(M,N)$ describing the perturbation of $f \in C^\infty(M,N)$ with respect to only the transversal direction of $\mathcal{F}$.

All arguments in this note are analogous to those for the jet-transversality theory on the ordinary Whitney C^∞-topology. Nevertheless, we write down the results because these contain some new concepts and these are the fundamental results for the study of singularities of Haefliger foliations.

2 TRANSVERSAL WHITNEY TOPOLOGY MODULO $\mathcal{F}$

In this section we introduce a new topology on $C^\infty(M, N)$ along the line of the description of Whitney C^∞-topology in [3]. First, we borrow the notion of jet spaces modulo $\mathcal{F}$ from Ikegami [4] for our purpose. Let $f, g : (M, x) \longrightarrow (N, y)$ be C^∞-map germs. Since $\mathcal{F}$ is a regular foliation on N, there exists a local chart $\eta_1 \times \eta_2 : U \longrightarrow D^q \times D^r \subset \mathbb{R}^q \times \mathbb{R}^r$ around $y \in N$ such that $U \cap \mathcal{F} = \{(\eta_1 \times \eta_2)^{-1}(\{c\} \times D^r) \mid c \in D^q\}$. For any $k \geq 1$, we say that f and g have k-th order contact at x modulo $\mathcal{F}$ if $\eta_1 \circ f$ and $\eta_1 \circ g$ have k-th order contact at x in the ordinary sense (cf., [3]). We can easily show that the above relation is an equivalence relation among the set of all C^∞-map germs $C^\infty(M, N)_{(x,y)}$. We denote the set of equivalence classes by $J^k(M, N; \mathcal{F})_{(x,y)}$ and we call it a k-jet space modulo $\mathcal{F}$. For $k = 0$, we consider that $J^0(M, N; \mathcal{F})_{(x,y)} = M \times N$. The element in $J^k(M, N; \mathcal{F})_{(x,y)}$ represented by a map-germ $f : (M, x) \longrightarrow (N, y)$ is denoted $j_{\mathcal{F}}^k f(x)$. We can define the k-jet bundle modulo $\mathcal{F}$ by

$$J^k(M, N; \mathcal{F}) = \cup_{(x,y) \in M \times N} J^k(M, N; \mathcal{F})_{(x,y)}.$$

The topology on $J^k(M, N; \mathcal{F})$ is defined like as the ordinary k-jet bundle. We also define a mapping $\alpha \times \beta : J^k(M, N; \mathcal{F}) \longrightarrow M \times N$ by $\alpha \times \beta(j_{\mathcal{F}}^k f(x)) = (x, f(x))$. We can summarise basic properties of $J^k(M, N; \mathcal{F})$. For the detailed descriptions, see Ikegami [4].

PROPOSITION 2.1 *Let M, N be smooth manifolds such that N is regularly foliated by $\mathcal{F}$ with* codim$\mathcal{F} = q$.
a) $J^k(M, N; \mathcal{F})$ *is a smooth manifold.*
b) *For any smooth mapping $f : M \longrightarrow N$, we can define a smooth mapping $j_{\mathcal{F}}^k f : M \longrightarrow J^k(M, N; \mathcal{F})$ by $(j_{\mathcal{F}}^k f)(x) = j_{\mathcal{F}}^k f(x)$. We call $j_{\mathcal{F}}^k f$ a k-jet extension modulo $\mathcal{F}$.*
c) $\alpha \times \beta : J^k(M, N; \mathcal{F}) \longrightarrow M \times N$ *is a smooth fibre bundle with the fibre*

$$B_{m,q}^k = \{P : (\mathbb{R}^m, 0) \longrightarrow (\mathbb{R}^q, 0) \mid P : \text{ polynomial map with degree } P \leq k\}.$$

We now define the notion of Whitney C^∞-topology modulo $\mathcal{F}$. The idea is very simple. We adopt $J^k(M, N; \mathcal{F})$ for the new topology instead of $J^k(M, N)$ for the ordinary Whitney C^∞-topology. For any non-negative integer k and a subset $U \subset J^k(M, N; \mathcal{F})$, we denote that

$$M_{\mathcal{F}}^k(U) = \{f \in C^\infty(M, N) \mid j_{\mathcal{F}}^k f(M) \subset U\}.$$

It is clear that $M_{\mathcal{F}}^k(U) \cap M_{\mathcal{F}}^k(V) = M_{\mathcal{F}}^k(U \cap V)$. It follows from these facts that $W_k^{\mathcal{F}} = \{M_{\mathcal{F}}^k(U) \mid U : \text{open subset}\}$ form a basis of a topology on $C^\infty(M, N)$. We can induce a topology on $C^\infty(M, N)$ such that the basis of the topology is given by $W_\infty^{\mathcal{F}} = \cup_{k=0}^\infty W_k^{\mathcal{F}}$. We denote $C^\infty(M, N; \mathcal{F})$ the space of smooth mappings with this new topology.

We call this topology *Whitney C^∞-topology modulo $\mathcal{F}$*. We can prove several basic facts on Whitney C^∞-topology modulo $\mathcal{F}$ like as the ordinary Whitney C^∞-topology.

PROPOSITION 2.2 *1) $C^\infty(M, N; \mathcal{F})$ is a Baire space.*
2) The mapping

$$j_{\mathcal{F}}^k : C^\infty(M, N; \mathcal{F}) \longrightarrow C^\infty(M, J^k(M, N; \mathcal{F}))$$

defined by $j_{\mathcal{F}}^k(f) = j_{\mathcal{F}}^k f$ *is continuous.*

Here we adopt the ordinary Whitney C^∞*-topology on* $C^\infty(M, J^k(M, N; \mathcal{F}))$.

Proof: Let d_k be a complete metric on $J^k(M, N; \mathcal{F})$. Let $U_1, U_2, \ldots$ be countably many open dense subsets of $C^\infty(M, N; \mathcal{F})$. For any open subset $V \subset C^\infty(M, N; \mathcal{F})$, we have to show that $V \cap \cap_{i=1}^\infty U_i \neq \emptyset$. Since $V \subset C^\infty(M, N; \mathcal{F})$ is an open subset with respect to Whitney C^∞ topology modulo $\mathcal{F}$, there exist a natural number k_0 and an open subset $W \subset J^{k_0}(M, N; \mathcal{F})$ such that $M_{\mathcal{F}}(\bar{W}) \subset V$ and $M_{\mathcal{F}}(W) \neq \emptyset$. It is enough to show that $M_{\mathcal{F}}(\bar{W}) \cap \cap_{i=1}^\infty U_i \neq \emptyset$.

For the purpose, we choose functions $f_1, f_2, \cdots \in C^\infty(M, N; \mathcal{F})$, non-negative integers $k_1, k_2, \cdots$ and open sets $W_i \subset J^{k_i}(M, N; \mathcal{F})$ with the following properties:

(A_i) $f_i \in M_{\mathcal{F}}(W) \cap \cap_{j=1}^{i-1} M_{\mathcal{F}}(W_j) \cap U_i$,

(B_i) $M_{\mathcal{F}}(\bar{W}) \subset U_i$ and $f_i \in M_{\mathcal{F}}(W_i)$,

(C_i) $d_s(j_{\mathcal{F}}^s f_i(x), j_{\mathcal{F}}^s f_{i-1}(x)) < \dfrac{1}{2^i}$ for $i > 1$ and $1 \le s \le i$.

(D_i) For any $x \in M$, there exists a local chart $\eta_1 \times \eta_2 : V \longrightarrow D^q \times D^r$ of N around $f_1(x)$ such that $V \cap \mathcal{F} = \{\eta^{-1}(c) \mid c \in D^q\}$, $f_i(x) \in V$ and $\eta_2 \circ f_i = \eta_2 \circ f_1$ on $(f_i)^{-1}(V) \cap (f_1)^{-1}(V)$ for any i.

The condition D_i is independent on the choice of local charts $(\eta_1 \times \eta_2, V)$ and it is the essential difference from the arguments on the ordinary Whitney C^∞-topology. We can choose the above subjects by the almost same arguments as those on the ordinary Whitney C^∞-topology. By the condition (C_i), $j_{\mathcal{F}}^s f_i(x)$ $(i = 1, 2, \ldots)$ is a Cauchy sequence in $J^s(M, N; \mathcal{F})$, so that we have the limit $g_s(x) = \lim_{i \to \infty} j_{\mathcal{F}}^s f_i(x)$. Since $J^0(M, N; \mathcal{F}) = M \times N$, we have $j_{\mathcal{F}}^0 f_i(x) = (x, f_i(x))$. Therefore, we define a mapping $g : M \longrightarrow N$ by $g_0(x) = (x, g(x))$. In order to prove that g is a smooth mapping, we need the condition (D_i). By exactly the same arguments as those on the ordinary Whitney C^∞-topology, we can show that $\eta_1 \circ g$ is smooth for a local chart $\eta_1 \times \eta_2 : V \longrightarrow D^q \times D^r$ of N around $g(x)$. By the condition (D_i), $\eta_2 \circ g = \eta_2 \circ f_1$. This means that g is smooth.

By the conditions (A_i), (B_i) and exactly the same arguments as those on the ordinary Whitney C^∞-topology, we can show that $g \in M_{\mathcal{F}}(\bar{W}) \cap \cap_{i=1}^\infty U_i$. This completes the proof of the assertion 1).

Let $U \subset J^\ell(M, J^k(M, N; \mathcal{F}))$ be an open subset. Since

$$M(U) = \{f \in C^\infty(M, J^k(M, N; \mathcal{F})) \mid j^\ell f(M) \subset U\}$$

is an element of the open basis for the ordinary Whitney C^∞-topology, it is enough to show that $(j_{\mathcal{F}}^k)^{-1}(M(U))$ is an open set in $C^\infty(M, N; \mathcal{F})$. We define a map

$$\alpha_{k,\ell} : J^{k+\ell}(M, N; \mathcal{F}) \longrightarrow J^\ell(M, J^k(M, J^k(M, N; \mathcal{F})))$$

by $\alpha_{k,\ell}(\sigma) = j^\ell(j_{\mathcal{F}}^k f)(x)$, where $\sigma = j_{\mathcal{F}}^{k+\ell} f(x)$. It is clear that $\alpha_{k,\ell}$ is a well-defined smooth map. Therefore, $\alpha_{k,\ell}^{-1}(U)$ is an open submanifold of $J^{k+\ell}(M, N; \mathcal{F})$. Since $\alpha_{k,\ell} \circ j_{\mathcal{F}}^{k+\ell} f = j^\ell \circ j_{\mathcal{F}}^k f$, we have $M_{\mathcal{F}}(\alpha_{k,\ell}^{-1}(U)) = (j_{\mathcal{F}}^k)^{-1}(M(U))$, so that it is open. This completes the proof.   □

3 A JET TRANSVERSALITY THEOREM MODULO $\mathcal{F}$

In this section we formulate the transversality theorem modulo $\mathcal{F}$ and give a proof. The proof of the theorem is a direct analogy of the proof of the original jet transversality theorem [3], so that we only give the sketch of the proof here.

THEOREM 3.1 *Let M, N be smooth manifolds and $\mathcal{F}$ be a regular foliation on N. For any submanifold $W \subset J^k(M, N; \mathcal{F})$, the set*

$$T_W^{\mathcal{F}} = \{f \in C^{\infty}(M, N; \mathcal{F}) \mid j_{\mathcal{F}}^k f \; : \; \text{transversal to } W \; \}$$

is a residual subset of $C^{\infty}(M, N; \mathcal{F})$. If W is a closed set, then $T_W^{\mathcal{F}}$ is an open set.

Proof: The proof is analogous to the proof of the ordinary jet transversality theorem, so that we only describe the different part from the ordinary one. We show that $T_W^{\mathcal{F}}$ is a intersection of countable open dense subsets of $C^{\infty}(M,N;\mathcal{F})$. Let $W_1, W_2, \cdots \subset W$ be an open covering of W with the following properties:

1) $\bar{W}_i \subset W$.

2) $\bar{W}_i$ is compact.

3) There exist coordinate neighbourhoods U_i of M and V_i of N such that $\alpha \times \beta(W_i) \subset U_i \times V_i$.

4) $\bar{U}_i$ is compact.

Since $\mathcal{F}$ is a regular foliation on N, we can choose the coordinate system $\eta_1 \times \eta_2 : V_i \longrightarrow D^q \times D^r \subset \mathbb{R}^q \times \mathbb{R}^r$ with $V_i \cap \mathcal{F} = \{(\eta_1 \times \eta_2)^{-1}(\{c\} \times D^r) \mid c \in D^q \}$. We define a set

$$T_{W_i}^{\mathcal{F}} = \{f \in C^{\infty}(M, N; \mathcal{F}) \mid j_{\mathcal{F}}^k f \; : \; \text{transversal to } W \text{ on } \bar{W}_i \; \}.$$

Since $T_W^{\mathcal{F}} = \cap_{i=1}^{\infty} T_{W_i}^{\mathcal{F}}$, it is enough to show that $T_{W_i}^{\mathcal{F}}$ is an open dense subset.
We denote by

$$T_i = \{g \in C^{\infty}(M, J^k(M, N; \mathcal{F})) \mid g \; : \; \text{transversal to } W \text{ on } \bar{W}_i \; \}.$$

Since we adopt the ordinary Whitney C^{∞}-topology on $C^{\infty}(M, J^k(M, N; \mathcal{F}))$, it is known that T_i is an open set ([3], page 57). It follows from the fact that $j_{\mathcal{F}}^k$ is continuous, we have an open subset $T_{W_i} = (j_{\mathcal{F}}^k)^{-1}(T_i)$. In order to prove the density, let $\psi : U_i \longrightarrow \mathbb{R}^n$ and $\eta_1 \times \eta_2 : V_i \longrightarrow D^q \times D^r \subset \mathbb{R}^q \times \mathbb{R}^r$ be coordinate charts. Let

$$\rho : \mathbb{R}^n \longrightarrow [0, 1], \quad \rho' : \mathbb{R}^r \longrightarrow [0, 1]$$

be continuous functions respectively defined by

$$\rho = \begin{cases} 1 & \text{on a neighbourhood of } \psi \circ \alpha(\bar{W}_i) \\ 0 & \text{off } \psi(U_i) \end{cases}$$

$$\rho' = \begin{cases} 1 & \text{on a neighbourhood of } \eta_1 \circ \alpha(\bar{W}_i) \\ 0 & \text{off } \eta_1(V_i). \end{cases}$$

Let $B(m, r : k)$ be a set of polynomial maps $p : \mathbb{R}^m \longrightarrow \mathbb{R}^r$ with degree $p \le k$ and $p(0) = 0$. Let $f : M \longrightarrow N$ be a smooth mapping. For any $b \in B(m, r : k)$, we define a smooth mapping $g_b : M \longrightarrow N$ by

$$g_b(x) = \begin{cases} f(x) \text{ if } x \notin U_i \text{ or } f(x) \notin V_i \\ (\eta_1 \times \eta_2)^{-1}(\rho(\psi(x))\rho'(\eta_1(f(x)))b(\psi(x)) + \eta_1(f(x)), \eta_2(f(x))) \text{ otherwise.} \end{cases}$$

We now define a smooth mapping $G : M \times B(m, r : k) \longrightarrow N$ by $G(x, b) = g_b(x)$. We also define a mapping $\Phi : M \times B(m, r : k) \longrightarrow J^k(M, N; \mathcal{F})$ by $\Phi(x, b) = j_{\mathcal{F}}^k g_b(x)$. By the similar arguments as those of in the proof of ordinary jet transversality theorem, we can easily show that there exists an open neighbourhood B of $0 \in B(m, r : k)$ such that $\Phi | M \times B$ is a local diffeomorphism at the point $(x, b) \in M \times B$ with $\Phi(x, b) \in \bar{W}_i$, so that $\Phi | M \times B$ is transversal to W on $\bar{W}_i$. By the fundamental transversality theorem of Thom (cf., [3]), we have a sequence $\{b_j\}_{j=1}^{\infty} \subset B$ which is uniformly convergent to $0 \in B$ such that $j_{\mathcal{F}}^k g_{b_j}$ is transversal to W on $\bar{W}_i$. Since $g_0 = f$ and $g_b = f$ outside of U_i, we have $\lim_{j \longrightarrow \infty} g_{b_j} = f$ in $C^\infty(M, N; \mathcal{F})$.

On the other hand, suppose that W is a closed set. Since we adopt the ordinary Whitney C^∞-topology on $C^\infty(M, J^k(M, N; \mathcal{F}))$, it has been known that the set

$$T_W^k = \{F \in C^\infty(M, J^k(M, N; \mathcal{F})) \mid F : \text{ transversal to } W \}$$

is an open set (cf., [3]). By the assertion 2) of Proposition 2.2,

$$j_{\mathcal{F}}^k : C^\infty(M, N; \mathcal{F}) \longrightarrow C^\infty(M, J^k(M, N; \mathcal{F}))$$

is continuous.

Hence, $T_W^{\mathcal{F}} = (j_{\mathcal{F}}^k)^{-1}(T_W^k)$ is an open subset of $C^\infty(M, N; \mathcal{F})$. $\square$

We remark that the multi-jet version of the above theorem also holds. Since we do not consider global situation here, we omit to describe the assertion and the proof.

4   APPLICATIONS

In this section we give some applications of the jet transversality theorem modulo $\mathcal{F}$.

1) Thom-Boardman Haefliger foliations Let M, N be smooth manifolds and $\mathcal{F}$ be a regular foliation on N with codim$\mathcal{F} = q$. We now consider *the (ordinary) Thom-Boardman singular set* $\Sigma^I(m, q) \subset J^k(m, q)$. Since $J^k(m, q)$ is considered to be the fibre of $J^k(M, N; \mathcal{F})$, we can define *a Thom-Boardman subbundle modulo $\mathcal{F}$* $\Sigma^I(M, N; \mathcal{F}) \subset J^k(M, N; \mathcal{F})$ as usual. By Theorem 3.1, the set

$$T_{\Sigma^I(M, N; \mathcal{F})} = \{f \in C^\infty(M, N; \mathcal{F}) \mid j_{\mathcal{F}}^k f : \text{ transversal to } \Sigma^I(M, N; \mathcal{F}) \}$$

is a residual subset of $C^\infty(M, N; \mathcal{F})$. For any $f \in T_{\Sigma^I(M, N; \mathcal{F})}$, we call $\mathcal{H} = f^{-1}\mathcal{F}$ a *Thom-Boardman Haefliger foliation* with I.

Suppose that the Boardman symbol is given by $I = (i_1, i_2, \ldots, i_k)$. For any $f \in T_{\Sigma^I(M, N; \mathcal{F})}$, we can choose local covering $\{U_j\}_{j \in J}$ of M and local C^∞-mappings $\phi_j : U_j \longrightarrow D^q \subset \mathbb{R}^q$ which define the Haefliger foliation $\mathcal{H} = f^{-1}\mathcal{F}$. By the construction of the Thom-Boardman singular set, the set

$$\Sigma^{i_1}(\mathcal{H}) = \{x \in M \mid \text{rank } d(\phi_j)_x = q - i_1 \text{ for } x \in U_j\}$$

is a smooth submanifold of M. Moreover,

$$\Sigma^{i_1,i_2}(\mathcal{H}) = \{x \in \Sigma^{i_1}(\mathcal{H}) \mid \operatorname{rank} d(\phi_j|\Sigma^{i_1}(\mathcal{H}))_x = \min(\dim \Sigma^{i_1}(\mathcal{H}), q-i_2) \text{ for } x \in U_j\}$$

is also a smooth submanifold of M. This procedure can be continued to k times. As a special case, we assume that $q = 1$, $k = 1$ and $I = (1)$. For any $f \in T_{\Sigma^1(M,N;\mathcal{F})}$, $\mathcal{H} = f^{-1}\mathcal{F}$ is called a *Morse type Haefliger foliation*. We have the following problem associated to Morse type Haefliger foliations:

PROBLEM: Suppose that $\mathcal{F}$ is a regular foliation on N with $\operatorname{codim}\mathcal{F} = 1$ and $\mathcal{H} = f^{-1}\mathcal{F}$ is a Morse type Haefliger foliation. Can we have a generalised Morse inequality? It might be related to the topology of the leaf space $N/\mathcal{F}$.

2) Stability of mappings modulo $\mathcal{F}$ For any $f, g \in C^\infty(M, N; \mathcal{F})$, we say that f, g are $\mathcal{A}_\mathcal{F}$-*equivalent* if there exist a diffeomorphism $h : M \longrightarrow M$ and a smooth mapping $k : N \longrightarrow N$ taking leaves of $\mathcal{F}$ to leaves such that the induced mapping $\tilde{k} : N/\mathcal{F} \longrightarrow N/\mathcal{F}$ is a homeomorphism and $\tilde{g} = \tilde{k} \circ \tilde{f} \circ h^{-1}$, where $\tilde{f} = \pi \circ f$ and $\pi : N \longrightarrow N/\mathcal{F}$ is the canonical projection. The local version of the above definition is as follows: For smooth map-germs $f : (M, x) \longrightarrow (N, y)$ and $g : (M, x') \longrightarrow (N, y')$, f, g are $\mathcal{A}_\mathcal{F}$-*equivalent* if there exist diffeomorphism-germs $h : (M, x) \longrightarrow (M, x')$ and $k_1 : (D^q, 0) \longrightarrow (D^q, 0)$ such that $k_1 \circ \pi_1 \circ (\eta_1 \times \eta_2) \circ f = \pi_1 \circ (\eta_1' \times \eta_2') \circ g \circ h$, where $\eta_1 \times \eta_2 : (N, y) \longrightarrow (D^q \times D^r, 0), \eta_1' \times \eta_2' : (N, y') \longrightarrow (D^q \times D^r, 0)$ are local charts and $\pi_1 : (D^q \times D^r, 0) \longrightarrow (D^q, 0)$ is the canonical projection. We remark that the above definition does not depend on the choice of local charts $\eta_1 \times \eta_2$, $\eta_1' \times \eta_2'$.

We now define the notion of stability corresponding to each definition. We say that $f \in C^\infty(M, N; \mathcal{F})$ is $\mathcal{A}_\mathcal{F}$-*stable modulo* $\mathcal{F}$ if there exists a neighbourhood V_f of f in $C^\infty(M, N; \mathcal{F})$ such that f and g are $\mathcal{A}_\mathcal{F}$-equivalent for any $g \in V_f$. We also say that a smooth map-germ $f : (M, x) \longrightarrow (N, y)$ is $\mathcal{A}_\mathcal{F}$-*stable modulo* $\mathcal{F}$ if for any representative $\tilde{f} : U \longrightarrow V$ of f with a local chart $\eta_1 \times \eta_2 : V \longrightarrow D^q \times D^r$, there exists a neighbourhood $V_{\tilde{f}}$ of $\tilde{f}$ in $C^\infty(U, V; \mathcal{F}|V)$ such that for any $g \in V_{\tilde{f}}$ there exists a point $x' \in U$ with $\tilde{f}$ and the germ $g : (M, x') \longrightarrow (N, g(x'))$ are $\mathcal{A}_\mathcal{F}$-equivalent. The notion of the $\mathcal{A}_\mathcal{F}$-stability modulo $\mathcal{F}$ for smooth map-germs is rather trivial because it is almost the same as the ordinary definition of the stability of smooth map-germs. However, the local stability of global mappings is important for the study of Haefliger foliations. We say that $f \in C^\infty(M, N; \mathcal{F})$ is *locally* $\mathcal{A}_\mathcal{F}$-*stable modulo* $\mathcal{F}$ if the map-germ $f : (M, x) \longrightarrow (N, f(x))$ at any point $x \in M$ represented by f is $\mathcal{A}_\mathcal{F}$-stable modulo $\mathcal{F}$ as a smooth map-germ.

We now define the infinitesimal version of the above notions of stability. We have the normal bundle $T\mathcal{F}^\perp = TN/T\mathcal{F}$ of the foliation $\mathcal{F}$. We say that $f \in C^\infty(M, N; \mathcal{F})$ is *infinitesimally* $\mathcal{A}_\mathcal{F}$-*stable modulo* $\mathcal{F}$ if for any $w \in \Gamma^\infty(f^*\mathcal{F}^\perp)$, there exist $u \in \Gamma^\infty(TM)$, $v \in \Gamma^\infty(T\mathcal{F}^\perp)$ such that $w = \pi_\mathcal{F} \circ df \circ u + v \circ f$, where $\pi_\mathcal{F} : TN \longrightarrow T\mathcal{F}^\perp$ is the canonical projection. We can also define the germ version of the infinitesimally $\mathcal{A}_\mathcal{F}$-stability modulo $\mathcal{F}$ as usual. A smooth map-germ $f : (M, x) \longrightarrow (N, y)$ is *infinitesimally* $\mathcal{A}_\mathcal{F}$-*stable modulo* $\mathcal{F}$ *as a germ* if for any $w \in \Gamma^\infty(f^*\mathcal{F}^\perp)_x$, there exist $u \in \Gamma^\infty(TM)_x$, $v \in \Gamma^\infty(T\mathcal{F}^\perp)_{f(x)}$ such that $w = \pi_\mathcal{F} \circ df \circ u + v \circ f$ for any $x \in M$, where we denote that $\Gamma^\infty(f^*\mathcal{F}^\perp)_x, u \in \Gamma^\infty(TM)_x, \Gamma^\infty(T\mathcal{F}^\perp)_{f(x)}$ are sets of germs of smooth sections of each vector bundle. Since the local $\mathcal{A}_\mathcal{F}$-group is the geometric subgroup of $\mathcal{A}$ and $\mathcal{K}$ in the sense of Damon [1], the infinitesimally $\mathcal{A}_\mathcal{F}$-stability modulo $\mathcal{F}$ implies the local $\mathcal{A}_\mathcal{F}$-stability modulo $\mathcal{F}$. The converse is also true by the jet transversality theorem modulo $\mathcal{F}$.

PROPOSITION 4.1 *For* $f \in C^\infty(M, N; \mathcal{F})$, *f is locally* $\mathcal{A}_\mathcal{F}$*-stable modulo* $\mathcal{F}$ *if and only if f is infinitesimally* $\mathcal{A}_\mathcal{F}$*-stable modulo* $\mathcal{F}$ *in the local sense.*

Moreover, we have the following characterisation of local stability: Since the fibre of the jet bundle modulo $\mathcal{F}$ $J^k(M, N; \mathcal{F})$ can be identified with the ordinary jet space $J^k(m, q)$, we can consider the $\mathcal{K}$-orbit in $J^k(m, q)$. Here $\mathcal{K}$ is the group corresponding to the contact equivalence in the sense of Mather (c.f., [3, 5]). For any $z = j_\mathcal{F}^k f(x) \in J^k(M, N; \mathcal{F})$, we denote the $\mathcal{K}$-orbit through z by $\mathcal{K}^k(z)$. By the characterisation theorem of Mather [5], we have the following proposition:

PROPOSITION 4.2 *Let* $f : (M, x) \longrightarrow (N, y)$ *be a smooth map-germ. Then f is* $\mathcal{A}_\mathcal{F}$*-stable modulo* $\mathcal{F}$ *if and only if* $j_\mathcal{F}^k f$ *is transversal to* $\mathcal{K}^k(z)$ *for* $k \geq q + 1$.

It follows from the transversality theorem modulo $\mathcal{F}$ and the celebrated results of Mather [5] that we have the following theorem.

THEOREM 4.3 *If* $(\dim M, \mathrm{codim}\, \mathcal{F})$ *is in the nice dimensions in the sense of* Mather [5], *the set of locally* $\mathcal{A}_\mathcal{F}$*-stable mappings modulo* $\mathcal{F}$ *is open and dense in* $C^\infty(M, N; \mathcal{F})$.

FINAL REMARK: We remark that Prof. J. P. Dufour pointed out that if we consider the foliation $\mathcal{F}$ given by the irrational flow on the torus $N = T^2$, there might be no global $\mathcal{A}_\mathcal{F}$-stable mappings modulo $\mathcal{F}$ from any compact manifold into N. This means that we need strong assumption on the foliation for the existence of global $\mathcal{A}_\mathcal{F}$-stable mappings modulo $\mathcal{F}$.

Apologies and Acknowledgments It was 1994 August, the first author proposed collaboration to Prof. L. A. Favaro on the study of smooth mappings along the above direction. We promised to proceed this project in that time. However, we have not been able to realise it because of his sorrowful death. The first author really would like to apologise to him that we did not start the project earlier.

On the other hand, the authors would like to thank the organisers of "The 6th Workshop On Real and Complex Singularities" for giving us an opportunity to talk about this incomplete project. The first author is also grateful to people at USP and UFSCAR for their kind and warm-hearted hospitality during his stay in São Carlos.

REFERENCES

[1] J. Damon. The unfolding and determinacy theorems for subgroups of $\mathcal{A}$ and $\mathcal{K}$. Memoirs Amer. Math. Soc. 50 306, 1984.

[2] L.A. Favaro. Differentiable mappings between foliated manifolds. Bol. Soc. Brasil Mat. 8:39–46, 1977.

[3] M Golubitsky, V. Guillemin. Stable Mappings and Their Singularities. Graduate Texts in Math. 14 Springer-Verlag, 1973.

[4] G. Ikegami. Vector fields tangent to foliations. Japan. J. Math. 112:95–120, 1986.

[5] J.N. Mather. I:Stability of C^∞-mappings I:The division theorem. Ann. of Math.. 87:89–104, 1968.

——————— II: Infinitesimally stability implies stability. Ann. of Math. 89:254–291, 1969.

——————— III: Finitely determined map-germs. Publi. Math. I.H.E.S. 35:127–156, 1968.

——————— IV: Classification of stable germs by $\mathbb{R}$ algebras. Pub. Math. I.H.E.S. 37:223–248, 1970.

——————— V:Transversality. Adv. Math., 4:301–336, 1970.

——————— VI:The nice dimensions. Lecture Notes in Math. 192:207–253, 1972.

On a Conjecture of Chisini for Coverings of the Plane with A-D-E-Singularities

VALENTINE S. KULIKOV Moscow State University of Printing, Department of mathematics 2A, Prjanishnikova str. Moscow 127550,
E-mail:valentin@masha.ips.ras.ru

1 INTRODUCTION

This is a report about our joint paper with my brother, to be published in Izvestiya RAN: Ser.Mat. 64:6 (2000) (see also math.AG/00021-74). We are interested in generic coverings $f : X \to \mathbb{P}^2$ of projective plane by surfaces X with A-D-E-singularities (in other terminology, rational double points, Du Val singularities and etc.). A classical result is that a generic projection of a non-singular surface $X \subset \mathbb{P}^r$ to $\mathbb{P}^2$ is a finite covering with at most folds and pleats as singularities, and the discriminant (= branch) curve $B \subset \mathbb{P}^2$ is cuspidal. A fold (resp., a pleat) is a singularity of a map locally equivalent to a projection of a surface, defined by equation $x = z^2$ (resp. $y = z^3 + xz$), to the x, y-plane. A cuspidal curve is a curve, which has at most ordinary nodes and cusps (locally defined by equation $xy = 0$ and $y^2 = x^3$). Over a node $b \in B$ there are two folds, and over a cusp — one pleat. A *generic* (or *simple*) *covering* is a covering, which possesses the same properties as a generic projection. First of all, I explain why it is important to consider coverings by surfaces with A-D-E-singularities.

A presentation of an algebraic variety as a finite covering of the projective space is one of the effective ways of studying projective varieties as well as their moduli. To compare we recall what such an approach gives in the case of curves. For a curve C of genus g a generic covering $f : C \to \mathbb{P}^1$ is a covering such that in every fibre there is at most one ramification point which is a double point (or a singular point of f of

This work was partially supported by grants RFBR 99-01-01133, NWO-RFBR 047-008-005 and INTAS-00-0259.

type A_1). Let $B \subset \mathbb{P}^1$ be the set of branch points, and $d = deg\ B$, i.e. $d = \sharp(B)$. Then according to the Hurwitz formula $d = 2N + 2g - 2$, where $N = deg\ f$. If $N \geq g + 1$, then any curve of genus g can be presented as a simple covering of $\mathbb{P}^1$ of degree N. The set of all simple coverings (up to equivalence) $f : C \to \mathbb{P}^1$ of degree N with d branch points is parametrized by the Hurwitz variety $H = H^{N,d}$. Let $\mathbb{P}^d \setminus \Delta$ (Δ is the discriminant) be the projective space parametrizing the sets of d different points of $\mathbb{P}^1$, and let M_g be the moduli space of curves of genus g. There are two maps: a map $h : H \to \mathbb{P}^d \setminus \Delta$ sending f to the set of branch points $B \subset \mathbb{P}^1$, and a map $\mu : H \to M_g$, sending f to the class of curves isomorphic to C. Hurwitz introduced and investigated the variety H in 1891. He proved that the variety H is connected, and h is a finite unramified covering. In modern functorial language H was studied also by W.Fulton (1969). The map μ is surjective and has fibres of dimension $N + (N - g + 1)$ ($f : C \to \mathbb{P}^1$ is a rational function on C; the first summand N is the number of points on C, chosen to be $f^{-1}(\infty)$ or the divisor of poles of function f; the second summand is the number of f with fixed poles defined by Riemann-Roch theorem). We obtain the number of moduli $2N + 2g - 2 - (2N - g + 1) = 3g - 3$ and the irreducibility of the moduli space M_g.

In the case of surfaces we also can consider an analog of Hurwitz variety H of all generic coverings (up to equivalence) $f : S \to \mathbb{P}^2$ of degree N and with discriminant curves B of degree d with given number n of nodes and given number c of cusps. Let $\mathbb{P}^\nu$, $\nu = \frac{d(d+3)}{2}$, be a projective space parametrizing curves of degree d, and $h : H \to \mathbb{P}^\nu$ be a mapping sending a covering f to its discriminant curve B. A so called Chisini conjecture concerns the fibres of h. It claims that if B is the discriminant curve of a generic covering f of degree $N \geq 5$, then f is defined by the curve B uniquely, up to equivalence. In other words, it means that the mapping h is injective (and, besides, $N = deg\ f$ is determined by B). In [6] it is proved that the Chisini conjecture is true for almost all generic coverings. In particular, it is true for generic coverings defined by a multiple canonical class. A construction of the moduli space of surfaces of general type uses pluricanonical maps. As it is known [3] , if S is a minimal surface of general type, then for $m \geq 5$ the linear system $|mK_S|$ blows down only (-2)-curves and gives a birational map of S to a surface $X \subset \mathbb{P}^r$ (the canonical model) with at most A-D-E-singularities. This requires a generalization of the notion of generic covering to the case of surfaces with A-D-E-singularities.

In our paper we:

1) generalize a classical result on singularities of generic projections of non-singular surfaces to the case of surfaces with A-D-E-singularities; this gives a notion of a generic covering;

2) compute numerical invariants of a generic covering in terms of intersection numbers on a minimal resolution of the surface;

3) obtain an inequality (the main inequality), which is the main tool in proving the Chisini conjecture; this inequality bounds the degree of a generic covering in the case of existence of another nonequivalent generic covering; our proof of the main inequality requires some investigations in singularity theory;

4) study local structure of fibre products of generic coverings with A-D-E-singularities and

5) investigate the canonical cycle of an A-D-E-singularity, with the help of which we compute numerical invariants of generic coverings; finally,

6) we apply the main inequality to prove the Chisini conjecture for generic m-canonical

coverings of surfaces of general type.

2   GENERIC PROJECTIONS OF SURFACES WITH A-D-E-SINGULARITIES

We consider a slightly more general situation, when a surface X may possess isolated singularities of multiplicity two, locally defined by equations $z^2 = h(x,y)$, which are called *double planes*. All A-D-E-singularities are double planes.

THEOREM 2.1 *Let $X \subset \mathbb{P}^r$ be a surface with at most isolated singularities of the form $z^2 = h(x,y)$ (= 'double planes'), $X \to \mathbb{P}^2$ be the restriction to X of a generic projection $\mathbb{P}^r \backslash L \to \mathbb{P}^2$, where L is a generic linear subspace of $\mathbb{P}^2$ of dimension $r - 3$. Then*

(i) f is a finite covering;

(ii) at non-singular points of X the covering f has as singularities at most either double points (folds), or singular points of cuspidal type (pleats); in a neighbourhood of these points f is equivalent to the projection of a surface $x = z^2$, respectively $y = z^3 + xz$, to the x, y-plane;

(iii) in a neighbourhood of a point $s \in Sing\ X$ the covering f is analytically equivalent to a projection of a surface $z^2 = h(x,y)$ to the x, y-plane; from (ii) and (iii) it follows that the ramification divisor is reduced, i.e. $f^*(B) = 2R + C$, where $B = f(R)$, and R and C are reduced curves;

(iv) except singular points $f(Sing\ X)$ the discriminant curve B is cuspidal;

(v) the restriction of f to R is of degree 1.

We prove a weakened version of Theorem 2.1, in which the projection goes after a Veronese embedding of the initial embedding. This is quite enough for the purposes described above. Actually, the main difficulty in the proof of this theorem lies in the classical case, when the surface X is non-singular. Unfortunately the authors do not know a complete (and mordern) proof of this theorem, and it seems that such a proof does not exist. Thus, its proof, even in the case of a non-singular surface, take interest.

We begin with a projection π_L of X onto a surface Y in $\mathbb{P}^3$ with ordinary singularities (double and triple points and pinches ('Whitney umbrellas')). Then we prove that for a generic point $\xi \in \mathbb{P}^3$ all lines $l \ni \xi$ are at most simple (i.e. $(Y \cdot l)_y = 2$) bitangents and simple stationary tangents (i.e. $(Y \cdot l)_y = 3$) with respect to Y, and l is transversal to tangent cones at the singular points. The projection $f = \pi_\xi \circ \pi_{L|x} : X \to \mathbb{P}^2$ at singular points is a covering of degree 2 or 3. To obtain normal forms of the projection at the singular points, we use a simple lemma [1]

LEMMA 2.2 *Let $(X, 0) \subset (\mathbb{C}^3, 0)$ be a non-singular surface, and $(\mathbb{C}^3, 0) \to (\mathbb{C}^2, 0)$ be a smooth morphism, the restriction of which $f : X \to \mathbb{C}^2$ is a finite covering of degree μ. Then one can choose local coordinates x, y in $\mathbb{C}^2$ and x, y, z in $\mathbb{C}^3$ such that the restriction to X is defined by the equation*

$$y = z^\mu + \lambda_1(x)z^{\mu-2} + \cdots + \lambda_{\mu-2}(x)z,$$

and f is the projection along z axis.

In the proof, the covering f is considered as a 2-parameter family of 0-dimensional hypersurface singularities of multiplicity μ, and, consequently, f is induced by a miniversal deformation of the singularity of type $A_{\mu-1}$. From this lemma it follows that for a non-singular X a covering $(X,0) \to (\mathbb{C}^2,0)$ of degree 2 is a fold, and a covering of degree 3 locally is a projection to the x,y-plane of one of the surfaces $y = z^3 + x^k z$, $k = 1,2,\ldots$, or $y = z^3$ ($k = \infty$). For a generic point $\xi \in \mathbb{P}^3$ the exponent $k = 1$, i.e. we have a pleat.

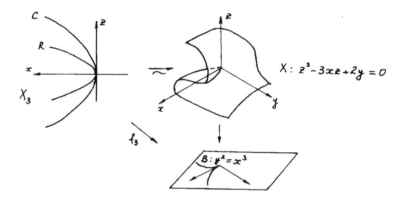

Figure 1

3 GENERIC COVERINGS

Let X be a surface with A-D-E-singularities, a_n singularities of type A_n, $n \geq 1$, d_n singularities of type D_n, $n \geq 4$, and e_n singularities of type E_n, $n = 6,7,8$. A finite morphism $f : X \to \mathbf{P}^2$ is called *generic* (or *simple*) *covering*, if it possesses the properties of a generic projection. The ramification divisor R of f is reduced, $f^*(B) = 2R + C$. The discriminant curve B has, firstly, 'the same' singularities as the surface X (and as the curve R), which are locally defined by the equation $h(x,y) = 0$, where $z^2 = h(x,y)$ is an equation of the corresponding singular point. These singularities on B we call *s-singularities*, in particular, *s-nodes* and *s-cusps*. Besides, there are nodes and cusps on B originated from singularities of the map f, which we call *p-nodes* and *p-cusps*.

Set $N = \deg f$. Let $\deg B = d$ and let B has n nodes and c cusps, $n_s = a_1$ and $c_s = a_2$ of which originates from $Sing\,X$, and n_p and c_p are p-nodes and p-cusps. The covering f is unbranched in $\mathbb{P}^2 \setminus B$. Over a point $b \in B \setminus Sing\,B$ the covering f has one fold $x = z^2$ and $N - 2$ 'étale sheets'. Over an s-point $b \in Sing\,B$ f has one singular point of X and $N - 2$ étale sheets; over a p-cusp f has one pleat and $N - 3$ étale sheets; over p-node f has two folds and $N - 4$ étale sheets.

Let $\pi : S \to X$ be a minimal resolution of X, and $\bar{f} = f \circ \pi : S \to \mathbb{P}^2$. Let $L \subset X$ be the preimage of a generic line $l \subset \mathbb{P}^2$. Denote by $\bar{R}$ and $\bar{L}$ the proper transforms of R and L on S. Then $\bar{R}$ is a normalization of the curve R, and $\bar{L} \simeq L$. The singularities of X being Gorenstein, the divisor R is locally principal, and

$$\pi^*(R) = \bar{R} + Z \,,$$

where $Z = \sum_{x \in Sing X} Z_x$ is the canonical cycle of S, Z_x are the canonical cycles of singularities $x \in Sing\, X$. We'll discuss the cycles Z_x later and compute the numbers $\delta_x = -\frac{1}{2}(Z_x^2)$, which we call *defects* of A-D-E-singularities. Set $\delta_X = \sum_{x \in Sing X} \delta_x$.

We compute intersection numbers

$$(\bar{L}^2) = N \,,\; \bar{L} \cdot \bar{R} = d \,,\; \bar{L} \cdot Z = 0 \,,\; \bar{R} \cdot Z = 2\delta_X \,,\; (Z^2) = -2\delta_X \,,$$

$$(\bar{R}^2) = 3\bar{d} + g - 1 - \delta_X \,,\; (\bar{R} + Z)^2 = 3\bar{d} + g - 1 + \delta_X = 3\bar{d} + p_a(R) - 1 \,,$$

where $\bar{d} = \frac{1}{2}d$, g is the geometric genus of B (or R), and $p_a(R) = g + \delta_X = \frac{(d-1)(d-2)}{2} - n_p - c_p$ is the arithmetical genus of R.

Besides, we compute the self-intersection number of the canonical class

$$(K_S^2) = 9N - 9\bar{d} + p_a(R) - 1 = 9N + \frac{1}{2}d(d-12) - n_p - c_p.$$

and the topological Euler characteristic

$$e(S) = 3N + 2g - 2 + 2\delta_X - c_p.$$

From Noether's formula $(K_S^2) + e(S) = 12p_a$, we have

$$p_a = 1 - q + p_g = N + \frac{\bar{d}(\bar{d}-3)}{2} - \frac{n_p}{4} - \frac{c_p}{3}.$$

and as in the case of a non-singular surface X, we obtain that $n_p \equiv 0 \;(mod\, 4)$, $c_p \equiv 0 \;(mod\, 3)$.

4   MAIN INEQUALITY

Let $f_1 : X_1 \to \mathbb{P}^2$ and $f_2 : X_2 \to \mathbb{P}^2$ be two generic coverings of surfaces with A-D-E-singularities and with the same discriminant curve B. Let $f_i^*(B) = 2R_i + C_i$, $i = 1, 2$. With respect to a pair of coverings f_1 and f_2 nodes and cusps of B are subdivided into four types: ss-, sp-, ps- and pp-nodes and cusps. For example, a sp-node $b \in B$ is a node, which is a s-node for f_1 and a p-node for f_2. The number of sp-nodes is denoted by n_{sp}. Then

$$n = n_{ss} + n_{sp} + n_{ps} + n_{pp} \,,\quad c = c_{ss} + c_{sp} + c_{ps} + c_{pp},$$

where $n_{\flat\sharp}$ and $c_{\flat\sharp}$ are numbers of $\flat\sharp$-nodes and $\flat\sharp$-cusps of . In particular, $n_{ss} + n_{sp} = a_1$ is the number of singularities of type A_1, and $_{ss} + _{sp} = a_2$ is the number of singularities of type A_2 on the surface X_1.

THEOREM 4.1 *If f_1 and f_2 are nonequivalent generic coverings of degrees N_1 and N_2, then*

$$N_2 \leq \frac{4(3\bar{d} + g_1 - 1)}{2(3\bar{d} + g_1 - 1) - \iota_1} \,,$$

where $g_1 = p_a(R_1)$ is the arithmetic genus of the curve R_1, and $\iota_1 = 2n_{sp} + 2c_{sp} + c_{pp}$.

We briefly sketch the proof of this theorem. To compare coverings f_1 and f_2, we consider a normalization X of the fiber product $X^\times = X_1 \times_{\mathbf{P}^2} X_2$ and the corresponding commutative diagram

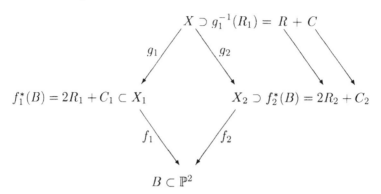

The surface X is a $N_1 N_2$-sheeted covering of $\mathbb{P}^2$ and it has at most A-D-E-singularities, which lie over $Sing\ B$.

LEMMA 4.2 *If coverings f_1 and f_2 are non equivalent, then the surface X is irreducible.*

For the proof it is shown that the group $\pi_1 = \pi_1(\mathbb{P}^2 \setminus B)$ acts transitively on a fibre of $f : X \to \mathbb{P}^2$. It is used that the monodromy homomorphisms $\varphi_i : \pi_1 \to S_{N_i}$, $i = 1, 2$, are epimorphisms.

The main idea in the proof of Theorem 4.1 is the same as in [6] in the case of non-singular surfaces: we partition the preimage $g_1^{-1}(R_1)$ of the curve R_1 into two parts

$$g_1^{-1}(R_1) = R + C,$$

where R is the curve mapped by g_2 onto R_2, and C is the curve mapped by g_2 onto C_2, and apply the Hodge index theorem to the pair of divisors R and C on X. More precisely, the surface X has A-D-E-singularitis, and so to use intersection theory we have to resolve the singularities. Let $\tilde{\pi} : S \to X$ be a minimal resolution of singularities of X, and denote by $\tilde{R}$ and $\tilde{\ }$ the proper transforms of R and C on S. Let $\tilde{\pi}^*(g_1^{-1}(R_1)) = \hat{R}_1$ be the total transform of R_1 on S. Then

$$\hat{R}_1 = \hat{R} + \hat{C}, \quad \text{where} \quad \hat{R} = \tilde{R} + Z_R, \quad \hat{C} = \tilde{C} + Z_C,$$

where Z_R and Z_C are the sums $\sum Z_x$, where Z_x is the canonical cycle of a point $x \in Sing\ X$, and the summation runs over $x \in R$, respectively $x \in C$. We apply the Hodge index theorem to divisors $\hat{R}$ and $\hat{C}$. Recall that this theorem states that for a non-singular projective surface S the cup product form on $H_{\mathbb{R}}^{1,1}(S)$ is non-degenerate of type $(1, h^{1,1} - 1)$. This gives that if $\hat{R}$ and $\hat{C}$ are divisors on S and $(\hat{R}^2) > 0$, then

$$\begin{vmatrix} (\hat{R}^2) & \hat{R} \cdot \hat{C} \\ \hat{R} \cdot \hat{C} & (\hat{C}^2) \end{vmatrix} \leq 0.$$

This is the desired main inequality. So our goal is to compute the mentioned intersection numbers and to show that $(\hat{R}^2) > 0$.

One more remark. To compute intersection numbers we need to make additional σ-processes on S (and on X_1) to disjoint the curves $\tilde{R}$ and $\tilde{C}$. For then we obtain a covering $\bar{X} \to \bar{X}_1$ and in neighbourhoods of proper transforms $\bar{R}$, $\bar{C}$ and $\bar{R}_1$ of $\tilde{R}$, $\tilde{C}$ and R_1, the restrictions $\bar{R} \to \bar{R}_1$ and $\bar{C} \to \bar{R}_1$ are finite coverings of degrees 2 and $N_2 - 2$ respectively. Therefore, $\left(\bar{R}^2\right) = 2\left(\bar{R}_1^2\right)$, $\left(\bar{C}^2\right) = (N_2 - 2)\left(\bar{R}_1^2\right)$, $\bar{R} \cdot \bar{C} = 0$. If we know exactly the performed σ-processes, we can compute the desired intersection numbers.

Thus, we have to examine carefully the normalization X of the fibre product, and the divisors R and C on X. This problem is local in its nature. Let $U \subset \mathbb{P}^2$ be a sufficiently small neighbourhood (in complex topology) of a point $b \in Sing\, B$. The preimage $f_1^{-1}(U)$ is a disjoint union of two parts, $f_1^{-1}(U) = V_1 \sqcup V_1'$, where V_1 is a part containing the ramification curve R_1, and V_1' is a part not containing R_1 and étale mapped to U. The map $V_1' \to U$ is an unramified covering of degree $N_1 - k$ ($k = 2, 3$ or 4 depending on the type of the singular point b). Analogously, $f_2^{-1}(U) = V_2 \sqcup V_2'$. Then $f^{-1}(U)$ is a disjoint union of four open sets – of normalizations of fibre products

$$W = \overline{V_1 \times_U V_2},\ W' = \overline{V_1 \times_U V_2'},\ \overline{V_1' \times_U V_2},\ \overline{V_1' \times_U V_2'}.$$

And only W and W' meet the curve $g_1^{-1}(R_1)$ and are essential for us. We can look over $g_1^{-1}(R_1) = R + C$ through the windows W and W'. W is the main window and it is studied in detail for different types of singular points $b \in B$. The window W' is auxiliary. It does not meet R. W' consists of $N_2 - k$ disjoint components isomorphically mapped to V_1.

Thus, to finish the proof we have to study the windows W.

5 LOCAL STRUCTURE OF FIBRE PRODUCTS

It is not difficult to describe the window W, when b is a non-singular or an ss-point of B. In this case V_1 and V_2 are double planes $z_1^2 = h(x, y)$, $z_2^2 = h(x, y)$, where $h(x, y) = 0$ locally defines B. Then $W^\times = V_1 \times_U V_2$ is defined in $\mathbb{C}^4 \ni (x, y, z_1, z_2)$ by two equations $z_1^2 = h(x, y)$, $z_2^2 = h(x, y)$. Hence, $z_1^2 = z_2^2$, and we see that $W^\times$ consists of two components, $W^\times = W_1^\times \cup W_2^\times$, lying in subspaces $z_1 = z_2$ and $z_1 = -z_2$. $W_1^\times$ and $W_2^\times$ are isomorphic to V_1 and V_2, and they meet along a curve $z_1 = z_2 = 0$, $h(x, y) = 0$. We obtain that $W = W_1^\times \sqcup W_2^\times$ is a disjoint union of two components isomorphic to V_1 and V_2. A visually-schematic picture is as follows

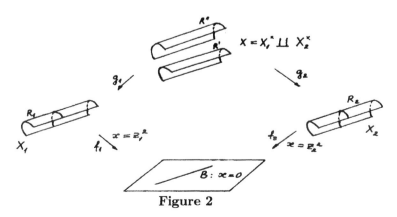

Figure 2

It remains to describe the window W for sp- and pp-points of type A_1 and A_2 of B. The most interesting is the case of an sp-point of type A_2. So let $f_2 : X_2 \to \mathbb{C}^2$ be a two-sheeted covering branched along a cusp $B : y^2 = x^3$, where f_2 is a projection of a surface $w^2 = x^3 - y^2$ onto x, y-plane. Here $0 \in B$ is a s-point of type A_2. Let $f_3 : X_3 \to \mathbb{C}^2$ be a projection of a surface $y = z^3 + xz$ onto x, y-plane. We can consider it as a miniversal deformation $z^3 + a_2 z + a_3 = 0$ $(a_2 = x, a_3 = -y)$ of a zero-dimensional singularity of type A_2. Its discriminant curve B is defined by equation $4a_2^3 + 27a_3^2 = 0$. To bring the equation of B to the form $y^2 = x^3$, we make a substitution

$$a_2 = -3x \,,\; a_3 = 2y \,.$$

We obtain a 3-sheeted covering $f_3 : X_3 \to \mathbb{C}^2$, where f_3 is a projection of a surface X_3, defined by equation $-2y = z^3 - 3xz$, to the x, y-plane. Or we can assume that $X_3 \simeq \mathbb{C}^2 \ni (x, z)$, and f_3 is defined by formulae $x = x$, $y = -\frac{1}{2}(z^3 - 3xz)$ (see Fig. 1). Then the ramification curve is defined by equation $x = z^2$, and $f_3^*(B) = 2R + C$, where C is defined by equation $x = \frac{1}{4}z^2$. Thus, we have a p-point $0 \in B$ of type A_2.

To obtain $X = \overline{X_2 \times_{\mathbb{C}^2} X_3}$ we must call to mind that in singularity theory there is a Viète map $f_6 : \mathbb{C}^2 \to \mathbb{C}^2$ of degree 6, branched along a cusp. It appears to be the desired normalization of $X_2 \times_{\mathbb{C}^2} X_3$. We recall that the Viète map sends the roots of a polinomial to its coefficients,

$$v : \mathbb{C}^3 \to \mathbb{C}^3 \,,\; (z_1, z_2, z_3) \longmapsto (a_1, a_2, a_3),$$

where $(z - z_1)(z - z_2)(z - z_3) = z^3 + a_1 z^2 + a_2 z + a_3$, i.e.

$$a_1 = -(z_1 + z_2 + z_3), \; a_2 = z_1 z_2 + z_2 z_3 + z_3 z_1, \; a_3 = -z_1 z_2 z_3.$$

The map v is ramified along diagonals $\Delta = \cup_{i \neq j} \{z_i = z_j\}$, and $v(\Delta) = D$ is defined by the discriminant of a polynomial of degree three. The map f_6 is obtained by restricting v to the invariant subspace $\mathbb{C}^2 = \{z_1 + z_2 + z_3 = 0\}$ (i.e. $a_1 = 0$). In coordinates x, y, where $a_2 = -3x$, $a_3 = 2y$, this map

$$\mathbb{C}^2 = \{z_1 + z_2 + z_3 = 0\} = X_6 \xrightarrow{f_6} \mathbb{C}^2 \ni (x, y)$$

is defined by formulae

$$f_6 : x = -\frac{1}{3}(z_1 z_2 + z_2 z_3 + z_3 z_1) \,,\; y = -\frac{1}{2} z_1 z_2 z_3.$$

The ramification divisor $R = L_1 + L_2 + L_3$ consists of three lines

$$L_i : z_j = z_k, \; z_1 + z_2 + z_3 = 0, \; \{i, j, k\} = \{1, 2, 3\} \,,$$

and the discriminant curve B has equation $y^2 = x^3$, and $f^*(B) = 2L_1 + 2L_2 + 2L_3$.

We obtain a commutative diagram

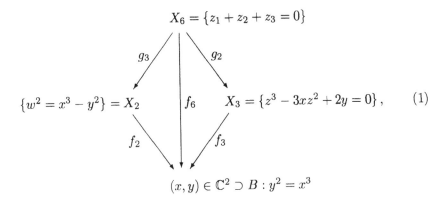

$$X_6 = \{z_1 + z_2 + z_3 = 0\}$$

g_3 g_2

$$\{w^2 = x^3 - y^2\} = X_2 \qquad f_6 \qquad X_3 = \{z^3 - 3xz^2 + 2y = 0\}\,, \qquad (1)$$

f_2 f_3

$$(x,y) \in \mathbb{C}^2 \supset B : y^2 = x^3$$

describing a normalization of the fibre product $W = \overline{V_1 \times_U V_2}$ in a neighbourhood of a sp-point of type A_2. It is easy to see that g_3 is a factorization under the action of a cyclic group $\mathbb{Z}_3 = \mathcal{A}_3 \subset S_3$, $X_2 = X_6/\mathcal{A}_3$, and g_2 is a factorization under the action of a cyclic group of order two $\mathbb{Z}_2 \simeq S_2 = \{(1),(2,3)\} \subset S_3$.

The diagram (1) can be visually-schematic presented as follows

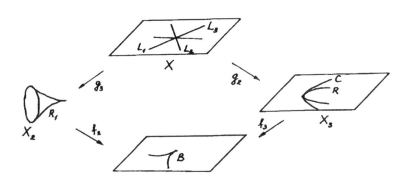

Figure 3

Direct computations show that $x - z^2 = \frac{1}{3}(z_2 - z_1)(z_1 - z_3)$, and $x - \frac{1}{4}z^2 = \frac{1}{12}(z_3 - z_2)^2$, i.e.

$$g_2^*(R) = L_2 + L_3 \,, \ g_2^*(C) = 2L_1.$$

It is interesting to note (Catanese F., [4]) that the constructed coverings f_2, f_3 and f_6 are characterized as the only coverings $f : (X,0) \to (\mathbb{C}^2,0)$ by a normal irreducible surface X, branched along an ordinary cusp $B \subset \mathbb{C}^2$, and the ramification curve of which is reduced, i.e. $f^*(B) = 2R + C$.

To finish the consideration of the case of sp-points of type A_2, we recall that for computing the intersection numbers we have to resolve the singular point of X_2 and to disjoint the curves $L_2 + L_3 (= R)$ and $L_1 (= C)$ on X_6. For this purpose we resolve the singularity of the cusp B by σ-processes,

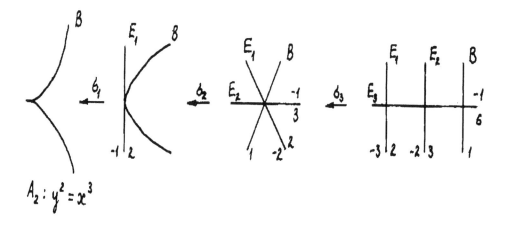

Figure 4

and obtain $\sigma : \bar{\mathbb{C}}^2 \to \mathbb{C}^2$, $\sigma = \sigma_3 \circ \sigma_2 \circ \sigma_1$. Then we lift (1) to $\bar{\mathbb{C}}^2$ and obtain a commutative diagram

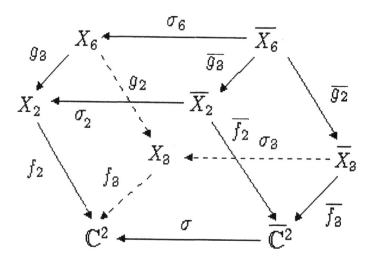

where $\bar{X}_i$ is a normalization of $X_i \times_{\mathbb{C}^2} \bar{\mathbb{C}}^2$. We obtain a detailed description of the right square of this diagram. Replacing the varieties at its vertices by the total transforms of the curve B, we obtain the following picture

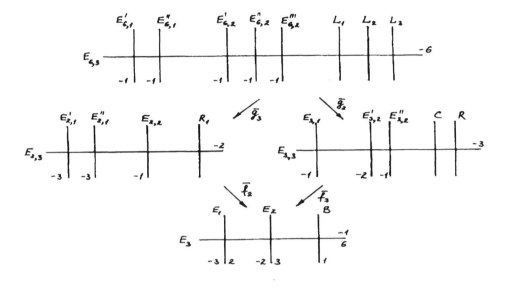

Figure 5

The rule of notation is as follows. The exceptional curves E_1, E_2, E_3 on $\bar{\mathbb{C}}^2$ are already denoted. Under double indexing $E_{i,j}$ the first index indicates the variety $\bar{X}_i$, where $E_{i,j}$ lies, and the second index indicates to what curve E_j the curve $E_{i,j}$ is mapped on $\bar{\mathbb{C}}^2$.

6   CANONICAL CYCLE OF A DU VAL SINGU-LARITY

Let (X, x) be a 2-dimensional A-D-E-singularity. Let $\pi : \bar{X} \to X$ be a minimal resolution, $L = \pi^{-1}(x)$ be the exceptional curve. As is known, the canonical class $K_{\bar{X}}$ is trivial in a neighbourhood of L, that is we can choose a divisor in $K_{\bar{X}}$ with a support not intersecting L. In other words, there is a differetial form ω on $\bar{X}$, which has neither poles nor zeroes in a neighbourhood of L, i.e. $(\omega) = 0$. On the other hand, (X, x) can be considered as a double plane, that is as a 2-sheeted covering $X \xrightarrow{f} Y$ of the plane $Y = \mathbb{C}^2$ (locally). Let $z^2 = h(x, y)$ be an equation of (X, x), $B : h(x, y) = 0$ be the discriminant curve, $f^{-1}(B) = R$, defined by the equation $z = 0$, be the ramification curve. We can consider the differential form $\omega = f^*(dx \wedge dy)$ lifted from Y. Then on $\bar{X}$ the divisor $(\omega) = (z) = \bar{R} + Z$, where $\bar{R} \subset \bar{X}$ is the proper transform of R , $Z = \sum \gamma_i L_i$ is a cycle on L. We shall say that Z is the *canonical cycle* of a 2-dimensional A-D-E-singularity. Thus, $-Z$ is a cycle on the exceptional curve L, which is equivalent to the ramification curve $\bar{R}$ in a neighbourhood of L. Let us calculate the canonical cycle for all A-D-E-singularities.

As for any double plane, a resolution of an A-D-E-singularity can be obtained by means of a resolution of the discriminant curve $B \subset Y = \mathbb{C}^2$, $B : h(x,y) = 0$. Let $\sigma : \bar{Y} \to Y$ be a composition of σ-processes, such that the total transform of B is a divisor with normal crossings. Let $\sigma^*(B) = \bar{B} + \sum_{i=1}^{r} \alpha_i l_i$, where $\bar{B}$ is the proper transform of B, $l_i \simeq \mathbb{P}^1$, $i = 1, \ldots, r$, are the exceptional curves, as well as their proper transforms, glued by σ-processes. Let $\bar{X}$ be the normalization of $\bar{Y} \times_Y X$, and $\bar{f}$ and π be induced by projections,

$$
\begin{array}{ccc}
\pi^{-1}(x) = L = L_1 \cup \ldots \cup L_r \subset \bar{X} & \xrightarrow{\ \pi\ } & X \supset R \ni x, \ R : z = 0 \\[2mm]
\bar{f} \downarrow & & \downarrow f \\[2mm]
\sigma^*(B) = \bar{B} + \sum_{i=1}^{r} \alpha_i l_i \subset \bar{Y} & \xrightarrow[\ \sigma\ ]{} & Y \supset B.
\end{array}
$$

Set $\bar{f}^{-1}(l_i) = L_i$. The curve L_i is either irreducible or consists of two components $L_i = L_i' + L_i''$, where $L_i' \simeq \mathbb{P}^1$, $L_i'' \simeq \mathbb{P}^1$. The mapping $\bar{f}$ is a 2-sheeted covering branched along the curve $\bar{B} + \sum_{\alpha_i-\text{odd}} l_i$. To be more graphic we denote the curves l_i, for which α_i are odd, also by $\bar{l}_i$, and L_i – respectively by $\bar{L}_i$. The surface $\bar{X}$ has singularities of type A_1 over nodes of the branch curve $\bar{B} + \sum \bar{l}_i$. If this curve is non-singular, that is, a disconnected union of components (one can reach this by performing one additional σ-processes for each node), then $\bar{X}$ is non-singular and is a resolution of the singularity (X, x). Let $\bar{R}$ be the proper transform of R w.r.t. π (= the proper transform of $\bar{B}$ w.r.t. $\bar{f}$). We have $\bar{f}^*(\bar{l}_i) = 2\bar{L}_i$, if α_i is odd, and $\bar{f}^*(l_i) = L_i$, if α_i is even. We have

$$
\big((\sigma \circ \bar{f})^* h(x,y)\big) = (z^2) = 2\bar{R} + \sum_{\alpha_i-\text{odd}} 2\alpha_i \bar{L}_i + \sum_{\alpha_i-\text{even}} \alpha_i L_i
$$

and, consequently, $(z) = \bar{R} + Z$, where

$$
Z = \sum_{\alpha_i-\text{odd}} \alpha_i \bar{L}_i + \sum_{\alpha_i-\text{even}} \frac{1}{2} \alpha_i L_i .
$$

For example, for the singularity $E_8 : x^3 + y^5$, these computations give

$$
Z = 3L_1 + 5L_2 + 9L_3 + 15L_4 + 10L_5 + 8L_6 + 12L_7 + 6L_8.
$$

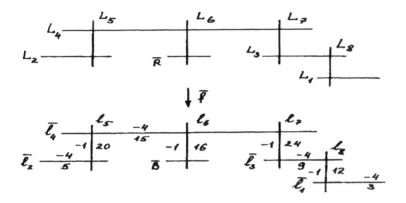

Figure 6

Define a *defect* δ of an A-D-E-singularity by the formula

$$\delta = \frac{1}{2}(\bar{R} \cdot Z).$$

Computing the canonical cycle for all A-D-E-singularities, we obtain the following

LEMMA 6.1 *For different types of A-D-E-singularities the defect equals*

$$\delta = \begin{cases} \left[\frac{n+1}{2}\right] & \text{for type } A_n; \\ \left[\frac{n}{2}\right] + 1 & \text{for type } D_n; \\ \left[\frac{n+1}{2}\right] & \text{for types } E_n, \ n = 6, 7, 8. \end{cases}$$

In particular, for the type A_1 (nodes) and A_2 (cusps) the defect $\delta = 1$.

Actually the defect δ is equal to the δ-invariant (genus) of the one-dimensional A-D-E-singularity.

7 APPLICATION OF THE MAIN INEQUALITY FOR THE PROOF OF THE CHISINI CONJECTURE FOR GENERIC m-CANONICAL COVERINGS

Let S be a minimal model of a surface of general type with numerical invariants $(K_S^2) = k$ and $e(S) = e$. Let X be a canonical model of the surface S, and $\pi : S \to X$ be the blowing down of (-2)-curves. Let $f : X \to \mathbb{P}^2$ be a *generic m-canonical covering*, i.e. a generic projection onto $\mathbb{P}^2$ of the surface $X = \varphi_m(S)$, where $\varphi_m : S \to \mathbb{P}^{p_m-1}$ is a m-canonical map, defined by the complete linear system $|mK_S|$, $p_m = \frac{1}{2}m(m-1)k + \chi(S)$. As is well known [3] , by a theorem of Bombieri $\varphi_m(S) \simeq X$ for $m \geq 5$, and φ_m gives the blowing down π.

For a m-canonical covering we can express $d = \deg B$, $N = \deg f$ and $p_a(R)$ by m and k,

$$d = m(3m+1)k, \ N = m^2 k, \ p_a(R) - 1 = \frac{1}{2}(3m+1)(3m+2)k \,.$$

THEOREM 7.1 *Let S_1 and S_2 be minimal models of surfaces of general type with the same (K_S^2) and $\chi(S)$, and let $f_1 : X_1 \to \mathbb{P}^2$, $f_2 : X_2 \to \mathbb{P}^2$ be generic m-canonical coverings with the same discriminant curve. Then for $m \geq 5$ the coverings f_1 and f_2 are equivalent.*

So let S_1 and S_2 be two surfaces of general type with numerical invariants k and e. Let $f_i : X_i \to \mathbb{P}^2$, $i = 1, 2$, be their m_i-canonical coverings having the same discriminant curve $B \subset \mathbf{P}^2$. In this situation we have $m_1 = m_2 = m$ and $\deg f_1 = \deg f_2 = N$. By the main inequality, for the proof that f_1 and f_2 are equivalent, it is sufficient to show that an inequality

$$N \left(2(3\bar{d} + p_a(R) - 1) - \iota \right) > 4(3\bar{d} + p_a(R) - 1)$$

holds (here R stands for R_1). Using a Hirzebruch-Miyaoka inequality ([3], p.215), $s \leq \frac{2}{9}\left(3e(S) - (K_S^2)\right)$, for the number s of disjoint (-2)-curves on S (to estimate ι), we obtain that it is enough to show that an equality of the form

$$3m^3(2m+1)k - 4(3m+1)^2 - (\frac{11}{9}k + 12)m^2 > 0$$

holds. It is easy to show that it is indeed so for $m \geq 3$ and for all k.

REFERENCES

[1] V.I. Arnol'd. Indices of singular points of 1-forms on a manifold with a boundary, convolution of invariants of groups generated by reflections, and singular projections of smooth surfaces. Usp. Math. Nauk. 34:2:3–38, 1979. (Engl. translation in Russ. Math. Surv., 34 2:1–42, 1979.)

[2] V.I. Arnol'd, S.M. Gusein-Zade, A.N. Varchenko. Singularities of differentiable maps, Vol. I, II. Birkhäuser, 1985.

[3] W. Barth, C. Peters, A. Van de Ven. Compact complex surfaces. Springer, 1984.

[4] F. Catanese. On a Problem of Chisini. Duke Math. J. 53 1:33–42 1986.

[5] Ph. Griffiths, J. Harris. Principles of algebraic geometry. John Wiley & Sons, New York, 1978.

[6] Vic. S. Kulikov. On a Chisini Conjecture. Izvestiya RAN: Ser.Mat., 63:683–116, 1999.

[7] B. Moishezon. Complex Surfaces and Connected Sums of Complex Projective Planes. LNM 603, Springer, 1977.

[8] O. Zariski. Algebraic surfaces. Berlin.Verlag von Julius Springer, 1935 Springer-Verlag, 1971.

Not All Codimension 1 Germs Have Good Real Pictures

DAVID MOND Mathematics Institute, University of Warwick, Coventry CV4 7AL, United Kingdon E-mail: mond@maths.warwick.ac.uk

ROBERTA G. WIK ATIQUE Departamento de Matemática, Instituto de Ciências Matemáticas e de Computação, Universidade de São Paulo - Campus de São Carlos, Caixa Postal 668, 15560-970, São Carlos SP, Brazil E-mail: rwik@icmc.sc.usp.br

1 INTRODUCTION

Some time ago the first author conjectured (in [9]) that every $\mathcal{A}_e$-codimension 1 equivalence class of map-germs $(\mathbb{C}^n, S) \to (\mathbb{C}^p, 0)$, with $n \geq p-1$ and (n, p) nice dimensions, should have a real form with a "good real perturbation" — that is, the $\mathcal{A}$-equivalence class should contain a real germ (one whose power-series expansion has purely real coefficients) which moreover should have a real perturbation whose real discriminant (if $n \geq p$) or real image (if $n = p-1$) carries the vanishing homology of its complexification. The purpose of this note is to give an example for which this does not hold — thus proving the conjecture false.

The example is not hard to describe: it is the simplest $\mathcal{A}_e$-codimension 1 class of map-germ $f : (\mathbb{C}^n, 0) \to (\mathbb{C}^{n+1}, 0)$ of corank 2 (i.e. such that $\ker(d_0 f)$ has dimension 2). Since the codimension in jet space $J^1(n, n+1)$ of the set Σ^2 of matrices of kernel rank 2 is equal to 6, such a singularity cannot occur if $n < 5$.

In fact one can easily construct map-germs $(\mathbb{C}^5, 0) \to (\mathbb{C}^6, 0)$ of $\mathcal{A}_e$-codimension 1 and corank 2. From the point of view of this paper, the most straightforward procedure is by transverse pull-back from the stable map-germ $F : (\mathbb{C}^6, 0) \to (\mathbb{C}^7, 0)$ defined by

$$F(A, B, C, D, x, y) = (A, B, C, D, x^2 + Ay, xy + Bx + Cy, y^2 + Dx).$$

That is, if $i : \mathbb{C}^6 \to \mathbb{C}^7$ is transverse to F, then the fibre square

$$
\begin{array}{ccc}
(\mathbb{C}^6, 0) & \xrightarrow{\;F\;} & (\mathbb{C}^7, 0) \\
\uparrow & & \uparrow i \\
(\mathbb{C}^6 \times_{\mathbb{C}^7} \mathbb{C}^6, 0) & \xrightarrow{\;i^*(F)\;} & (\mathbb{C}^6, 0)
\end{array}
$$

gives us a map-germ $i^*(F)$ from the smooth 5-dimensional complex manifold germ $(\mathbb{C}^6 \times_{\mathbb{C}^7} \mathbb{C}^6, 0)$ to $(\mathbb{C}^6, 0)$. If i is suitably generic (in a sense we will make precise below) then $i^*(F)$ has $\mathcal{A}_e$-codimension 1. Every map-germ $(\mathbb{C}^5, 0) \to (\mathbb{C}^6, 0)$ with local algebra isomorphic to that of F (which is the simplest local algebra of type Σ^2, isomorphic to $\mathcal{O}_{\mathbb{C}^2,0}/(m_{\mathbb{C}^2,0})^2$), can be obtained by transverse pull-back from F.

To give a concrete example, one can compute that the germ

$$
f(u, v, w, x, y) = (u, v, w, x^2 + uy, xy + vx + wy, y^2 - ux)
$$

has $\mathcal{A}_e$-codimension 1. This germ is obtained as $i^*(F)$ by taking

$$
i(u, v, w, x, y, z) = (u, v, w, -u, x, y, z).
$$

Of course there is no canonical choice of coordinates on the fibre product $(\mathbb{C}^6 \times_{\mathbb{C}^7} \mathbb{C}^6, 0)$, and we could equally well have written $i^*(F)$ in a different form. In fact the $\mathcal{A}$-equivalence class of $i^*(F)$ is determined by the image L of i, and so a more economical way of studying the codimension 1 map-germs obtained in this way is to consider the space of smooth hypersurface-germs transverse to F. For the same reason we also adopt the notation $F_{|L}$ in place of $i^*(F)$, although the reader will of course appreciate that since L is contained in the *target* of F, this denotes base-change rather than restriction.

In Section 2 we show first that to obtain every $\mathcal{A}_e$-equivalence class we need consider only hyperplanes, and among hyperplanes only those of the form $kA + lB + mC + nD = 0$. We then show that within the space $\mathcal{L}$ of these hyperplanes there is a discriminant Δ, itself a hypersurface, such that if $L \notin \Delta$ then $F_{|L}$ has $\mathcal{A}_e$-codimension 1.

In Section 3 we study the topology of the complement in $\mathcal{L}_{\mathbb{R}}$ of $\Delta_{\mathbb{R}}$; we do not determine its homology, but we do locate at least one point in each connected component. If L_1 and L_2 are in the same connected component of $\mathcal{L}_{\mathbb{R}} \setminus \Delta_{\mathbb{R}}$, the two germs $F_{|L_1}$ and $F_{|L_2}$ are $\mathcal{A}$-equivalent (over $\mathbb{R}$) (Remark 2.5 below) and thus we do not need to examine both of them.

In Section 4, we compute the homotopy type of the image of a stable perturbation of map-germs $F_{|L}$, for each hyperplane L in the list that we determine in Section 3. Up to homeomorphism each real codimension 1 map-germ has two real stable perturbations, corresponding to the sign of the deformation parameter; however here it turns out that the two are isomorphic. This implies a rather strong version of triviality of the monodromy of the family of complex stable perturbations, over $\mathbb{C}^*$.

Using Morse theory, we show that each of these complex images has the homotopy type of a 5-sphere; in consequence, the algebraic closure of the real image of each stable perturbation has the homotopy type of an n-sphere for some n between 0 and 5. For none of the examples in our list does n equal 5.

It turns out that $\Delta \subset \mathcal{L}$ is a free divisor, with some curious properties; in Section 5 we record some of these.

We are grateful to Jan Stevens for a geometrical observation concerning Δ, and to the organisers of the VIth Workshop for running such an enjoyable meeting.

2 $\mathcal{A}_e$-CODIMENSION   1   GERMS $(\mathbb{C}^5, 0) \to (\mathbb{C}^6, 0)$ OF TYPE Σ^2

LEMMA 2.1 *Let $i \colon (\mathbb{C}^6, 0) \to (\mathbb{C}^7, 0)$ be a map germ such that $i^*(F)$ has $\mathcal{A}_e$-codimension 1.*

Then

1. *i is an immersion, and*

2. *$i^*(F)$ is $\mathcal{A}$-equivalent to $i_\ell^*(F)$, where i_ℓ is the linear part of i.*

Proof: Write $f = i^*(F)$. (1) By a theorem of J. Damon ([6], but see also [10], Section 8),

$$\frac{\theta(f)}{T\mathcal{A}_e f} \simeq N\mathcal{K}_{E,e}i := \frac{\theta(i)}{ti(\theta_{\mathbb{C}^6}) + i^*(\mathrm{Der}(\log E))}$$

where $E = F(\mathbb{C}^6)$.

As F is not a trivial unfolding of any lower-dimensional germ, $\mathrm{Der}(\log E) \subset m_{\mathbb{C}^7, 0}\theta_{\mathbb{C}^7}$;
hence

$$1 = \mathcal{A}_e\text{-codimension } f \geq \dim_{\mathbb{C}} \frac{\theta(i)}{ti(\theta_{\mathbb{C}^6}) + i^*(\mathrm{Der}(\log E)) + m_{\mathbb{C}^7,0}\theta(i)} = \dim_{\mathbb{C}} \frac{\mathbb{C}^7}{i_\ell(\mathbb{C}^6)}.$$

It follows that the inequality in the centre is an equality, and i is an immersion. (2) is essentially Proposition 2.6 of [1], and can also be found, in this formulation, as of [13]. For completeness we outline the proof.

Since $i^*(F)$ has $\mathcal{A}_e$-codimension 1, we must have

$$ti(m_{\mathbb{C}^6,0}\theta_{\mathbb{C}^6}) + i^*(\mathrm{Der}(\log E)) = m_{\mathbb{C}^7,0}\theta(i);$$

by Nakayama's lemma this holds if and only if

$$ti(m_{\mathbb{C}^6,0}\theta_{\mathbb{C}^6}) + i^*(\mathrm{Der}(\log E)) + m_{\mathbb{C}^7,0}^2\theta(i) = m_{\mathbb{C}^7,0}\theta(i);$$

hence it depends only on the linear part of i. Let $i_\lambda = i_\ell + \lambda(i - i_\ell)$ be the linear interpolation between i_ℓ and i.

For each value of λ, $T\mathcal{K}_E i_\lambda = m_{\mathbb{C}^7,0}\theta(i_\lambda)$, and a standard argument using Mather's Lemma (of [7]) and finite $\mathcal{K}_E$-determinacy now shows that the family i_λ is $\mathcal{K}_E$-trivial (over $\mathbb{R}$).

It follows that the family $i_\lambda^*(F)$ is $\mathcal{A}$-trivial, and $i^*(F)$ is $\mathcal{A}$-equivalent to $i_\ell^*(F)$.
$\square$

LEMMA 2.2 *If the $\mathcal{A}_e$-codimension of $i_\ell^*(F)$ is equal to 1 then i_ℓ is $\mathcal{K}_E$- equivalent to a linear immersion whose image has equation of the form $kA + lB + mC + nD = 0$.*

Proof: Let p define the image of i_ℓ. By the preceding lemma we may assume that p is linear. We can express the $\mathcal{K}_{E,e}$-normal space of i_ℓ (and hence the $\mathcal{A}_e$-normal space of $i_\ell^*(F)$) in terms of p:

$$NK_{E,e}(i_\ell) \simeq \frac{\theta(p)}{tp(\mathrm{Der}(\log E)) + (p)\theta(p)}.$$

For if we write $\mathbb{C}^7 = L \times \mathbb{C}$ and identify p with projection to $\mathbb{C}$, then

$$\frac{\theta(p)}{tp(\mathrm{Der}(\log E)) + (p)\theta(p)} \simeq \frac{\theta_{L \times \mathbb{C}}}{\theta_{L \times \mathbb{C}/\mathbb{C}} + \mathrm{Der}(\log E) + m_{\mathbb{C},0}\theta_{L \times \mathbb{C}}}.$$

Now

$$\theta(i_\ell) = (\theta_{L \times \mathbb{C}})|_L = \theta_{L \times \mathbb{C}}/m_{\mathbb{C},0}\theta_{L \times \mathbb{C}}$$

and within $\theta_{L \times \mathbb{C}}/m_{\mathbb{C},0}\theta_{L \times \mathbb{C}}$,

$$\theta_{L \times \mathbb{C}/\mathbb{C}} = ti_\ell(\theta_L),$$

so

$$\frac{\theta_{L \times \mathbb{C}}}{\theta_{L \times \mathbb{C}/\mathbb{C}} + \mathrm{Der}(\log E) + m_{\mathbb{C},0}\theta_{L \times \mathbb{C}}} \simeq \frac{\theta(i_\ell)}{ti_\ell(\theta_L) + i^*(\mathrm{Der}(\log E))}.$$

If $i_\ell^*(F)$ has $\mathcal{A}_e$-codimension 1, then $tp(\mathrm{Der}(\log E)) + (p)\theta(p)$ must be equal to $m_{\mathbb{C}^7,0}$. Now the module $tp(\mathrm{Der}(\log E)) + (p)\theta(p)$ is the tangent space of p under the action of the group $\mathcal{K}(E)$, the semi-direct product of the group $\mathcal{R}_E$ of diffeomorphisms of $(\mathbb{C}^7, 0)$ preserving E, with the group $\mathcal{C}$. This is a geometric group in the sense of Damon [5]; it follows that p is finitely $\mathcal{K}(E)$-determined.

There is a 7-dimensional family of linear maps p,

$$\mathcal{L}_1 := \{p = kA + lB + mC + nD + pU + qV + rW : k, l, m, n, p, q, r \in \mathbb{C}\},$$

and within $\mathcal{L}_1$ a discriminant, Δ_1, in principle possibly equal to all of $\mathcal{L}_1$, or reduced to the zero map, consisting of linear maps for which $tp(\mathrm{Der}(\log E)) + (p)\theta(p)$ fails to be equal to $m_{\mathbb{C}^7,0}$.

A calculation with *Macaulay* shows that the annihilator in $\mathbb{C}[A, \ldots, W, k, \ldots, r]$ of

$$\frac{(A, \ldots, W, k, \ldots, r)}{T\mathcal{K}(E)p}$$

is generated by

$$f(k, l, m, n) := 8kl^3 + 3l^2m^2 + 24klmn + 8m^3n - 16k^2n^2$$

together with polynomials in the ideal $(A, \ldots, W)$. Thus, the equation of the discriminant is $f = 0$, and in particular is independent of the coefficients p, q, r. Suppose $L \in \mathcal{L}_1 \setminus \Delta_1$ has equation ℓ, and let ℓ_0 be obtained from ℓ by setting to zero the coefficients of U, V and W.

Then for each t, the hyperplane $L_t = \{(1 - t)\ell_0 + t\ell = 0\}$ also lies outside Δ_1, and it follows from Mather's Lemma (and finite determinacy) that the linear interpolation between ℓ and the map ℓ_0 is $\mathcal{K}(E)$-trivial. Hence the corresponding family of immersions $\mathbb{C}^6 \to \mathbb{C}^7$ is $\mathcal{K}_E$-trivial. This completes the proof. $\square$

The Lemma tells us we can forget the family $\mathcal{L}_1$ of all linear maps $\mathbb{C}^7 \to \mathbb{C}$ and consider only those of the form $kA + lB + mC + nD$; we denote the family of these by $\mathcal{L}$.

REMARK 2.3 The equation for the discriminant Δ may also be obtained by the following alternative method, which also relies on *Macaulay*, but which is primarily topological rather than deformation-theoretic.

Suppose that L is a hyperplane in $\mathbb{C}^7$, with defining equation $h_L(A, \dots, W) = \mathbf{u_L} \cdot (A, \dots, W) = 0$, and let G be a reduced defining equation for the image E of F.

If the $\mathcal{A}_e$-codimension of $F_{|L}$ is 1, then the family of normal translations of L, $L_t := \{t\mathbf{u_L}\} + L$, $t \in (\mathbb{C}, 0)$, induces an $\mathcal{A}_e$-versal deformation of $F_{|L}$. This follows from the fact that for a deformation of a codimension 1 germ, versality is equivalent to infinitesimal non-triviality.

By a Morse-theoretic argument of Dirk Siersma ([12]), the image of $F_{|L_t}$, i.e. $E \cap L_t$, has the homotopy type of a wedge of spheres, whose number, the *image Milnor number* of $F_{|L}$, $\mu_I(F_{|L})$), is equal to the sum of the Milnor numbers of the (isolated) singular points of $G_{|L}$ which move off the zero locus $E \cap L_t$ as t moves off 0. Let us call these points *indicator points*. For each value of t, the number of indicator points is finite; as t varies, they form a curve. In fact this curve is simply the polar curve of G with respect to the linear form h_L; its equations can easily be obtained as the generators of the transporter ideal $J_L : G$, where J_L is the ideal of maximal minors of the 2×7 jacobian matrix of the map (G, h_L).

It is worth noting that $V(J_L : G)$ evidently *contains* all the indicator points. However, *a priori* it is not clear that every zero of $J_L : G$ (off $\{t = 0\}$) is an indicator point. Nevertheless, this is the case. For provided $t \neq 0$, each germ of $F_{|L_t}$ is right-left stable. In Mather's nice dimensions (which contain $(5,6)$, cf [8]), every stable germ is weighted homogeneous in appropriate coordinates, and so its image has a weighted homogeneous defining equation. It follows that $E \cap L_t$ is locally weighted homogeneous at each point, and thus $G_{|L_t} \in J_{|L_t}$ at any point where $G = 0$.

Since, for $t \neq 0$, the deformation induced by varying t is right-left trivial, it follows that $G \in J_L$ at any point where $G = 0$ and $t \neq 0$. Hence the set of indicator points coincides with $V(J_L : G)$.

All of this can be done simultaneously for the family of all hyperplanes $L = \{kA + lB + mC + nD = 0\}$. We work in $\mathbb{C}[A, \dots, W, k, \dots, n]$. Let $h(A, \dots, n) = kA + lB + mC + nD$, let J be the ideal of maximal minors of the jacobian of (G, h) with respect to the variables $A, \dots, W$, and let

$$\mathcal{I}_0 := J : G.$$

Evidently $\mathcal{I}_0$ has a (k, l, m, n)-primary component, corresponding to the absent hyperplane for which $k = l = m = n = 0$, which we remove by using a transporter ideal. We find that

$$\mathcal{I} = \mathcal{I}_0 : (k, l, m, n)$$

has no (k, l, m, n)-primary component, and is thus the ideal we require. It is equal to

$(Ak - 1/3Bl + 1/3Cm - Dn, \, 4Bm - 4Ck - 3Dl, \, 3Am + 4Bn - 4Cl, B^2 - 3W, \, C^2 - 3U, 8BC - 3AD + 12V).$

We claim that $V(\mathcal{I}) \to \mathcal{L}$ restricts to a locally trivial fibration over $\mathcal{L} \setminus \Delta$, with fibre over $L \in \mathcal{L} \setminus \Delta$ equal to the set of indicator points of the family $F_{|L_t}$, where $L_t = \{t\mathbf{u_L}\} + L$ is the normal translation of L.

In fact this follows immediately from what we showed in the preceding paragraph: for $t \neq 0$ and $L \notin \Delta$, the deformation induced by varying L is locally trivial, and so the fibrewise equality

$$\{\text{indicator points}\} = V(J_L : G)$$

gives a relative equality

$$\{\text{relative indicator points}\} = V(J : G) = V(\mathcal{I}).$$

Now $F_{|L}$ has image Milnor number 1 if and only if the curve of indicator points in the deformation $F_{|L_t}$ is non-singular and transverse to L. We find the ideal of the set of points where this fails, by taking the determinant of the jacobian matrix of the pair $(g_1, \dots, g_6, h)$, where $g_1, \dots, g_6$ generate $\mathcal{I}$.

Up to a scalar factor, this (thanks to *Macaulay*) gives the equation f that we found above,

$$f(k, l, m, n) = 8kl^3 + 3l^2m^2 + 24klmn + 8m^3n - 16k^2n^2.$$

We note as soon as we assign numerical values to the parameters k, l, m and n for which this equation does not vanish, thus determining a hyperplane $L \notin \Delta$, then the specialisation $\mathcal{I}_L$ of the ideal $\mathcal{I}$ defines a smooth curve $\mathbb{C}^7$ (the curve of indicator points), transverse to L and to its normal translates $L_t = \{t\mathbf{u_L}\} + L$. At any point P of this curve,

$$\mathcal{O}_{\mathbb{C}^7,P}/\mathcal{I}_L + (h_L - t) \simeq \mathcal{O}_{L_t,P}/J_{G|L_t};$$

non-singularity of the curve implies, by Siersma's result, that $\mu_I(F_{|L}) = 1$. This conclusion retrospectively justifies the approach we adopt in this remark.

We record this conclusion as:

PROPOSITION 2.4 *Any map-germ* $(\mathbb{C}^5, 0) \to (\mathbb{C}^6, 0)$ *with* $\mathcal{A}_e$-*codimension equal to 1 and Boardman type* Σ^2 *has image Milnor number* μ_I *equal to 1.*   □

REMARK 2.5 If L_λ is any path in $\mathcal{L} \setminus \Delta$ then the argument of Lemma 2.2 shows that the family $F_{|L_\lambda}$ is $\mathcal{A}$-trivial. The ideas just used prove also

PROPOSITION 2.6 *If* $F_{|L}$ *has* $\mathcal{A}_e$-*codimension 1 then it is 2-determined for* $\mathcal{A}$-*equivalence.*

Proof: Since $f := F_{|L}$ has $\mathcal{A}_e$-codimension 1, and there is no stable germ in the same $\mathcal{K}$-orbit, its $\mathcal{A}$-orbit (in any jet space) must be open in its $\mathcal{K}$-orbit, and the tangent spaces to the two orbits coincide at each point of the former. The latter tangent space is easily shown to contain $m_{\mathbb{C}^5,0}^2 \, \theta(f)$. The result now follows by Mather's Lemma.□

3   THE COMPLEMENT OF THE REAL DISCRIMINANT $\Delta_{\mathbb{R}}$

To determine a list of normal forms for $\mathcal{A}_e$-codimension 1 germs $(\mathbb{R}^5, 0) \to (\mathbb{R}^6, 0)$ of type Σ^2, we have to select one real point from each component of $\mathbb{P}(\mathcal{L} \setminus \Delta)$. This can be done as follows:

within each component of $S^3 \setminus \Delta_{\mathbb{R}}$, the function $f_{| S^3}$ achieves at least one local maximum or local minimum. The critical points of $f_{| S^3}$ are the zeros of the ideal I of 2×2 minors of the jacobian matrix of (f, r^2), where r^2 is the usual distance-squared function. To eliminate the components lying in Δ we calculate the transporter ideal $I : f$. Once again by a *Macaulay* calculation, we find that this is equal to

$$(km - 4/7lm, lm - 7/4ln, k^2 - 1/3l^2 + 1/3m^2 - n^2).$$

This ideal defines a curve consisting of the eight lines through $(0, 0, 0, 0)$, shown in the left hand column of table 1. The second column shows the signs of the eigenvalues of the Hessian of $f_{| S^3}$ at the points where these lines meet the sphere.

The third column shows information obtained in the next section: it is the dimension of the sphere to which the real section $L_t \cap E \cap \mathbb{R}^6$ is homotopy-equivalent for $t \neq 0$.

By a Morse-theoretic argument given in the next section, this is one less than the index displayed in the third column of Table 2.

Line	Eigenvalues	Homotopy dimension of $L_t \cap E \cap \mathbb{R}^6$
$\text{Sp}\{(1, \sqrt{3}, 0, 0)\} \cap S^3$	$-\ -\ -$	4
$\text{Sp}\{(1, -\sqrt{3}, 0, 0)\} \cap S^3$	$+\ +\ -$	3
$\text{Sp}\{(0, 0, \sqrt{3}, 1)\} \cap S^3$	$-\ -\ -$	4
$\text{Sp}\{(0, 0, -\sqrt{3}, 1)\} \cap S^3$	$+\ +\ -$	3
$\text{Sp}\{(1, 0, 0, 1)\} \cap S^3$	$+\ +\ +$	3
$\text{Sp}(1, 0, 0, -1)\} \cap S^3$	$+\ +\ +$	3
$\text{Sp}\{(4, 7, 7, 4)\} \cap S^3$	$+\ -\ -$	4
$\text{Sp}\{(4, 7, -7, -4)\} \cap S^3$	$+\ -\ -$	4

Table 1: Critical points of $f_{| S^3}$

To obtain a representative of each $\mathcal{A}$- equivalence class we need take only those $F_{|L}$ for which L corresponds to one of the lines at which $f_{| S^3}$ has a local maximum or minimum: $L = \{A + \sqrt{3}B = 0\}$, $L = \{\sqrt{3}C + D = 0\}$, $L = \{A + D = 0\}$ and $L = \{A - D = 0\}$. In fact F has an obvious symmetry permuting x and y in the source and U and W in the target, and permuting also A and D, and B and C in both source and target. This symmetry interchanges the first and second of the hyperplanes just listed, and therefore we do not need to consider both.

4   THE TOPOLOGY OF REAL PERTURBATIONS

Let $L_{\mathbb{R}} \subset \mathbb{R}^7$ be a hyperplane (containing 0), let $\mathbf{u_L} \neq 0$ be orthogonal to $L_{\mathbb{R}}$, and let $L_{\mathbb{R},t}$ be the affine translate of $L_{\mathbb{R}}$ by $t\mathbf{u_L}$. If $F_{|L_{\mathbb{R}}}$ has $\mathcal{A}_e$-codimension 1, then the family $F_{\mathbb{R}|L_{\mathbb{R},t}}$ is an $\mathcal{A}_e$- versal deformation of $F_{\mathbb{R}|L_{\mathbb{R}}}$.

From 2.4 it follows (*cf* [3]) that $L_{\mathbb{R},t} \cap E_{\mathbb{R}}$ is homotopy-equivalent to a d-sphere, for some d between 0 and 5. For by Siersma's result, $F_{|L_t}$ has just one indicator point; this must therefore be real (otherwise its complex conjugate would also be an indicator point), and now $L_{\mathbb{R},t} \cap E_{\mathbb{R}}$ is homotopy-equivalent to the boundary of a k-cell, where k is the index of $G_{\mathbb{R}|L_{\mathbb{R},t}}$ at the indicator point.

The indicator point can be determined by substituting the values of k, l, m and n into the expressions for the generators of the ideal $\mathcal{I}$ listed in Remark 2.3.

The Hessian of $G_{|L_t}$ at the indicator point can be computed as $P^t H_G P$, where H_G is the Hessian of G and P is a 7×6 matrix whose columns form an orthonormal basis for L.

The results of this computation are displayed in Table 2. In principle we should display the signs of the eigenvalues of the Hessian of $G_{|L_t}$ at the indicator point for two values of t, one positive and one negative. However, there is an obvious symmetry which renders this unnecessary: if (A, B, C, D, U, V, W) is the indicator point for L_t then $(-A, -B, -C, -D, U, V, W)$ is the indicator point for L_{-t}, and the Hessian matrices of G at these two indicator points are the same.

Equation of L	Ideal of indicator point	Index
$A + \sqrt{3}B$	$(\sqrt{3}A - B, C, D, U, V, 3W - B^2)$	5
$A - \sqrt{3}B$	$(\sqrt{3}A + B, C, D, U, V, 3W - B^2)$	4
$\sqrt{3}C + D$	$(\sqrt{3}D - C, A, B, W, V, 3U - C^2)$	5
$\sqrt{3}C - D$	$(\sqrt{3}D + C, A, B, W, V, 3U - C^2)$	4
$A + D$	$(B, C, A - D, D^2 - 4V, W, U)$	4
$A - D$	$(B, C, A + D, D^2 + 4V, W, U)$	4
$4A+7B+7C+4D$	$(B - \frac{7}{4}D, A - D, C - B, D^2 - \frac{48}{49}W, U - W, V + \frac{86}{49}W)$	5
$4A + 7B - 7C - 4D$	$(B + \frac{7}{4}D, A + D, C + B, D^2 - \frac{48}{49}W, U - W, V - \frac{86}{49}W)$	5

Table 2: Indicator points and Morse indices

5   CONCLUDING REMARKS

This counterexample reveals a sharp difference between singularities of corank 1 and of higher corank. The main theorem of [4] states that *every* $\mathcal{A}_e$-codimension 1 multi-germ $(\mathbb{C}^n, S) \to (\mathbb{C}^p, 0)$ $(n \geq p - 1, (n, p)$ nice dimensions), in which each component has corank ≤ 1, has a real form with a good real perturbation, and thus a "good real picture".

It is clear from our example here that good real pictures cannot be expected for germs of corank greater than 1.

6 APPENDIX: GEOMETRY OF THE DISCRIM-INANT Δ

Jan Stevens has pointed out to us that the discriminant Δ of Section 3 is the affine cone over the tangent developable surface of the rational normal curve (a twisted cubic) in $\mathbb{P}^3$ parametrised by

$$(u, v) \mapsto (u^3, -2uv^2, 2u^2v, -v^3).$$

For it is easy to check that the singular locus of Δ is the cone over this curve, and then to check that Δ coincides with the tangent developable.

The tangent developable has a cuspidal edge along the curve, and is otherwise smooth. The map $\mathbb{RP}^1 \times \mathbb{RP}^1 \to \mathbb{RP}^3$ parametrising the real part of the tangent developable is therefore a homeomorphism onto its image, and from this it follows by a straightforward duality argument that the complement of $\Delta_{\mathbb{R}}$ has just two connected components. Thus, it is no surprise that the eight lines in the complement of $\Delta_{\mathbb{R}}$ that we listed in Section 3, and the eight hyperplanes in $L \subset \mathbb{R}^7$ that they correspond to, should give rise to just two inequivalent real codimension 1 map-germs $F_{|L}$.

It also turns out that Δ is a free divisor. This is easy to check, using Saito's criterion (cf [11]): a *Macaulay* calculation shows that the module $\mathrm{Der}(\log \Delta)$ of vector fields on $\mathcal{L}$ which are tangent to Δ is generated by the vector fields $\chi_1, \ldots, \chi_4$ whose coefficients in the basis $\partial/\partial k, \partial/\partial l, \partial/\partial m, \partial/\partial n$ form the columns of the matrix

$$M = \begin{pmatrix} k & -3/4k & 0 & -3/4m \\ l & 1/4l & -m & -n \\ m & -1/4m & k & l \\ n & 3/4n & 3/4l & 0 \end{pmatrix},$$

and the determinant of this matrix is a reduced defining equation of Δ. We note for future reference that $\chi_1 \cdot f = 4f$ and $\chi_i \cdot f = 0$ for $i \neq 1$.

The free divisor Δ is unusual in that all of the generating vector fields are linear. This makes it particularly easy to calculate the cohomology of the complement: from its geometrical description, Δ is clearly locally quasihomogeneous, and thus we can use the theorem of [2]:

THEOREM 6.1 *If $D \subset \mathbb{C}^n$ is a locally quasihomogenenous free divisor, then integration of forms along cycles defines an isomorphism*

$$h^q(\Gamma(\mathbb{C}^n, \Omega^{\bullet}(\log \Delta))) \simeq H^q(\mathbb{C}^n \setminus \Delta; \mathbb{C}).$$

$\square$

Via a contracting homotopy defined by the Lie derivative with respect to the Euler vector field, the complex $\Omega^{\bullet}(\log \Delta)_0$ is quasi-isomorphic to its subcomplex of weight zero. As all the generating vector fields are linear, they have weight zero, and so dually the generators $\omega_1, \omega_2, \omega_3, \omega_4$ of $\Omega^1(\log \Delta)$ (determined by the property that $\omega_i(\chi_j) = \delta_{i,j}$) also have weight zero. Thus the weight zero subalgebra of $\Omega^\bullet(\log \Delta)$ is the free exterior algebra over $\mathbb{C}$ on generators $\omega_1, \ldots, \omega_4$. As soon as we determine the coefficients $\Gamma^i_{j,k}$ in

$$d\omega_i = \sum_{j<k} \Gamma^i_{j,k} \omega_j \wedge \omega_k$$

which must be simply complex numbers, determination of $H^*(\mathbb{C}^4 \setminus \Delta)$ is reduced to linear algebra.

In fact we have

$$\begin{aligned}
d\omega_1 &= 0 \\
d\omega_2 &= -\omega_3 \wedge \omega_4 \\
d\omega_3 &= \tfrac{1}{2}\,\omega_2 \wedge \omega_3 \\
d\omega_4 &= -\tfrac{1}{2}\,\omega_2 \wedge \omega_4
\end{aligned}$$

and thus

$$\begin{aligned}
d(\omega_1 \wedge \omega_2) &= \omega_1 \wedge \omega_3 \wedge \omega_4 & d(\omega_2 \wedge \omega_3) &= 0 \\
d(\omega_1 \wedge \omega_3) &= -\tfrac{1}{2}\omega_1 \wedge \omega_2 \wedge \omega_3 & d(\omega_2 \wedge \omega_4) &= 0 \\
d(\omega_1 \wedge \omega_4) &= \tfrac{1}{2}\omega_1 \wedge \omega_2 \wedge \omega_4 & d(\omega_3 \wedge \omega_4) &= 0
\end{aligned}$$

and

$$d(\omega_i \wedge \omega_j \wedge \omega_k) = 0$$

for all i, j, k.

Hence

PROPOSITION 6.2

$$H^q(\mathcal{L} \setminus \Delta; \mathbb{C}) = \begin{cases}
\mathbb{C} & \textit{if } q = 0 \\
\mathbb{C}, \textit{ generated by } \omega_1 & \textit{if } q = 1 \\
0 & \textit{if } q = 2 \\
\mathbb{C}, \textit{ generated by } \omega_2 \wedge \omega_3 \wedge \omega_4 & \textit{if } q = 3 \\
\mathbb{C}, \textit{ generated by } \omega_1 \wedge \omega_2 \wedge \omega_3 \wedge \omega_4 & \textit{if } q = 4
\end{cases}$$

The equality $d\omega_2 = -\omega_3 \wedge \omega_4$ shows that the family $f : \mathbb{C}^4 \to \mathbb{C}$ is a *logarithmic contact bundle*. For, as with any free divisor $D = \{f = 0\}$, the wedge product of all the generators of $\Omega^1(\log D)$ is equal to the generator of the top-dimensional module of logarithmic differential forms. Since $\omega_1 = df/f$ can be taken as one of these generators, this shows that the wedge product of the remaining generators of $\Omega^1(\log D)$ generates the relative dualising module ω_f (for it can be written as the "wedge division" $dz_1 \wedge \cdots \wedge dz_n/df$) and thus its restriction to the regular fibres is a holomorphic volume form.

Since in our example, $\omega_2 \wedge d\omega_2 = -\omega_2 \wedge \omega_3 \wedge \omega_4$, we have

PROPOSITION 6.3 *The restriction of ω_2 to each fibre of f is a contact form, and so $f : \mathbb{C}^4 \to \mathbb{C}$ is a logarithmic contact bundle.* $\square$

We note that a similar phenomenon can never occur for a function with isolated singularity, in any dimension. For if $f : (\mathbb{C}^{2n}, 0) \to (\mathbb{C}, 0)$ has isolated singularity at 0 and $D = f^{-1}(0)$, then

$$\Omega^1(\log D) = \Omega^1_{\mathbb{C}^{2n}} + \mathcal{O}_{\mathbb{C}^{2n}} \cdot df/f,$$

and so the restriction to the fibre of any logarithmic 1-form ω is actually the restriction of a regular form. Hence for any logarithmic 1-form ω, $df \wedge \omega \wedge (d\omega)^{\wedge n-1} \subset df \wedge \Omega^{2n-2}_{\mathbb{C}^{2n}}$; since $\omega_f / df \wedge \Omega^{2n}_{\mathbb{C}^{2n+1}}$ has positive dimension $\tau(D)$ if f is singular, no element of $df \wedge \Omega^{2n}$ can generate ω_f.

We do not know whether this property of Δ, or the fact that it is a free divisor, can be deduced from the singularity theory in which it is embedded.

REFERENCES

[1] J.W. Bruce, R.M. Roberts. Critical points of functions on analytic varities. Topology 27 1:57–90 1988.

[2] F. Castro, D. Mond, L. Narváez. Cohomology of the complement of a free divisor. Trans. Amer. Math. Soc. 348 8:3037–3049, 1996.

[3] T. Cooper, D. Mond. Complex monodromy and changing real pictures. J. London Math. Soc. 57 2:599–608, 1998.

[4] T. Cooper, D. Mond, R. Wik Atique. Vanishing topology of codimension 1 germs over $\mathbb{R}$ and $\mathbb{C}$. Compositio Math. to appear.

[5] J.N. Damon. The unfolding and determinacy theorems for subgroups of $\mathcal{A}$ and $\mathcal{K}$. A.M.S. Memoirs 306, Providence R.I. 1984.

[6] J.N. Damon. $\mathcal{A}$-equivalence and equivalence of sections of images and discriminants. In: Singularity Theory and Applications, eds. D. Mond, J. Montaldi. Warwick 1989, Part I. Lecture Notes in Maths. 1462, Springer Verlag, 1991, pp. 93–121.

[7] J.N. Mather. Stability of C^∞ mappings IV, Classification of stable germs by $\mathbb{R}$-algebras. Publ. Math. I.H.E.S. 37:223–248, 1969.

[8] J.N. Mather, Stability of C^∞ mappings VI, The nice dimensions. In: Singularities, ed. C.T.C. Wall. Liverpool 1971, Lecture Notes in Maths. 192, Springer Verlag, 1972, pp. 207–253.

[9] D. Mond. How good are real pictures? In: Algebraic and Analytic Geometry, La Rábida, eds. A. Campillo, L.Narváez, Progress in Math. 134, Birkhauser Verlag, 1996, pp.259–276.

[10] D. Mond. Differential forms on free and almost free divisors. Proc. London Math. Soc. 81 3:587–617, 2000.

[11] K. Saito. Theory of logarithmic differential forms and logarithmic vector fields. J. Fac. Sci. Univ. Tokyo Sect. Math. 27:265–291, 1980.

[12] D. Siersma. Vanishing cycles and special fibres, In: Singularity Theory and Applications, eds. D. Mond, J. Montaldi, Warwick 1989, Part I. Lecture Notes in Math 1462, Springer Verlag 1991, pp.292–301.

[13] R. Wik Atique. On the classification of multi-germs of maps from $\mathbb{C}^2 \to \mathbb{C}^3$ under $\mathcal{A}$-equivalence, In: Real and Complex Singularities, eds. J.W. Bruce, F. Tari. Research Notes in Math. Chapman and Hall/CRC 412, 2000, pp.119–133.

On the Topology of
Hypersurface Singularities

JOSÉ SEADE   Instituto de Matemáticas, Unidad Cuernavaca, Universidad Nacional Autónoma de México, Apartado Postal 273-3, Cuernavaca, Morelos, 62251, México. E-mail: jseade@math.unam.mx

1   INTRODUCTION

The topology of complex isolated hypersurface singularities has been long studied by many authors, as for instance Milnor [3], Lê Dũng Tráng [2] and many more. There is a beautiful and well developed theory to this respect, though there are still many things to be understood. This note is concerned with the real counterpart of this theory and it was inspired by [3, 4]. The motivation was to consider a class of real hypersurface singularities that arise naturally from complex singularities. We study the topology of the real hypersurfaces that arise when we consider a $\mathbb{C}$-valued holomorphic function H on an open set $\mathcal{U}$ in $\mathbb{C}^n$ and we compose this function with the projection onto a real line through the origin in $\mathbb{C}$. We show that, for a fixed function H, all these hypersurfaces are homeomorphic and the link of the singularity is the double of the Milnor fibre of the holomorphic function H. It would be interesting to make a similar study for isolated complete intersection complex singularities with codimension higher than 1. This general case is certainly more interesting, and more difficult, than the one envisaged in this note, since the type of the real singularities so obtained, by composing the complex map with the projection onto a real line, will vary as the corresponding line intersects the discriminant of the original function. So this article should be regarded as the first approach to that general, and hopefully interesting, problem.

2   THE THEOREMS

Let $\mathcal{U} \subset \mathbb{C}^{n+1}$ be a connected open set, $0 \in \mathcal{U}$, and let

$$H : (\mathcal{U}, 0) \to (\mathbb{C}, 0) ,$$

be a continuous function. For each real line $\mathcal{L}_\theta \subset \mathbb{C}$ passing through the origin with an angle θ, $\theta \in [0, \pi[$, we let $\pi_\theta : \mathbb{C} \to \mathcal{L}_\theta$ be the orthogonal projection; set $h_\theta = \pi_\theta \circ H$, so that h_0 and $h_{\frac{\pi}{2}}$ are, respectively, the real and the imaginary parts of H. We set $M_\theta = h_\theta^{-1}(0)$ and $\mathcal{V} = H^{-1}(0)$. We define the map $\tilde{\phi} : \mathcal{U} - \mathcal{V} \to S^1$ by $\tilde{\phi}(z) = \frac{iH(z)}{\|H(z)\|}$. For each $e^{i\theta} \in S^1$, we set $E_\theta = \tilde{\phi}^{-1}(e^{i\theta})$.

We prove the following results:

LEMMA 2.1 *One has :*

$$\mathcal{U} = \cup M_\theta \ \text{and} \ \mathcal{V} = \cap M_\theta ,$$

for $\theta \in [0, \pi[$. Also, for each $\theta \in [0, \pi[$ one has :

$$M_\theta = E_\theta \cup \mathcal{V} \cup E_{\theta + \pi} .$$

Similarly, if S_ε is a sphere embedded in $\mathcal{U}$ with centre at 0, one has,

$$S_\varepsilon = \cup(M_\theta \cap S_\varepsilon) , \ \mathcal{V} \cap S_\varepsilon = \cap(M_\theta \cap S_\varepsilon),$$

and

$$(M \cap S_\varepsilon) = (E_\theta \cap S_\varepsilon) \cup (\mathcal{V} \cap S_\varepsilon) \cup (E_{\theta + \pi} \cap S_\varepsilon) .$$

THEOREM 2.2 *If H is holomorpic, let $h_\theta = \pi_\theta \circ H$ be as before and let*

$$\phi = \frac{H}{\|H\|} : S_\varepsilon - K \to S^1 ,$$

be the usual Milnor fibration [3], where $K = \mathcal{V} \cap S_\varepsilon$ is the link of 0. Then :
i) Each $M_\theta = h_\theta^{-1}(0)$ is a real analytic hypersurface in $\mathcal{U}$, whose singular set is the singular set of $\mathcal{V}$.
ii) Each pair of antipodal fibers F_θ and $F_{\theta + \pi}$ of ϕ are naturally glued together along K, forming a real analytic variety isomorphic to $N_\theta = M_\theta \cap S_\varepsilon$.

This theorem is telling us that the link N_θ of M_θ is, in some sense, the double of the Milnor fibre of H, but N_θ is singular if K is singular. We notice that even in this case, there is also a fibration [2, 3],

$$H : H^{-1}(S_\delta^1) \cap B_\varepsilon \to S_\delta^1 ,$$

where S_δ^1 is a small circle centered at $0 \in \mathbb{C}$ and B_ε is a small open ball in $\mathbb{C}^{n+1}$. This fibration is equivalent to the Milnor fibration. The closure of each fibre $E_\theta = H^{-1}(e^{i\theta})$ in this fibration is a compact non-singular variety with boundary $\tilde{K}$, while the closure of the Milnor fibre F is the union of F with the link K, which may be singular. I thank Lê Dũng Tráng for explaining me that there is a natural contraction function from $\tilde{K}$ onto K; it would be interesting to study the relation of this function with our construction. Equivalently, one may study the relation between the double of the

Milnor fibre, which is a closed manifold, and the link N_θ of the real analytic variety $M_\theta = h_\theta^{-1}(0)$.

We notice that M_0 and $M_{\frac{\pi}{2}}$ are, respectively, the sets of points where $Re\,H(z) = 0$ and $Im\,H(z) = 0$.

If the singular set of $\mathcal{V}$ consists of an isolated point, one has the following theorem:

THEOREM 2.3 *If* $0 \in \mathcal{U}$ *is an isolated critical point of* H, *the* M_θ'*s are non-singular away from* 0, *and they are all homeomorphic. The link* $N_\theta = M_\theta \cap S_\varepsilon$ *is diffeomorphic to the double of the Milnor fibre* F *of* H, *hence* N_θ *is* $(n-1)$-*connected;* N_θ *is always stably parallelizable, and it is actually parallelizable if and only if* n *is odd and the Milnor number* μ *of* H *is* 1.

We recall that a manifold M is parallelizable if its tangent bundle TM is trivial; M is stably parallelizable if $TM \oplus (k)$ is trivial, where (k) is a trivial bundle over M, or equivalently, if M can be embedded in some sphere S^N with trivial normal bundle.

Since the topology of the Milnor fibre of H and the link K are well understood, these theorems determine the topology of the real hypersurfaces M_θ.

EXAMPLE 2.4 Consider the complex polynomial $f(z_1, z_2) = z_1^p + z_2^q$, with $p, q > 1$. The topology of this singularity is well understood. Its link K is a torus link (or knot if p, q are relative prime) and its Milnor number μ is $(p-1)(q-1)$; we recall that the Milnor fibre F is in this case an oriented surface in the 3-sphere, with boundary K and the homotopy type of μ circles. Now consider the real and the imaginary parts of f,

$$f_1 = Re\,f = z_1^p + z_2^q + \bar{z}_1^p + \bar{z}_2^q\,,$$

$$f_2 = Re\,f = z_1^p + z_2^q - \bar{z}_1^p - \bar{z}_2^q\,.$$

Both define real hypersurfaces in $\mathbb{C}^2$, which are cones over their link L_1, L_2. By the theorem above, L_1 and L_2 are homeomorphic, actually diffeomorphic, and they are the double of the Milnor fibre F. Hence L_1 and L_2 are closed, oriented surfaces of genus $(p-1)(q-1)$ in the 3-sphere $S^3 \subset \mathbb{C}^2$. This can be interesting because it provides explicit analytic embeddings of surfaces of all genera in $S^3 \subset \mathbb{C}^2$. For instance, if $p = 2 = q$, then K is the Hopf link, F is a cylinder $S^1 \times I$ and L_1, L_2 are tori $S^1 \times S^1$, obtained by taking two copies of F and glueing them along their boundary. The link K is the intersection of the tori L_1, L_2.

3   THE PROOFS

Proof of The Lemma 2.1:   If $z \in M_\theta$, then $H(z)$ is contained in the line passing through 0 with an angle $\theta \pm \frac{\pi}{2}$; Hence $z \in M_{\theta_1} \cap M_{\theta_2}$ if and only if $z \in \mathcal{V}$ where $\theta_1 = \theta_2 + k\pi$, $k \in \mathbb{Z}$; We prove that $\mathcal{V} = \cap M_\theta$ for each $\theta \in [0, \pi[$. It is clear that each point in $\mathcal{U}$ is either in $\mathcal{V}$ itself or else it is in a certain E_θ. So the claim $\mathcal{U} = \cup M_\theta$ follows from the claim $M_\theta = E_\theta \cup \mathcal{V} \cup E_{\theta+\pi}$. We prove this last claim. For this we notice that M_θ is the set of points $z \in \mathbb{C}^{n+1}$ such that $e^{-i\theta}H(z)$ is a purely imaginarian number. One has :

$$M_\theta = \{z \in \mathbb{C}^{n+1} \mid Re\,e^{-i\theta}H(z) = 0\}\,.$$

It is now clear that $\mathcal{V} \subset M_\theta$. Let us prove that $E_\theta \subset M_\theta$. If $z \in E_\theta$ then

$$\frac{iH(z)}{\|H(z)\|} = e^{i\theta},$$

which implies

$$\frac{e^{i\theta}iH(z)}{\|H(z)\|} = 1,$$

hence,

$$Re\, e^{-i\theta}\frac{H(z)}{\|H(z)\|} = 0.$$

Thus one has $z \in M_\theta$. Similarly , if $z \in E_{\theta+\pi}$, then

$$\frac{iH(z)}{\|H(z)\|} = e^{i\theta+\pi} = -e^{i\theta},$$

so

$$e^{-i\theta}\frac{iH(z)}{\|H(z)\|} = -1,$$

which implies $Re\, e^{-i\theta}H(z) = 0$. Hence one has $E_{\theta+\pi} \subset M_\theta$. Conversely, if $z \in M_\theta$, then $Re\, e^{-i\theta}H(z) = 0$; If $H(z) = 0$, one has $z \in \mathcal{V}$ and there is nothing to prove. If $H(z) \neq 0$, then

$$e^{-i\theta}\frac{iH(z)}{\|H(z)\|} = \pm 1,$$

so z is in E_θ or in $E_{\theta+\pi}$, according to its sign, and we arrive to the formula above. To prove the second statement in the lemma we just restrict the previous discussion to the sphere S_ε.

Proof of The Theorem 2.2:   The second statement in Theorem 2.2 is a consequence of the lemma above, because the map $\tilde{\phi}$ is ϕ followed by the diffeomorphism $z \to iz$ of $\mathbb{C}$.   Hence the intersection of each M_θ with each sphere S_ε is a real analytic variety of dimension $2n$; since M_θ is a cone near 0, by [3], we know that M_θ is a real hypersurface. Its singular set consists of the critical points of h_θ, i.e. the points where all the partial derivatives of h_θ vanish. This set does not change if we multiply h_θ by the number $e^{-i\theta}$. Thus, to prove the claim about the singularities of M_θ, it is enough to consider the case $\theta = 0$, i.e. for the real part of H, $f = ReH$. One has,

$$f = \frac{1}{2}(H + \overline{H}),$$

where $\overline{H}$ is the complexe conjugate of H. Therefore,

$$2\Delta f = \left(\frac{\partial H}{\partial z_1}, \frac{\overline{\partial H}}{\partial z_1}, ..., \frac{\partial H}{\partial z_{n+1}}, \frac{\overline{\partial H}}{\partial z_{n+1}}\right),$$

because the partial derivatives of H with respect to the $\overline{z}_i$ are all 0, the partial derivatives of $\overline{H}$ with respect to the z_i are all 0 and $\frac{\partial \overline{H}}{\partial \overline{z}_i} = \overline{\frac{\partial H}{\partial z_i}}$. Thus, the critical points of H are the critical points of f.

Proof of The Theorem 2.3. The claims that the M_θ are regular away from 0, that they are all homeomorphic (actually diffeomorphic away from 0) and the link is diffeomorphic to the double of the Milnor fibre of f, are all consequences of Theorem 2.2, N_θ is connected because F is connected, by [3]. Furthermore, if $n > 1$, then K is connected and F is simply connected, by [3]. Hence Van Kampen's Theorem ([6], p. 151) implies that N_θ is simply connected. Moreover, by [3], F is a wedge of n-spheres and K is $(n-2)$-connected, thus Mayer-Vietoris (reduced if n=2) implies that one has

$$H_1(N_\theta; \mathbb{Z}) \cong \ldots \cong H_{n-1}(N_\theta; \mathbb{Z}) \cong 0 \,.$$

Therefore Hurewicz's isomorphism ([6],p. 397) implies that N_θ is $(n-1)$-connected. Finally, N_θ is stably parallelizable because it is a codimension 1, oriented submanifold of S_ε. Thus, ([1], Th.IX) , N_θ is parallelizable if and only if its Euler-Poincaré characteristic $\chi(N_\theta)$ vanishes. One has:

$$\chi(N_\theta) = 2\chi(F) = 2 + 2(-1)^n \, \mu \,,$$

where $\mu > 0$ is the Milnor number of H, since N_T is the double of F, and F is a wedge of μ n-sphères. So $\chi(N_\theta) = 0$ if and only if n is odd and $\mu = 1$.

REFERENCES

[1] M. Kervaire. Courbure intégrale généralisée et homotopie. Math. Ann. 131:219–252, 1956.

[2] Lê Dũng Tráng. Some remarks on the relative monodromy. Real and complex singularities, Sijthoff and Noordhoff, Oslo, 1976, pp. 397-403

[3] J. Milnor. Singular points of complex hypersurfaces, Annals of Maths. Studies, 61, Princeton Univ. Press, 1968.

[4] L. Paunescu. The topology of the real part of a holomorphic function. Math. Nachr. 174:265-272, 1995.

[5] J.A. Seade. Fibred links and a construction of real singularities via complex geometry. Bol. Soc. Bras. Mat. 27:199-215, 1996.

[6] E. Spanier. Algebraic Topology. Springer Verlag, 1966.

Polar Multiplicities and Equisingularity of Map Germs from $\mathbb{C}^3$ to $\mathbb{C}^4$

VICTOR HUGO JORGE PÉREZ   Universidade Estadual de Maringá, Departamento de Mate- mática, Av. Colombo 5790, Campus Universitário, CEP 87020-900, Maringá (PR), Brazil. E-mail: vhjperez@uem.br

1   INTRODUCTION

In this work we consider the following problem: given a 1-parameter family of map germs, find analytic invariants whose constancy in the family implies the family is topologically trivial or Whitney equisingular.

Gaffney in [6] describes this problem for families of finitely determined map germs of discrete stable type, in which the topological triviality of a 1-parameter family of a class of map germs is shown to be consequence of the constancy of finitely many analytic invariants associate to a member of that family, the number of invariants depending upon the particular dimensions of source and of target space being considered. The necessary invariants are the zero-stable invariants and the polar multiplicities defined by Teissier [17] and Gaffney in [6].

Gaffney in [6] and [3] uses this result to analyze mapping from the plane to plane and the plane to space. The author in [8] applies the results of Gaffney for mapping from 3-space to 3-space, in this case the number of invariants necessary has been reduced from 21 to 7. More recently Gaffney's approach has been used by Vohra [18], to study map germs from n-space ($n \geq 3$) to the plane.

In particular, the number of invariants required for Whitney equisingularity in each of these situations was shown to be smaller than the a priori number given by the general result of Gaffney [6].

In this paper we deal with the case of map germs from 3-space to 4-space. According to Gaffney's result, for a family $f_t : \mathbb{C}^3, 0 \to \mathbb{C}^4, 0$ to be Whitney equisingular it is needed the constancy of 20 invariants. We reduce this number to 8 for corank

1 germs. We do this by finding relations among the invariants and using the fact that these are upper semi-continuous. In section 2 we recall some results of [6] and prove our main result in section 3. We apply this result in section 4 to compute some invariants for normal forms of map germs given in [2].

2   BASIC DEFINITIONS AND RESULTS

We denote a holomorphic map germ $f : (\mathbb{C}^n, 0) \to (\mathbb{C}^p, 0)$ by writing $f \in \mathcal{O}(n, p)$. Our interest is primarily in finitely determined map-germs; a standard reference in this context is [19], although the material in section 1 of [6] is sufficient for our purposes. We denote by $\mathcal{R}$ the group of diffeomorphisms of the source $(\mathbb{C}^n, 0)$, and by $\mathcal{L}$ the group of diffeomorphisms of the target $(\mathbb{C}^p, 0)$. The action of the product $\mathcal{A} := \mathcal{R} \times \mathcal{L}$ leads to $\mathcal{A}$-equivalence of map germs: $f, g \in \mathcal{O}(n, p)$ are $\mathcal{A}$-equivalent if they are equivalent by smooth coordinate changes at source and target. Similarly the action of the semi direct product $\mathcal{K} := (\mathcal{R}.\mathcal{C})$ gives rise to $\mathcal{K}$-equivalence of map germs.

A germ is said to be k-$\mathcal{A}$-determined if any $g \in \mathcal{O}(n, p)$ with the same k-jet as f, i.e. $j^k g = j^k f$, is $\mathcal{A}$-equivalent to f. The germ f is said to be finitely $\mathcal{A}$-determined if it is k-$\mathcal{A}$-determined for some k.

A map-germ $f : \mathbb{C}^n, 0 \to \mathbb{C}^p, 0$ **is stable** if, up families of (bianalytic) diffeomorphisms in source and target, every deformation is trivial. That is, if f_t is a 1-parameter family with $f_0 = f$, then there should exist 1-parameter families φ_t and ψ_t of diffeomorphisms of source and target such that and $\psi_t \circ f \circ \varphi_t = f_t$. **Stable type** is the $\mathcal{A}$-equivalence class of stable germs. A finitely determined germ f has **discrete stable type** if there exist a versal unfolding of f in which only a finite number of stable types occur. If the numbers (n, p) are in Mather's "nice dimensions" (which is true for (3,4), our focus here) or on the boundary thereof, then every finitely determined germ $f \in \mathcal{O}(n, p)$ has discrete stable type.

2.1   Colength and multiplicities of ideals

The **colength** of a given ideal I in a complex analytic ring R, is defined as $\dim_{\mathbb{C}}(R/I)$; it may or not be finite. The **multiplicity** of an ideal I is an integer invariant denoted by $e(I)$ that is defined whenever I has finite colength. If $R = < R, m >$ is local and **Cohen-Macaulay**, if I is m-primary and a complete intersection, then the multiplicity of I is just its colength.

2.2   Finite maps and degree

A smooth map germ $f : (X, x) \to (Y, y)$ is said to be **finite** if the dimension of its local algebra is finite, i.e. if the number $m(f) := \dim_{\mathbb{C}} \frac{\mathcal{O}_{(X,x)}}{f^*(m_y)\mathcal{O}_{(X,x)}} < \infty$. Note that for f to be finite, it is necessary that $\dim X \leq \dim Y$. In the context of complex analytic geometry, we have the following important result for finite maps.

Let $f : (X, x) \to (Y, y)$ be an analytic map of analytic spaces of the same dimension, such that $f(X)$ is Zariski-dense in Y. Suppose $f(x)$ is a smooth point of Y, and $\{x\}$ a component of the fiber $f^{-1}(f(x))$. For open neighborhoods $U \subset X$ of x, and $V \subset Y$ and a closed analytic subset $B \subset Y$ such that:

(i) $V \setminus B$ is connected;

(ii) $f(U) \subset V$, $f|_U$ is proper, $f^{-1}(f(x)) = \{x\}$; and

(iii) $f|_{U \setminus f^{-1}(B)}$ smooth,

then the number of pre-images in U, counted with multiplicity, of any point $y \in V \setminus B$, is called the **degree** of f at x, denoted $\deg(f)$, for a proof see 3.12 in [15].

In particular if $X \subset (\mathbb{C}^n, 0)$ is an analytic space germ defined as the zero set of germs $g_1, ..., g_t$ with $\dim_0(X) = d$, and if $\mathcal{O}_X$ is Cohen-Macaulay, then we can often use a projection $\pi : (\mathbb{C}^n, 0) \to (\mathbb{C}^d, 0)$ such that $\deg(\pi|_X)$ is the colength of the ideal $(\pi_1, ..., \pi_d)$ in $\mathcal{O}_X$, i.e. $\dim_\mathbb{C} \frac{\mathcal{O}_n}{(\pi_1,...,\pi_d,g_1,...,g_t)}$.

2.3 Unfoldings

A 1-parameter unfolding F of f is a **good unfolding** if there exist neighborhoods U and W of the origin in $\mathbb{C} \times \mathbb{C}^n$ and in $\mathbb{C} \times \mathbb{C}^p$, respectively, such that $F^{-1}(W) = U$, F maps $(U \cap \Sigma(f)) \setminus \mathbb{C} \times 0$ to $W \setminus \mathbb{C} \times 0$ ($\Sigma(f)$ is the critical set of f), and if $(t_0, y_0) \in W \setminus \mathbb{C} \times 0$, with $S := F^{-1}(t_0, y_0) \cap \Sigma(F)$, then the germ $f_{t_0} : (\mathbb{C}^n, S) \to (\mathbb{C}^p, y_0)$ is a stable germ. We say that a good unfolding F of f is **excellent** if f is a finitely determined germ of discrete stable type, and the 0-stable invariants, see section 3 of [6], are constant in F, if $n = p$, then an additional requirement for excellence is that the degree of f be constant in the unfolding.

2.4 Whitney equisingularity and stratification of maps

A **Whitney stratification** of a given space is a stratification such that for any pair of strata S, S', with $S' \subset S$, the big stratum $\overline{S}$ is Whitney regular along S'. The local topological type remains constant along each stratum of a Whitney stratification of a given space. Note that a Whitney stratification always exists in the local complex analytic space, see section 1 of [5]. We say a given space X is **Whitney equisingular** along Y if there is a Whitney stratification of X with Y as a stratum.

Recall that if $F : \mathbb{C}^n \to \mathbb{C}^p$ is a morphism, and $A \subset \mathbb{C}^n$, $A' \subset \mathbb{C}^p$ subsets such that $F(A) \subset A'$, then a **stratification** of $F : A \to A'$ is a pair $(\mathcal{A}, \mathcal{A}')$ of stratifications of A and A' respectively, such that F maps strata submersively to strata. A given stratification $(\mathcal{A}, \mathcal{A}')$ of F is a regular stratification if $\mathcal{A}$, $\mathcal{A}'$ satisfy the Whitney regularity conditions, and all pairs of incident strata in the source satisfy Thom's A_f condition: *Let U be an open subset of some affine space, $f : U \to \mathbb{C}$ be an analytic function and M be a submanifold of U. Thom's A_f condition is satisfied between $U - \Sigma(f)$ and M if, whenever $p_i \in U - \Sigma(f)$, $p_i \to p \in M$, and $T_{p_i} V(f - f(p_i)) \to T$, then $T_p M \subseteq T$.*

If $F = f_t$ is an unfolding with parameter axis T, then a regular stratification $(\mathcal{A}, \mathcal{A}')$ of F is said to be a **Whitney equisingular** along T if T is a stratum of $\mathcal{A}$ and of $\mathcal{A}'$, ($\mathcal{A}$ and $\mathcal{A}'$ are Whitney equisingular along T). We also say in this case that F is a **Whitney equisingular map**.

The polar multiplicities of the polar varieties (defined by Teissier in [17]) of the stable types are the invariants needed to show the Whitney equisingularity of unfoldings.

DEFINITION 2.1 *Suppose $f : (X, 0) \to (S, 0)$ is a flat map smooth which fibers at every point of $X - sing(X)$. Let $p : \mathbb{C}^n \to \mathbb{C}^{d-k+1}$ be a linear projection such that $\ker p = D_{d-k+1}$ where D_{d-k+1} is a linear subspace of $(\mathbb{C}^n, 0)$ of dimension k. For*

$x \in X - sing(X)$, the fiber $X(f(x))$ is non-singular at x contained in $\{f(x)\} \times \mathbb{C}^n$ and one denotes by $\pi_x : X(f(x)) \to \mathbb{C}^{d-k+1}$ the restriction of p to $X(f(x))$. Let $P_k(f,p)$ be the closure of points $x \in X - sing(X)$ such that $x \in \Sigma(\pi_x)$, one calls the closed analytic subspace $P_k(f,p)$ of X, the relative polar variety of codimension k associate to D_{d-k+1}. If f is the constant map, we denote this by $P_k(X)$, called absolute polar variety.

The key invariant of $P_k(f,p)$ is its polar multiplicity which we denote by $m_k(X,f)$, if f is the constant map, we denote this by $m_0(P_k(X))$ or $m_k(X)$.

Gaffney in [6] page 195, defines a new invariant as following: Take a versal unfolding $F : \mathbb{C}^n \times \mathbb{C}^s \to \mathbb{C}^s \times \mathbb{C}^{n+1}$ of f. Specify a stable singularity type or stratum $\mathcal{D}(f)$ in source or target such that $\dim \overline{\mathcal{D}(f)} \geq 1$. Select D_1 a linear subspace of $(\mathbb{C}^{n+1}, 0)$ of dimension 1 and form $P_d(\mathcal{D}(F))$ the polar variety on $\mathcal{D}(F)$ with the projection $(p, \pi_s) : \mathbb{C}^n \times \mathbb{C}^s \to \mathbb{C} \times \mathbb{C}^s$ where $d = \dim(\overline{\mathcal{D}(F)}) - s$. The d-th stable multiplicity of f of type $\mathcal{D}(f)$, denoted $m_d(\mathcal{D}(f))$, is the multiplicity of $m_s \mathcal{O}_{\overline{\mathcal{D}(F)},(0,0)}$ in $\mathcal{O}_{\overline{\mathcal{D}(F)},(0,0)}$.

Using the polar multiplicities of the stable types and Thom's condition A_f, Gaffney showed the following principal result.

THEOREM 2.2 ([6] pp. 206-207) Suppose that $F : \mathbb{C} \times \mathbb{C}^n, (0,0) \to \mathbb{C} \times \mathbb{C}^p, (0,0)$ is an excellent unfolding of a finitely determined germ $f \in \mathcal{O}(n,p)$. Also suppose that the polar invariants of all the stable types defined in the discriminant $\Delta(f)$, in $\Sigma(f)$ and in $f^{-1}(\Delta(f)) - \Sigma(f)$ are constant at the origin for f_t. Then the unfolding is Whitney equisingular.

REMARK 2.3 1. The theorem also implies that such unfolding is topologicaly trivial; for the proof of this result Gaffney uses Thom's second isotopy lemma for complex analytic mappings, see [6] page 204.

2. ([18]) The theorem remains valid if we replace the term "an excellent unfolding" in the hypothesis by "a 1-parameter unfolding which, when stratified by stable types and by the parameter axis T, has only the parameter axis T as 1-dimensional stratum at the origin".

In the next section we shall need the following definition and results, these results are a key tool in finding relations among our invariants.

THEOREM 2.4 (Lê-Greuel, [10],[13]) Let X_1 be a complete intersection with isolated singularity at $0 \in \mathbb{C}^n$ (an ICIS). Let X be an ICIS defined in X_1 by $f_k = 0$, and let $f_1, ..., f_{k-1}$ be the generators of the ideal that defines X_1 at 0 in $\mathbb{C}^n$. Then

$$\mu(X_1, 0) + \mu(X, 0) = \dim_{\mathbb{C}} \frac{\mathcal{O}_n}{(f_1, ..., f_{k-1}, J(f_1, ..., f_k))}$$

REMARK 2.5 In the extreme case of a zero-dimensional ICIS we can appeal to the simpler formula: Let $f : (\mathbb{C}^k, 0) \to (\mathbb{C}^k, 0)$ be a germ such that $X = f^{-1}(0)$ is an ICIS. Then $\mu(X, 0) = \delta(f) - 1$, where $\delta(f) = \dim_{\mathbb{C}} \frac{\mathcal{O}_n}{f^*(m_n)\mathcal{O}_n}$, see [12] page 78.

Other elementary result that we appeal is: Let $f : (\mathbb{C}^n, 0) \to (\mathbb{C}^{n+1}, 0)$ be a finitely determined germ. Then $(f : \mathbb{C}^n, 0) \to f(\mathbb{C}^n) \subset (\mathbb{C}^{n+1}, 0)$ is bimeromorphic; see[4] page 138.

3   EQUISINGULARITY OF MAP GERMS IN $\mathcal{O}(3,4)$

3.1   The stable types in $\mathcal{O}(3,4)$

As highlighted in the introduction, our aim is to minimize the number of invariants defined in the stable types of f whose constancy in the family f_t implies the family is Whitney equisingular (therefore topologically trivial).

The strategy is to apply Theorem 2.2 and the techniques used by Gaffney in [6], that is, stratify the source and the target by the stable types and establish relations among the invariants on the strata. As these invariants are upper semi-continuous, the relations will allow us to reduce the number of invariants required in Gaffney's theorem.

For this we first give the following preliminary definition. Given a continuous mapping $f : X \to Y$ on analytic spaces, we define the k^{tk} multiple point space of f as

$$D^k(f) = \text{closure}\{(x_1, x_2, ..., x_k) \in X^k : f(x_1) = ... = f(x_k) \text{ for } x_i \neq x_j, i \neq j\}.$$

If $g : \mathbb{C}^n \to \mathbb{C}$ is a function then we define $V_i^k(g) : \mathbb{C}^{n+k-1} \to \mathbb{C}$ to be

$$
\begin{vmatrix}
1 & z_1 & \cdots & z_1^{i-1} & g(x, z_1) & z_1^{i+1} & \cdots & z_1^{k-1} \\
\vdots & \vdots & & \vdots & \vdots & \vdots & & \vdots \\
1 & z_k & \cdots & z_k^{i-1} & g(x, z_k) & z_k^{i+1} & \cdots & z_k^{k-1}
\end{vmatrix}
\Bigg/
\begin{vmatrix}
1 & z_1 & \cdots & z_1^{k-1} \\
\vdots & \vdots & & \vdots \\
1 & z_k & \cdots & z_k^{k-1}
\end{vmatrix}
$$

Suppose $f : \mathbb{C}^n, 0 \to \mathbb{C}^p, 0$, with $p \geq n$, is of corank 1 and is given in the form $f(x_1, ..., x_{n-1}, z) = (x_1, ..., x_{n-1}, h_1(x, z), ..., h_{p-n+1}(x, z))$.

THEOREM 3.1 ([14]) $D^k(f)$ is defined in $\mathbb{C}^{n+k-1}$ by the ideal $\mathcal{I}^k(f)$ generated by $V_i^k(h_j(x, z))$ for all $i = 1, ..., k - 1$ and $j = 1, ..., p - n + 1$.

In what follows we will take coordinates on $\mathbb{C}^{n+k-1} = \mathbb{C}^{n-1} \times \mathbb{C}^k$ to be $(x, z) = (x_1, ..., x_{n-1}, z_1, ..., z_k)$.

EXAMPLE 3.2 For a corank 1 map-germ $f : \mathbb{C}^n, 0 \to \mathbb{C}^p, 0$, $D^2(f)$ is defined by the ideal generated by the system $\left\{ \frac{h_i(x,z_1) - h_i(x,z_2)}{z_1 - z_2}, \ i = 1, ..., p - n + 1 \right\}$.

The main result of [14] is the theorem 2.14 where a description of $D^k(f)$ is obtained for a finitely $\mathcal{A}$-determined corank 1 map germ. There, it is shown that the multiple point spaces of f are ICIS. More precisely, f is finitely-determined if and only if for each k, with $p - k(p - n) \geq 0$, $D^k(f)$ is ICIS of dimension $p - k(p - n)$ or empty, and for those k with $p - k(p - n) < 0$, $D^k(f) = \{0\}$. Furthermore, f is stable if and only if the spaces are non-singular or empty.

DEFINITION 3.3 Let $\mathcal{P} = (r_1, r_2, ..., r_m)$ be a partition of k, i.e., $r_1 + r_2 + ... + r_m = k$, with $k = 1, ..., 4$. Let $\mathcal{I}(\mathcal{P})$ be the ideal in $\mathcal{O}_{n-1+k}$ generated by the $k - m$ elements $y_i - y_{i+1}$ for $r_1 + r_2 + ... + r_{j-1} + 1 \leq i \leq r_1 + r_2 + ... + r_j - 1, 1 \leq j \leq m$, and let $\Delta(\mathcal{P}) = V(\mathcal{I}(\mathcal{P}))$.

If $\mathcal{P}, \Re$ are two partitions of k, we say $\mathcal{P} < \Re$ if $\mathcal{I}(\mathcal{P}) \subsetneq \mathcal{I}(\Re)$. We define a generic point of $\Delta(\mathcal{P})$ for any partition $\Re$ of k with $\mathcal{P} < \Re$.

Define

$$\mathcal{I}^k(f,\mathcal{P}) = \mathcal{I}^k(f) + \mathcal{I}(\mathcal{P}), \quad D^k(f,\mathcal{P}) = V(\mathcal{I}^k(f,\mathcal{P})),$$

equipped with the sheaf structure $\mathcal{O}_{n-1+k}/\mathcal{I}^k(f,\mathcal{P})$.

Given a partition $\mathcal{P} = (r_1,...,r_m)$ of k, define projections $\pi_i(\mathcal{P}) : \mathbb{C}^{n-1+k} \to \mathbb{C}^n$, for $1 \le i \le m$, by $\pi_i(\mathcal{P})(x, z_1, ..., z_k) = (x, z_{r_1+...+r_{i-1}+1})$.

The geometric significance of $D^k(f,\mathcal{P})$ is given in Lemma 2.7 by Marar and Mond [14] page 559.

LEMMA 3.4 *Let* $\mathcal{P} = (r_1,...,r_m)$ *be a partition of* k*; at a generic point* (x,z) *of* $\Delta(\mathcal{P})$ *we have:*

$$\mathcal{I}_k(f,\mathcal{P}) = \mathcal{I}(\mathcal{P}) + \left\{ \frac{\partial^s f_j}{\partial z^s} \circ \pi_i(\mathcal{P}) \mid j = 1,...,p-n+1, 1 \le s \le r_i - 1, 1 \le i \le m \right\} +$$

$$\{f_j \circ \pi_1(\mathcal{P}) - f_j \circ \pi_i(\mathcal{P}) \mid j = 1,....p-n+1, 2 \le i \le m\} \text{ in } \mathcal{O}_{n-1+k}, (x,z).$$

In the corollary 2.15 of [14] page 562 it is shown the following result. If f is finitely determined then for each partition $\mathcal{P} = (r_1,...,r_m)$ of k satisfying $p-k(p-n+1)+m \ge 0$, the germ of $D^k(f,\mathcal{P})$ at 0 is either an ICIS of dimension $p-k(p-n+1)+m$, or is empty. Moreover, those $D^k(f,\mathcal{P})$ for $\mathcal{P}$ not satisfying the inequality consist at most of the single point 0.

With the definitions above, for any finitely determined germ $f \in \mathcal{O}(3,4)$ we denote by $D_1^2(f,\mathcal{P})$ and $D_1^k(f)$, with $k = 2,3$, the projections of $D^2(f,\mathcal{P})$ and $D^k(f)$ respectively, to the (x,y,z_1)-space. The stratifications in the source and target are as follows.

In the source: The set of critical points $\Sigma(f) = \mathbb{C}^3$, the set of double points $D_1^2(f)$, the set of triple points $D_1^3(f)$, the family of cross caps $D_1^2(f,2)$ and the zero-dimensional set $D_1^2(f,(2,1))$.

In the target: The discriminant of f, $\Delta(f) = f(\mathbb{C}^3)$, the image of the double points set $f(D^2(f))$, the image of the triples points set $f(D^3(f))$, the image $f(D^2(f,(2)))$ and the zero-dimensional stable types. These are, normal crossing of a hyperplane with the family of cross caps, denoted by (A_1C) and the set of quadruple points (Q).

There exist $k+1$ polar invariants associated to a k-dimensional variety. As $f(\mathbb{C}^3)$ is of dimension 3, the dimension of $D^3(f)$, $f(D^3(f))$, $D^2(f,(2))$ and $f(D^2(f,(2)))$ is 1, and the dimension of $D^2(f)$ and $f(D^2(f))$ is 2, there are 18 polar invariants defined on these sets. We also have 2 multiplicities of the zero-dimensional stable types. These are, the number of normal crossing of a hyperplane with the family of cross caps, denoted by $(\sharp A_1C(f))$ and the number of quadruple points $(\sharp Q(f))$. Therefore to apply Theorem 2.2 to germs in $\mathcal{O}(3,4)$ we needed the constancy of 20 invariants, we reduce this number to 8 invariants.

3.2 Relations among the invariants of the stable types in the target

We start analyzing the discriminant $f(\mathbb{C}^3)$. We have the following general result.

THEOREM 3.5 *Let $f \in \mathcal{O}(3,4)$ be a finitely determined germ. Then*

$$-m_3(f(\mathbb{C}^3)) + m_2(f(\mathbb{C}^3)) - m_1(f(\mathbb{C}^3)) + m_0(f(\mathbb{C}^3)) = 1.$$

Proof: We choose a generic linear projection $p_3 : \mathbb{C}^4 \to \mathbb{C}^3$ such that the degree of $p_3|f(\mathbb{C}^3)$ is equal to the multiplicity of $f(\mathbb{C}^3)$ at the origin, and the multiplicity of the polar variety $P_1(f(\mathbb{C}^3)) = \overline{\Sigma(p_3|f(\mathbb{C}^3)^0)}$ is (by definition) $m_1(f(\mathbb{C}^3))$. By the result of Teissier [16], $P_1(f(\mathbb{C}^3))$ has dimension 2. Then let $p_2 : \mathbb{C}^3 \to \mathbb{C}^2$ be a generic linear projection such that the degree of $p_2 \circ p_3|P_1(f(\mathbb{C}^3))$ is $m_1(f(\mathbb{C}^3))$. We can suppose that the projection $p_2 \circ p_3$ is also generic and defines the polar variety $P_2(f(\mathbb{C}^3)) = \overline{\Sigma(p_2 \circ p_3|f(\mathbb{C}^3)^0)}$. Choose $p_1 : \mathbb{C}^2 \to \mathbb{C}$ generic linear projection so that the degree of $p_1 \circ p_2 \circ p_3|P_2(f(\mathbb{C}^3))$ is $m_2(f(\mathbb{C}^3))$. Suppose that the projection $p_1 \circ p_2 \circ p_3$ is also generic and defines the multiplicity polar $m_3(f(\mathbb{C}^3))$. Let

$$X_1 = V(p_2 \circ p_3 \circ f),$$

$$X = V(p_3 \circ f).$$

As these varieties are ICIS, we apply Lê-Greuel theorem and obtain

$$\mu(X_1) + \mu(X) = \dim_{\mathbb{C}} \frac{\mathcal{O}_3}{(p_2 \circ p_3 \circ f, J[p_3 \circ f])}.$$

Applying Lê-Greuel again to $X_2 = V(p_1 \circ p_2 \circ p_3 \circ f)$ and X_1, we have

$$\mu(X_1) + \mu(X_2) = \dim_{\mathbb{C}} \frac{\mathcal{O}_3}{(p_1 \circ p_2 \circ p_3 \circ f, J(p_2 \circ p_3 \circ f))} = m_2(f(\mathbb{C}^3)).$$

As X is a 0-dimensional ICIS , we have $\mu(X) = \deg(p_3 \circ f) - 1$. As $f : \mathbb{C}^3 \to \mathbb{C}^4$ is bimeromorphic onto its image, $\mu(X) = m_0(f(\mathbb{C}^3)) - 1$ and as f is bimeromorphic we have that

$$\dim_{\mathbb{C}} \frac{\mathcal{O}_3}{(p_2 \circ p_3 \circ f, J(p_2 \circ p_3 \circ f))} = \deg(p_2 \circ p_3|P_2(f(\mathbb{C}^3))) = m_1(f(\mathbb{C}^3)),$$

$$\dim_{\mathbb{C}} \frac{\mathcal{O}_3}{(p_1 \circ p_2 \circ p_3 \circ f, J(p_2 \circ p_3 \circ f))} = \deg(p_1 \circ p_2 \circ p_3|P_2(f(\mathbb{C}^3))) = m_2(f(\mathbb{C}^3)).$$

Therefore

$$m_2(f(\mathbb{C}^3)) - \mu(X_2) + m_0(f(\mathbb{C}^3)) - 1 = m_1(f(\mathbb{C}^3)).$$

But $\mu(X_2) = m_3(f(\mathbb{C}^3))$. We take a versal unfolding $F : \mathbb{C}^3 \times \mathbb{C}^s \to \mathbb{C}^4 \times \mathbb{C}^s$ of f and consider $(p, \pi_s) : \mathbb{C}^4 \times \mathbb{C}^s \to \mathbb{C} \times \mathbb{C}^s$ a projection in target such that, $F : V(J(p_1 \circ p_2 \circ p_3 \circ \overline{f})) \subset \mathbb{C}^3 \times \mathbb{C}^s \to P_3(F(\mathbb{C}^3 \times \mathbb{C}^s)) \subset \mathbb{C}^3 \times \mathbb{C}^s$ is bimeromorphic, we have that the variety $V(J(p_1 \circ p_2 \circ p_3 \circ \overline{f}))$ has dimension s, since u is a generic parameter value, the $\deg(\pi_s|P_3(F(\mathbb{C}^3 \times \mathbb{C}^s))) = \deg(\pi_s|V(J(p_1 \circ p_2 \circ p_3 \circ \overline{f}))) = \mu(X_2) = m_3(f(\mathbb{C}^3))$.

$\square$

REMARK 3.6 (1). Note that in the kernel rank 1 case, $m_2(f(\mathbb{C}^3)) = 0$, since $\mu(X_2) = 0$, $m_3(f(\mathbb{C}^3)) = 0$, then $m_0(f(\mathbb{C}^3)) = \delta(f) - 1$.

(2). In the kernel rank 2 case $m_3(f(\mathbb{C}^3)) = 0$, so the constancy of $m_1(f(\mathbb{C}^3))$ implies the constancy of $m_2(f(\mathbb{C}^3)), m_0(f(\mathbb{C}^3))$ since all these invariants are upper semi-continuous.

(3). The alternating sum of polar multiplicities in the statement is reminiscent of an Euler obstruction for singular complex analytic spaces, there is a relationship between the Euler obstruction on stable types $f : \mathbb{C}^n \to \mathbb{C}^p$ and the the polar multiplicities; this result is in preparation by the author [9] .

The other strata in the target are $f(D^k(f))$, $k = 2, 3$. These have dimensions 2 and 1. Their polar multiplicities are $m_0(f(D^2(f)))$, $m_1(f(D^2(f)))$, $m_2(f(D^2(f)))$, $m_3(f(D^2(f)))$ and $m_0(f(D^3(f)))$, $m_1(f(D^3(f)))$.

To compute $m_2(f(D^2(f)))$, we take a versal unfolding $F : \mathbb{C}^3 \times \mathbb{C}^s \to \mathbb{C}^4 \times \mathbb{C}^s$ of f and consider $(p, \pi_s) : \mathbb{C}^4 \times \mathbb{C}^s \to \mathbb{C}^3 \times \mathbb{C}^s$ a projection in the source such that $F \circ (p, \pi_s) : D^2(F) \subset \mathbb{C}^4 \times \mathbb{C}^s \to F(D^2(F)) \subset \mathbb{C}^4 \times \mathbb{C}^s$. We know that $F \circ (p, \pi_s)|D^2(F)^0$ is a 2 fold cover of $F(D^2(F))^0$.

Choose $p_1 \circ p_2 : \mathbb{C}^4 \to \mathbb{C}$, a generic linear projection for $(F(D^2(F)), (p_1 \circ p_2, \pi_s))$. To work directly with

$$P_2(F(D^2(F)), (p_1 \circ p_2, \pi_s), D_1 \times \mathbb{C}^s) = \overline{\Sigma((p_1 \circ p_2, \pi_s)|F(D^2(F))^0)}$$

where $D_1 \times \mathbb{C}^s$ is the kernel of $(p_1 \circ p_2, \pi_s)$ we must work with

$$\overline{\Sigma((p_1 \circ p_2, \pi_s)|F(D^2(F))^0)}.$$

However, it is much easier to work with $\Sigma((p_1 \circ p_2, \pi_s) \circ F \circ (p, \pi_s)|D^2(F))$ which is

$$V(\mathcal{I}^2(F), J_x((p_1 \circ p_2, \pi_s) \circ F \circ (p, \pi_s), \mathcal{I}^2(F))),$$

where $\mathcal{I}^2(F)$ is the ideal that defines $D^2(F)$, we include the singular set of $D^2(F)$ in the critical set of $(p_1 \circ p_2, \pi_s) \circ F \circ (p, \pi_s)|D^2(F)$. This set has two advantages. It is in the source, and its equations are computable. Our strategy is to extract an invariant from it which will be simply related to $m_2(f(D^2(f)))$ and in fact which will control it.

The variety V has dimension s, since u is a generic parameter value, the degree of $\pi_s|V(\mathcal{I}^2(F), J_x((p_1 \circ p_2, \pi_s) \circ F \circ (p, \pi_s), \mathcal{I}^2(F)))$ is just the colength of m_s in the local ring of the source at $(0,0)$ and this is just

$$e_{D^2(f)}(f) = \dim_{\mathbb{C}} \frac{\mathcal{O}_4}{(\mathcal{I}^2(f), J_x((p_1 \circ p_2 \circ f \circ p, \mathcal{I}^2(f))))} \tag{3.1}$$

where $\mathcal{I}^2(f)$ defines $D^2(f)$.

We then have the following relation between $e_{D^2(f)}(f)$ and other invariants.

THEOREM 3.7 *Let $f \in \mathcal{O}(3,4)$ be a finitely determined germ of kernel rank 1. Then*

$$e_{D^2(f)}(f) = m_2(f(D^2(f))) + c_1 \sharp Q + c_2 \sharp(f),$$

where $c_1, c_2 \in \mathbb{Z}^+$

Proof: The components of the variety $V(\mathcal{I}^2(F), J_x((p_1 \circ p_2, \pi_s) \circ F \circ (p, \pi_s), \mathcal{I}^2(F)))$ are the closure of the set $F^{-1}(P_2(D^2(F)))$, the set $F^{-1}(Q)$, and the set $F^{-1}(A_1C)$. These have dimension at least s.

For a generic projection p_1, the dimension of V is s. As the multiplicity of V is the sum of the multiplicities of these components, it is enough to calculate the contribution of the degree of π_s restricted to each component, where π_s is the projection of $\mathbb{C}^4 \times \mathbb{C}^s$ to $\mathbb{C}^s$.

We choose neighborhoods U_1 of 0 in $\mathbb{C}^s$ and U_2 of 0 in $\mathbb{C}^4 \times \mathbb{C}^s$ such that at each point in U_1, π_s has $e_{D^2(f)}(f)$ pre-images in $V \cap U_2$ counting multiplicity. If $u \in \mathbb{C}^s$ is a generic parameter close to the origin we have

$$e_{D^2(f)}(f) = \sum_{x \in S} \dim_{\mathbb{C}} \frac{\mathcal{O}_{s+4,x}}{(m_s, \mathcal{I}^2(F), J_x((p_1, \pi_s) \circ F \circ (p, \pi_s), \mathcal{I}^2(F)))}$$

$$= \sum_{x \in S} \dim_{\mathbb{C}} \frac{\mathcal{O}_{4,x}}{(\mathcal{I}^2(f_s), J_x(p_1 \circ f_s \circ p, \mathcal{I}^2(f_s)))}$$

where $S = \pi_s^{-1}(0) \cap V$.

We can compute this last number using normal forms. Hence the singularities of type Q and A_1C contributes with $c_1 \sharp Q(f)$ and $c_2 \sharp A_1 C(f)$ to $e_{D^2(f)}(f)$, where $c_1, c_2 \in \mathbb{Z}^+$. A similar argument shows that $F^{-1}(P_2(F(D^2(F))))$ contributes with $m_2(f(D^2(f)))$. Since F restricted to each type of component is bimeromorphic and finite, and the generic point of each component is reduced, we have that

$$\deg(\pi_{|V}) = m_2(f(D^2(f))) + c_1 \sharp Q(f) + c_2 \sharp A_1 C(f)$$

□

PROPOSITION 3.8 *Let $f \in \mathcal{O}(3,4)$ be a finitely determined germ of corank 1. Then*

$$e_{D^2(f)}(f) = -2m_0(f(D^2(f))) + 2m_1(f(D^2(f))) + \mu(D^2(f)) + 1$$

Proof: The stratum in the target is $f(D^2(f))$, it has dimension 2 so there are 3 polar invariants $m_0(f(D^2(f)))$, $m_1(f(D^2(f)))$ and $m_2(f(D^2(f)))$. Since $f|D_1^2(f) - \{0\}$ is a 2-fold cover of $f(D^2(f)) - \{0\}$ it follows that $e((f^*(m_4))\mathcal{O}_{D^2(f),0}) = 2m_0(f(D^2(f)))$.

To compute $m_1(f(D^2(f)))$ choose $p_2 : \mathbb{C}^4 \to \mathbb{C}^2$, a generic linear projection for $f(D^2(f))$ such that the polar variety $P_1(f(D^2(f))) = \overline{\Sigma(p_2|f(D^2(f))^0)}$ and the multiplicity of $P_1(f(D^2(f)))$ is $m_1(f(D^2(f)))$.

Let $p_1 : \mathbb{C}^2 \to \mathbb{C}$ be a linear projection such that the degree of $p_1 \circ p_2|P_1(f(D^2(f)))$ is $m_1(f(D^2(f)))$. We also require that $p_1 \circ p_2$ be a generic projection that defines $m_2(f(D^2(f)))$. Let

$$X_1 = V(p_1 \circ p_2 \circ f \circ p, \mathcal{I}^2(f)),$$

$$X = V(p_2 \circ f \circ p, \mathcal{I}^2(f)).$$

These spaces are ICIS, where $\mathcal{I}^2(f)$ is the ideal that defined $D^2(f)$ and $p : D^2(f) \subset \mathbb{C}^4 \to \mathbb{C}^3$ a projection of $D^2(f)$ to the source of f.

Then by Lê-Greuel theorem we have

$$\mu(X_1) + \mu(X) = \dim_{\mathbb{C}} \frac{\mathcal{O}_4}{(p_1 \circ p_2 \circ f \circ p, \mathcal{I}^2(f), J(p_2 \circ f \circ p, \mathcal{I}^2(f)))} \qquad (3.2)$$

By our genericity assumption on p_1 and p_2, the right-hand side of (3.2) is just $2m_1(f(D^2(f)))$. Applying Lê-Greuel theorem again to the following spaces that are ICIS

$$X_1 = V(p_1 \circ p_2 \circ f \circ p, \mathcal{I}^2(f)), \text{ and } X_2 = D^2(f)$$

We get

$$\mu(X_1) + \mu(X_2) = \dim_{\mathbb{C}} \frac{\mathcal{O}_4}{(\mathcal{I}^2(f), J(p_1 \circ p_2 \circ f \circ p, \mathcal{I}^2(f)))}.$$

Then $\mu(X_1) = \dim_{\mathbb{C}} \frac{\mathcal{O}_4}{(\mathcal{I}^2(f), J(p_1 \circ p_2 \circ f \circ p, \mathcal{I}^2(f)))} - \mu(D^2(f))$.

In (3.2), we obtain

$$\dim_{\mathbb{C}} \frac{\mathcal{O}_4}{(\mathcal{I}^2(f), J(p_1 \circ p_2 \circ f \circ p, \mathcal{I}^2(f)))} = 2m_1(f(D^2(f))) + \mu(D^2(f)) - \mu(X)$$

Since X is ICIS of dimension zero we have that $\mu(X) = \deg(p_2 \circ f \circ p, \mathcal{I}^2(f)) - 1$. As $\mathcal{O}_{D^2(f)}$ is a Cohen Macaulay ring

$$\mu(X) = \dim_{\mathbb{C}} \frac{\mathcal{O}_4}{(p_2 \circ f \circ p, \mathcal{I}^k(f))} - 1 = 2m_0(f(D^2(f))) - 1.$$

Since $f : \mathbb{C}^3 \to \mathbb{C}^4$ is bimeromorphic

$$e_{D^2(f)}(f) = m_1(f(D^2(f))) + \mu(D^2(f)) - 2m_0(f(D^2(f))) + 1.$$

□

We now deal with the stratum $f(D^3(f))$. This has dimension 1 so there are two invariants $m_0(f(D^3(f)))$ and $m_1(f(D^3(f)))$. Since $p \circ f|D^3(f)^0$ is a 3-fold cover of $f(D^3(f))^0$ it follows that $e(f^*(m_4)\mathcal{O}_{D^3(f),0}) = 3!m_0(f(D^3(f)))$.

To compute $m_1(f(D^3(f)))$, we take a versal unfolding $F : \mathbb{C}^3 \times \mathbb{C}^s \to \mathbb{C}^4 \times \mathbb{C}^s$ and consider $(p, \pi_s) : \mathbb{C}^5 \times \mathbb{C}^s \to \mathbb{C}^3 \times \mathbb{C}^s$ a projection in the source, such that $F \circ (p, \pi_s) : D^3(F) \subset \mathbb{C}^5 \times \mathbb{C}^s \to F(D^3(F)) \subset \mathbb{C}^4 \times \mathbb{C}^s$. We know also that $F \circ (p, \pi_s)|D^3(F)^0$ is a 3 fold cover $F(D^3(F))^0$.

Choose $p_1 : \mathbb{C}^4 \to \mathbb{C}$, a generic linear projection for $(F(D^3(F)), (p_1, \pi_s))$. To work with

$$P_1(F(D^3(F)), (p_1, \pi_s), D_1 \times \mathbb{C}^s) = \overline{\Sigma((p_1, \pi_s)|F(D^3(F))^0)}$$

where $D_1 \times \mathbb{C}^s$ is the kernel of (p_1, π_s) we must work with

$$\overline{\Sigma((p_1, \pi_s)|F(D^3(F))^0)}.$$

However, it is much easier to work with $\Sigma((p_1, \pi_s) \circ F \circ (p, \pi_s)|D^3(F))$ which is

$$V(\mathcal{I}^3(F), J_x((p_1, \pi_s) \circ F \circ (p, \pi_s), \mathcal{I}^3(F)))$$

where $\mathcal{I}^3(F)$ is the ideal that defines $D^3(F)$, we include the singular set of $D^3(F)$ in the critical set of $(p_1, \pi_s) \circ F \circ (p, \pi_s)|D^3(F)$.

So V has dimension s, since u is a generic parameter value, the degree of

$$\pi_s|V(\mathcal{I}^3(F), J_x((p_1, \pi_s) \circ F \circ (p, \pi_s), \mathcal{I}^3(F)))$$

is just the colength of m_s in the local ring of the source at $(0,0)$ and this is

$$e_{D^3(f)}(f) = \dim_\mathbb{C} \frac{\mathcal{O}_5}{(\mathcal{I}^3(f), J_x((p_1 \circ f \circ p, \mathcal{I}^3(f))))} \qquad (3.3)$$

where $\mathcal{I}^3(f)$ defines $D^3(f)$.

We then have the following relation between $e_{D^3(f)}(f)$ and the invariants of the zero stable types.

THEOREM 3.9 *Let $f \in \mathcal{O}(3,4)$ be a finitely determined germ of corank* 1. *Then*

$$e_{D^3(f)}(f) = 4!b_1 \sharp Q(f) + b_2 \sharp A_1 C(f) + 3!m_1(f(D^3(f)))$$

where $b_1, b_2 \in \mathbb{Z}^+$

Proof: Using (3.3) and as in theorem above, we obtain the result. □

THEOREM 3.10 *Let $f \in \mathcal{O}(3,4)$ be a finitely determined germ of corank* 1. *Then*

$$\mu(D^3(f)) + 3!m_0(f(D^3(f))) - 1 = e_{D^3(f)}(f).$$

Proof: Since f is finitely determined, the ideal $\mathcal{I}^3(F)$ of a versal unfolding F of f, has the property that $\mathcal{I}^3(f_0) = 0$ is the defining equation of $D^3(f) \subset \mathbb{C}^5$ with the reduced structure. Hence Lê-Greuel theorem applies to the following varieties

$$X_1 = D^3(f)$$

$$X = V(p_1 \circ f \circ p, \mathcal{I}^3(f))$$

that are ICIS, where $p_1 : \mathbb{C}^4 \to \mathbb{C}$ is a generic projection that defines $e_{D^3(f)}(f)$, and $m_1(f(D^3(f)))$ and $p : \mathbb{C}^5 \to \mathbb{C}^3$ is a projection of $D^3(f)$ to the source of f.

So we have

$$\mu(X_1) + \mu(X) = \dim_\mathbb{C} \frac{\mathcal{O}_5}{(\mathcal{I}^3(f), J(\mathcal{I}^3(f), p_1 \circ f \circ p))}.$$

As X has dimension zero, it follows by 2.5 that

$$\mu(X) = \deg(p_1 \circ f \circ p, \mathcal{I}^3(f)) - 1.$$

Since $\mathcal{O}_{D^3(f)}$ is a Cohen Macaulay ring, we obtain

$$\deg(p_1 \circ f \circ p, \mathcal{I}^3(f)) = \dim_\mathbb{C} \frac{\mathcal{O}_5}{(\mathcal{I}^3(f), p_1 \circ f \circ p)}$$

and as f is bimeromorphic and finite

$$\dim_\mathbb{C} \frac{\mathcal{O}_5}{(\mathcal{I}^3(f), p_1 \circ f \circ p)} = 3!m_0(f(D^3(f))),$$

and the result follows. □

By the definition of $D^2(f, 2)$ we consider it as a subset of $\mathbb{C}^3$, then the stratum in the target is $f(D^2(f, 2))$. This has dimension 1 so there are two invariants $m_0(f(D^2(f, 2)))$ and $m_1(f(D^2(f, 2)))$. Since $f|D^2(f, 2)^0$ is a 1-fold cover of $f(D^2(f, 2))^0$ it follows that $e(f^*(m_4)\mathcal{O}_{D^2(f,2),0}) = m_0(f(D^2(f, 2)))$.

To compute $m_1(f(D^2(f, 2)))$, we take a versal unfolding $F : \mathbb{C}^3 \times \mathbb{C}^s \to \mathbb{C}^4 \times \mathbb{C}^s$ of f, We know also that $f|D^2(F, 2)^0$ is a 1-fold cover $F(D^2(F, 2))^0$.

Choose $p_1 : \mathbb{C}^4 \to \mathbb{C}$, a generic linear projection for $(F(D^2(F, 2)), (p_1, \pi_s))$. To work with

$$P_1(F(D^2(F, 2)), (p_1, \pi_s), D_1 \times \mathbb{C}^s) = \overline{\Sigma((p_1, \pi_s)|F(D^2(F, 2))^0)}$$

where $D_1 \times \mathbb{C}^s$ is the kernel of (p_1, π_s) we must work with

$$\overline{\Sigma((p_1, \pi_s)|F(D^2(F, 2))^0)}.$$

However, it is much easier to work with $\Sigma((p_1, \pi_s) \circ f|D^2(F, 2))$ which is

$$V(\mathcal{I}^2(F, 2), J_x((p_1, \pi_s) \circ F, \mathcal{I}^2(F, 2)))$$

where $\mathcal{I}^2(F, 2)$ is the ideal that defines $D^2(F, 2)$, we include the singular set of $D^2(F, 2)$ in the critical set of $(p_1, \pi_s) \circ f|D^2(F, 2)$. This set is in the source and its equations are computable. We follow the same strategy in the case of the set of double points to extract $m_1(f(D^2(f, 2)))$.

The variety V has dimension s, since s is a generic parameter value, the degree of $\pi_s|V(\mathcal{I}^2(F, 2), J_x((p_1, \pi_s) \circ F, \mathcal{I}^2(F, 2)))$ is just the colength of m_s in the local ring of the source at $(0, 0)$ and this is

$$e_{D^2(f,2)}(f) = \dim_{\mathbb{C}} \frac{\mathcal{O}_3}{(\mathcal{I}^2(f, 2), J_x((p_1 \circ f, \mathcal{I}^2(f, 2)))} \tag{3.4}$$

where $\mathcal{I}^2(f, 2)$ defines $D^2(f, 2)$.

We then have the following relation between $e_{D^2(f,2)}(f)$ and the invariants of the zero stables types.

THEOREM 3.11 *Let $f \in \mathcal{O}(3, 4)$ be a finitely determined germ of corank 1. Then $e_{D^2(f,2)}(f) = m_1(f(D^2(f, 2)))$*

Proof: To understand the geometric meaning of the right hand side of the equality (3.4) we choose a versal unfolding $F = (s, \overline{f}(s, x))$ of f and consider the variety $V(\mathcal{I}^2(F, 2), J_x(p_1 \circ F, \mathcal{I}^2(F, 2)))$. The components of this variety are the closure of the sets $F^{-1}(P_1(D^2(F, 2)))$, the set $F^{-1}(Q)$ and the set $F^{-1}(A_1 C)$. These have dimension at least s.

For a generic projection p_1, the dimension of V is s. As the multiplicity of V is the sum of the multiplicities of these components, it is enough to calculate the contribution of the degree of π_s restricted to each component, where π_s is the projection from $\mathbb{C}^3 \times \mathbb{C}^s$ to $\mathbb{C}^s$.

We choose neighborhoods U_1 of 0 in $\mathbb{C}^s$ and U_2 of 0 in $\mathbb{C}^3 \times \mathbb{C}^s$ such that at each point in U_1, π_s has $e_{D^2(f,2)}(f)$ pre-images in $V \cap U_2$ counting multiplicity. If $u \in \mathbb{C}^s$ is a generic parameter close to the origin we have

$$\begin{aligned}
e_{D^2(f,2)}(f) &= \sum_{x \in S} \dim_{\mathbb{C}} \frac{\mathcal{O}_{s+3,x}}{(m_s, \mathcal{I}^2(F,2), J_x((p_1, \pi_s) \circ F, \mathcal{I}^2(F,2)))} \\
&= \sum_{x \in S} \dim_{\mathbb{C}} \frac{\mathcal{O}_{3,x}}{(\mathcal{I}^2(f_s, 2), J_x(p_1 \circ f_s, \mathcal{I}^2(f_s, 2)))} \tag{3.5}
\end{aligned}$$

where $S = \pi_s^{-1}(0) \cap V$. To count the contribution of singularities of type A_1C, we take $f_s = \{(x, y, z, 0); (x, y, z^2, xz)\}$.

Then, to calculate the contribution of the singularities of type A_1C in $V(\mathcal{I}^2(F, 2), J_x((p_1, \pi_s) \circ F, \mathcal{I}^2(F, 2)))$ it is enough to use the ideal $\mathcal{I}^2(f, 2) = (2z, x)$. Using the formula (3.5) we have

$$e_{D^2(f,2)}(f) = \sum_{x \in S} \dim_{\mathbb{C}} \frac{\mathcal{O}_{3,x}}{(\mathcal{I}^2(f_s, 2), J_x(p_1 \circ f_s \circ p, \mathcal{I}^2(f_s, 2)))} = 0,$$

where $p(x, y, , z, z_1, z_2) = ax + by + cz$ is a generic projection ($b \neq 0$). This means that the points of type A_1C appear in V with a contribution equals to 0.

In the same way, to count the contribution of the points Q it is enough to choose the normal form $f_s = \{(x, y, z, 0), (0, x, y, z), (x, 0, y, z), (x, y, 0, z)\}$. Then the contribution of Q in V is also 0.

Since F restricted to each component is finite and bimeromorphic, we have

$$\deg(\pi_s | V) = m_1(f(D^2(f, 2))).$$

The theorem follows now by joining all the above equalities. $\qquad\qquad\square$

THEOREM 3.12 *Let* $f \in \mathcal{O}(3, 4)$ *be a finitely determined germ of corank* 1. *Then*

$$\mu(D^2(f, 2)) + m_0(f(D^2(f, 2))) - 1 = e_{D^2(f,2)}(f).$$

Proof: Since f is finitely determined of corank 1, the ideal $\mathcal{I}^2(F, 2)$ of a versal unfolding F of f has the property that $\mathcal{I}^2(f_0, 2) = 0$ is the defining equation of $D^2(f, 2)$ and we can consider $\mathcal{I}^2(f_0, 2) \subset \mathcal{O}_3$, therefore $D^2(f, 2) \subset \mathbb{C}^3$ with the reduced structure. Hence Lê-Greuel theorem applies for

$$X_1 = D^2(f, 2)$$

$$X = V(p_1 \circ f \circ p, \mathcal{I}^2(f, 2))$$

which are ICIS, where $p_1 : \mathbb{C}^4 \to \mathbb{C}$ a generic projection.

So we have

$$\mu(X_1) + \mu(X) = \dim_{\mathbb{C}} \frac{\mathcal{O}_3}{(\mathcal{I}^2(f, 2), J(\mathcal{I}^2(f, 2), p_1 \circ f))}.$$

As X has dimension zero, it follows that

$$\mu(X) = \deg(p_1 \circ f, \mathcal{I}^2(f, 2)) - 1.$$

Since $\mathcal{O}_{D^2(f,2)}$ is a Cohen Macaulay ring, we obtain

$$\deg(p_1 \circ f, \mathcal{I}^2(f, 2)) = \dim_{\mathbb{C}} \frac{\mathcal{O}_3}{(\mathcal{I}^2(f, 2), p_1 \circ f)}$$

and as f is bimeromorphic and finite $\dim_{\mathbb{C}} \frac{\mathcal{O}_3}{(\mathcal{I}^2(f,2), p_1 \circ f)} = m_0(f(D^2(f, 2)))$, and the result follows. $\qquad\qquad\square$

3.3   Relations among the invariants of the stable types in the source

In this section we stablish relations among the invariants of the stable types in the source. The strata in this case are the set of double points, the curve of triple points, the 0-dimensional stable types and the critical points set ($\mathbb{C}^3$), minus the previous strata. The situation is less difficult than in the case of the target because the set of double points and the curve of the triple points are ICIS of corank 1 germs .

We know from [16] that the absolute polar multiplicities of a hypersurface X with isolated singularity are related to the Milnor numbers $\mu^{(k)}$ of the plane sections ($\mu^{(k)}(X) = \mu(X \cap H^k)$) by the following equalities

$$m_k(X) = \mu^{(k+1)}(X) + \mu^{(k)}(X),$$

for $0 \leq k \leq d - 1$, where $d = \dim(X)$. This result is also valid for ICIS (see [10], [7]). The absolute polar multiplicities are defined when the dimension of X is ≥ 1. The multiplicity $m_d(X)$ cannot be defined directly like the other m_k, $0 \leq k \leq d - 1$, because the singularities of $p_1|X$ are isolated points. However, Gaffney [7] defines this multiplicity for spaces that are ICIS as follows.

DEFINITION 3.13 *The d-th polar multiplicity of $(X^d, 0)$ (X^d is ICIS of dimension d), denoted by $m_d(X^d)$, is defined by*

$$m_d(X^d) = \dim_{\mathbb{C}} \frac{\mathcal{O}_X}{J(p_1, f)}$$

where $f : (\mathbb{C}^n, 0) \to (\mathbb{C}^{n-d}, 0)$, $f^{-1}(0) = X^d$ and $p_1 : \mathbb{C}^n \to \mathbb{C}$ is a generic linear projection.

REMARK 3.14 As $V(p_1, f)$ is ICIS, then by Lê-Greuel theorem, we have

$$m_d(X^d) = \mu(X^d) + \mu(X^d \cap p_1^{-1}(0)).$$

When $f \in \mathcal{O}(3, 4)$ is finitely determined and of corank 1, the double points set $D^2(f)$ and the triple points curve $D^3(f)$ are ICIS. Therefore we can apply Definition 3.13 and all the properties above to obtain the following.

PROPOSITION 3.15 *Let $f \in \mathcal{O}(3, 4)$ be a finitely determined germ of corank 1. Then*

(i)

$$m_2(D^2(f)) - m_1(D^2(f)) + m_0(D^2(f)) = \mu^{(3)}(D^2(f)) + 1$$

(ii)

$$m_1(D^3(f)) - m_0(D^3(f)) = \mu^{(2)}(D^3(f)) - 1$$

(iii)

$$m_1(D^2(f, 2)) - m_0(D^2(f, 2)) = \mu^{(2)}(D^2(f, 2)) - 1$$

These equalities in the above proposition are the same for $D_1^2(f)$, $D_1^3(f)$, and $D_1^2(f, (2))$, therefore we can now deduce our main theorem. Using the results of subsections 3.2 and 3.3, we reduce the number of invariants in Gaffney's theorem 2.2 from 20 to 8 in the corank 1 case.

THEOREM 3.16 *Suppose that $f \in \mathcal{O}(3,4)$ is a finitely determined germ of corank 1 and $F = (t, f_t)$ is a good 1-parameter unfolding. Then F is Whitney equisingular along $T = \mathbb{C} \times \{0\}$ if, and only if, $m_0(f_t(\mathbb{C}^3))$, $e_{D^2(f_t)(f_t)}$, $m_0(f(D^2(f_t)))$, $e_{D^3(f_t)(f_t)}$, $m_1(D^2(f_t))$, $m_0(f_t(D^2(f_t, 2)))$, $m_0(f_t(D^3(f_t)))$, $m_0(D^2(f_t, 2))$ are constant for t close to the origin.*

THEOREM 3.17 *Suppose that $f \in \mathcal{O}(3,4)$ is a finitely determined germ and F an excellent unfolding of f. Then F is Whitney equisingular along $T = \mathbb{C} \times \{0\}$ if, and only if, $m_1(f(\mathbb{C}^3))$, the polar multiplicities of the 1-dimensional and the 2-dimensional stable types in the source and in the target are constant for t close to the origin.*

4   POLAR MULTIPLICITIES OF GERMS IN $\mathcal{O}(3,4)$

In [2] Houston and Kirk classified germs from $\mathbb{C}^3, 0 \to \mathbb{C}^4, 0$. The models for these germs are the following:

Label	Singularity
–	$(x, y, z, 0)$
E_7	$(x, y, z^2, z(z^2 + x^3 + xy^3))$
A_0	(x, y, z^2, xz)
E_8	$(x, y, z^2, z(z^2 + x^3 + y^5))$
A_k	$(x, y, z^2, z(z^2 + x^2 + y^{k+1})$
B_k	$(x, y, z^2, z(x^2 + y^3 + z^{2k}))$
D_k	$(x, y, z^2, z(z^2 + x^2y + y^{k-1})$
C_k	$(x, y, z^2, z(x^2 + yz^2 + y^k))$
E_6	$(x, y, z^2, z(z^2 + x^3 + y^4))$
F_4	$(x, y, z^2, z(x^2 + y^3 + z^4))$

We apply in this section the results of the previous sections to calculate the polar multiplicities of the stable types $f(\mathbb{C}^3)$ and $f(D^2(f))$.

PROPOSITION 4.1 *Let $f \colon \mathbb{C}^3, 0 \to \mathbb{C}^4, 0$ be one of the normal forms given above. Then we have the following:*

Label	$m_0(f(\mathbb{C}^3))$	$m_1(f(\mathbb{C}^3))$	$e_{D^2(f)}$	$m_0(f(D^2(f)))$	$m_1(f(D^2(f)))$	$\mu(D^2(f))$
A_0	2	1	–	–	–	–
A_k	2	1	$k+1$	1	2	k
D_k	2	1	$k+2$	1	3	k
E_6	2	1	9	1	4	6
E_7	2	1	9	1	3	7
E_8	2	1	10	1	3	8
B_k	2	1	$4k+1$	2	$2k+1$	$2k$
C_k	2	1	$4k-2$	2	$3k$	$k+1$
F_4	2	1	9	2	6	6

Proof: Let $f(x, y, z) = (x, y, z^2, z(z^2 + x^2 + y^{k+1}))$ be a germ of type A_k. To calculate $m_i(f(\mathbb{C}^3))$, $i = 0, 1, 2, 3$ according to Theorem 3.16 and Remark 3.6 we have to calculate only $m_0(f(\mathbb{C}^3))$ and $m_1(f(\mathbb{C}^3))$. We calculate first the polar multiplicity $m_0(f(\mathbb{C}^3))$:

$$m_0(f(\mathbb{C}^3)) = \dim_{\mathbb{C}} \frac{\mathcal{O}_3}{(p_3 \circ f)}$$

where p_3 is a generic projection. We take $p_3(x,y,z,w) = (x,y,z)$. Then

$$
\begin{aligned}
m_0(f(\mathbb{C}^3)) &= \dim_{\mathbb{C}} \frac{\mathcal{O}_3}{(x,y,z^2)} \\
&= 2
\end{aligned}
$$

We compute now the multiplicity $m_1(f(\mathbb{C}^3))$. As p_3 is generic, we choose the generic linear projection $p_2(x,y,z) = (x,y)$. Then

$$
\begin{aligned}
m_1(f(\mathbb{C}^3)) &= \dim_{\mathbb{C}} \frac{\mathcal{O}_3}{(p_2 \circ p_3 \circ f, J[p_3 \circ f])} \\
&= \dim_{\mathbb{C}} \frac{\mathcal{O}_3}{(x,y,z)} \\
&= 1
\end{aligned}
$$

In the same way, we can calculate $m_0(f(\mathbb{C}^3))$ and $m_1(f(\mathbb{C}^3))$ for the other normal forms above.

To calculate the multiplicity $m_i(f(D^2(f)))$, $i = 0,1$, and $e_{D^2(f)}(f)$ we need to find the ideal that defines the curve $D^2(f)$.

As $2m_0(f(D^2(f))) = \dim_{\mathbb{C}} \frac{\mathcal{O}_3}{(p_2 \circ f, \mathcal{I}_1^2(f))}$ where $p_2(x,y,z,w) = (x,y)$ and $\mathcal{I}_1^2(f) = z^2 + x^2 + y^{k+1}$, then $2m_0(f(D^2(f))) = \dim_{\mathbb{C}} \frac{\mathcal{O}_3}{(x,y,z^2+x^2+y^{k+1})} = 2$.

We now calculate $m_1(f(D_1^2(f)))$. We have

$$
\begin{aligned}
m_1(f(D_1^2(f))) &= \dim_{\mathbb{C}} \frac{\mathcal{O}_3}{(p_1 \circ p_2 \circ f, \mathcal{I}_1^2(f), J[p_2 \circ f, \mathcal{I}_1^2(f)])} \\
&= \dim_{\mathbb{C}} \frac{\mathcal{O}_3}{(x, z^2+x^2+y^{k+1}, J[x,y,z^2+x^2+y^{k+1}])} \\
&= 2.
\end{aligned}
$$

As the Milnor number of $D_1^2(f) = z^2 + x^2 + y^{k+1}$ is k, it follows from Proposition 3.8, that $e_{D^2(f)}(f) = k+1$.

Let $f(x,y,z) = (x,y,z^2, z(z^2 + x^2y + y^{k-1}))$. We compute the multiplicities $m_i(f(D^2(f)))$, $i = 0,1$, and $e_{D^2(f)}(f)$, for this we need to find the ideal that defines the curve $D^2(f)$.

As $2m_0(f(D^2(f))) = \dim_{\mathbb{C}} \frac{\mathcal{O}_3}{(p_2 \circ f, \mathcal{I}_1^2(f))}$ where $p_2(x,y,z,w) = (x+y, x-y)$ and $\mathcal{I}_1^2(f) = z^2 + x^2y + y^{k-1}$, then $2m_0(f(D^2(f))) = \dim_{\mathbb{C}} \frac{\mathcal{O}_3}{(x+y,x-y,z^2+x^2y+y^{k-1})} = 2$.

We now calculate $m_1(f(D_1^2(f)))$. Choosing $p_1(x,y,z) = x + y$ which is generic, we have

$$
\begin{aligned}
m_1(f(D_1^2(f))) &= \dim_{\mathbb{C}} \frac{\mathcal{O}_3}{(p_1 \circ p_2 \circ f, \mathcal{I}_1^2(f), J[p_2 \circ f, \mathcal{I}_1^2(f)])} \\
&= \dim_{\mathbb{C}} \frac{\mathcal{O}_3}{(x+y, z^2+x^2y+y^{k-1}, J[x+y,x-y,z^2+x^2y+y^{k-1}])} \\
&= 3.
\end{aligned}
$$

As the Milnor number of $D_1^2(f) = z^2 + x^2y + y^{k-1}$ is k, it follows from the Proposition 3.8 that $e_{D^2(f)}(f) = k+2$.

The invariants of the other normal forms are computed in the same way as above.

$\square$

5   APPLICATION

EXAMPLE 5.1 Let $F(t, x, y, z) = (t, x, y, z^3 + xz, yz^2 + z^5 + z^6 + tz^7)$, unfolding of the germ, $f(x, y, z) = (x, y, z^3 + xz, yz^2 + z^5 + z^6)$. Then the family $f_t(x, y, z) = F(t, x, y, z)$ is Whitney equisingular for $0 < t < 1/2$.

According to Theorem 3.16 we have to calculate 7 invariants. We calculate first the multiplicity of the discriminant $m_0(f_t(\mathbb{C}^3))$:

$$m_0(f_t(\mathbb{C}^3)) = \dim_\mathbb{C} \frac{\mathcal{O}_3}{(p_3 \circ f_t)} = \dim_\mathbb{C} \frac{\mathcal{O}_3}{(x, y, xz + xz + z^3)} = 3, \forall t,$$

where $p_3(x, y, z) = (x, y)$ is a generic projection.

Now we compute the invariants $e_{D^2(f_t)}(f_t)$, $m_0(f_t(D^2(f_t)))$ and $m_1(D^2(f_t))$, for this we need the ideal that defines $D^2(f_t)$, which is

$$I^2(f_t) = (x + z_1^2 + z_1 z_2 + z_2^2, y(z_1 + z_2) + P_4(z_1, z_2) + P_5(z_1, z_2) + tP_6(z_1, z_2)),$$

with $P_4(z_1, z_2) = \Sigma_{k=0}^4 z_1^{4-k} z_2^k$, $P_5(z_1, z_2) = \Sigma_{k=0}^5 z_1^{5-k} z_2^k$ and $P_6(z_1, z_2) = \Sigma_{k=0}^6 z_1^{6-k} z_2^k$.

We compute now the multiplicity $m_1(D^2(f_t))$. As $D^2(f_t)$ is ICIS and isomorphic to $y(z_1 + z_2) + P_4(z_1, z_2) + P_5(z_1, z_2) + tP_6(z_1, z_2))$, then for the relation

$$m_1(D^2(f_t)) = \mu^{(2)}(D^2(f_t)) + \mu^{(1)}(D^2(f_t))$$

we have $\mu^{(2)}(D^2(f_t)) = 1$ and $\mu^{(1)}(D^2(f_t)) = 3$ therefore, $m_1(D^2(f_t)) = 4 \; \forall t$.

Now to calculate $m_0(f_t(D^2(f_t)))$ we apply the formula

$$2m_0(f_t(D^2(f_t))) = \dim_\mathbb{C} \frac{\mathcal{O}_4}{(p_2 \circ f \circ p, I^2(f_t))},$$

choosing the generic projections $p(x, y, z_1, z_2) = (x, y, z_1)$ and $p_2(w_1, w_2, w_3, w_4) = (w_1, w_2)$, we have

$$
\begin{aligned}
2m_0(f_t(D^2(f_t))) &= \dim_\mathbb{C} \frac{\mathcal{O}_4}{(x, y, x+z_1^2+z_1 z_2+z_2^2, y(z_1+z_2)+P_4(z_1,z_2)+P_5(z_1,z_2)+tP_6(z_1,z_2))} \\
&= \dim_\mathbb{C} \frac{\mathcal{O}_2}{(z_1^2+z_1 z_2+z_2^2, P_4(z_1,z_2)+P_5(z_1,z_2)+tP_6(z_1,z_2))} \\
&= 8,
\end{aligned}
$$

therefore $m_0(f_t(D^2(f_t))) = 4$. To calculate $e_{D^2(f_t)(f_t)}$, we choose the generic projection $(p_2 \circ f \circ p)(x, y, z_1, z_2) = y$ and obtain

$$
\begin{aligned}
e_{D^2(f_t)(f_t)} &= \dim_\mathbb{C} \frac{\mathcal{O}_4}{(I^2(f_t), J(p_2 \circ f \circ p, I^2(f_t)))} \\
&= \mu(I^2(f_t)) + \mu(p_2 \circ f \circ p, I^2(f_t)) \\
&= 4 + \mu(p_2 \circ f \circ p, I^2(f_t)) \\
&= 13.
\end{aligned}
$$

In the same way, to calculate $e_{D^3_{(f_t)}}(f_t)$, $m_0(D^3(f_t))$ and $m_0(f_t(D^3(f_t)))$, the ideal $\mathcal{I}^3(f_t)$ that defines $D^3(f_t)$ is given by $V_i^3(g_j)$ where $i = 1, 2$ and $j = 1, 2$. The $V_i^k(g)$ are easily calculated using Maple. Then by theorem 3.10 it is sufficient to compute $m_0(f_t(D^3(f_t)))$ and $\mu(D^3(f_t))$, and $D^3(f_t)$ is defined by

$$(x - (z_1 z_2 + z_2 z_3 + z_1 z_3), z_1 + z_2 + z_3, V_1^3(yz^2 + z^5 + z^6 + tz^7), V_2^3(yz^2 + z^5 + z^6 + tz^7)).$$

So eliminating x and z_3 we have, that $D^3(f_t)$ is isomorphic to the hypersurface defined by $-tz_1 z_2^2 + z_1^4 + 3z_1^2 z_2^2 + 2z_1 z_2^3 + 2z_1^3 z_2 - tz_1^2 z_2 + z_2^4$ in $\mathbb{C}^3$, therefore $\mu(D^3(f_t)) = 7$, $\forall\ t \neq 0, 1, 1/2, 5/4, 3/2$.

Now, as $3! m_0(f_t(D^3(f_t))) = \dim_\mathbb{C} \frac{\mathcal{O}_5}{(\mathcal{I}^3(f_t), p_1 \circ f_t \circ p)}$, we choose generic projections $p(x, y, z_1, z_2, z_3) = (x, y, z_1)$ and $p(w_1, w_2, w_3, w_4) = w_2$, thus $m_0(f_t(D^3(f_t))) = 1$, $\forall t \neq 0, 1, 1/2$, therefore, $e_{D^3(f_t)}(f_t) = 12$, $t \neq 0, 1, 1/2, 5/4, 3/2$.

The calculation for $m_0(f_t(D^2(f_t, 2))) = 4$ and $m_0(D^2(f_t, 2)) = 2$ are similar.

REFERENCES

[1] J.W. Bruce, A.A. du Plessis, C.T.C. Wall. Determinacy and unipotency. Invent. Math. 88:521–554, 1987.

[2] K. Houston, N.P. Kirk. On the classification and geometry of map-germs from 3-space to 4-space. In: Proccedings of the European Singularities Conference in honour of C.T.C. Wall on the occasion of his 60th Birthday, eds. J.W. Bruce, D. Mond, Singularity Theory, London Math. Soc. Lecture Note Series 263, Cambridge University Press. pp. 325-351, 1999.

[3] T. Gaffney, D. Mond. Cusps and double folds germs of analytic maps $\mathbb{C}^2 \to \mathbb{C}^2$. J. London Math. Soc. 43 2:185-192, 1991.

[4] T. Gaffney. Properties of finitely determined germs. Ph.D. Thesis, Brandeis University, 1975.

[5] T. Gaffney, D. Massey. Trends in Equisingularity theory. In: Singularity Theory, ed. J.W. Bruce, D. Mond. Cambridge University Press, 1999.

[6] T. Gaffney. Polar multiplicities and equisingularity of map germs. Topology 32 1:185-223, 1993.

[7] T. Gaffney. Multiplicities and equisingularity of ICIS germs. Invent. Math. 123:209-220, 1996.

[8] V.H. Jorge Pérez. Polar multiplicities and equisingularity of map germs from $\mathbb{C}^3$ to $\mathbb{C}^3$, Cadernos de Matemática ICMC 1 1:261–280, 1999.

[9] V.H. Jorge Pérez, D. Levcovitz. Euler obstruction and Polar multiplicities on stable types of map germs from $\mathbb{C}^n$ to $\mathbb{C}^p$, $n \leq p$. In Preparation.

[10] L.D. Tráng. Calculation of Milnor number of isolated singularity of complete intersection. Funktsional'ny: Analizi Ego Prilozheniya. 8 2:45–49, 1974.

[11] L.D. Tráng, C.P. Ramanujam. The invariance of Milnor's number implies the invariance of the topological type. Amer. J. Math. 98:67-78, 1976.

[12] E.J.N. Looijenga. Isolated singular points on complete intersections. London Mathematical Soc. Lecture Note Series 77, 1984.

[13] G.M. Greuel. Der Gauss-Manin Zusammenhang isolierter Singularitäten von vollständigen Durchschnitten. Dissertation. Göttingen, 1973.

[14] W.L. Marar, D. Mond. Multiple point schemes for corank 1 maps. J. London Math. Soc. 39 2:553–567, 1989.

[15] D. Mumford. Introduction to Algebraic Geometry. Springer Verlag, 1976.

[16] B. Teissier. Cycles évanescents, sections planes et conditions de Whitney. Singularités à Cargèse 1972. Asterisque, 7–8, 1973.

[17] B. Teissier. Variétés polaires 2: Multiplicités polaires, sections planes, et conditions de Whitney. Actes de la conference de géometrie algébique à la Rábida, Springer Lecture Notes, 961 pp. 314–491, 1981.

[18] R. Vohra. Equisingularity of map germs from n-space,$(n \geq 3)$, to the plane, PhD. Thesis, Northeastern University, 2000.

[19] C.T.C. Wall. Finite determinacy of smooth map-germs. Bull. London Math. Soc. 13:481-539, 1981.

Topological Invariants of Stable Maps from a Surface to the Plane from a Global Viewpoint

D. HACON Dep. de Matemática, PUC-RIO, Rio de Janeiro-RJ, Brazil
E-mail: derek@mat.puc-rio.br

C. MENDES de JESUS* Dep. de Matemática, PUC-RIO, Rio de Janeiro-RJ, Brazil
E-mail: catarina@mat.puc-rio.br

M. C. ROMERO FUSTER** Dep. Geometría y Topología, Facultad de
Matemáticas, Universitat de València, Burjasot (Valencia), Spain
E-mail: carmen.romero@uv.es

We associate to any stable map of a surface M to the plane certain global invariants: the first is a weighted graph which describes the pair $M, \Sigma f$ (where Σf is the singular set of f). The other is a Vassiliev type invariant related to the number of connected components of Σf. We describe this invariant in terms of Chíncaro's classification singularities of codimension one of maps between surfaces.

1 INTRODUCTION

Vassiliev introduced the concept of finite order invariants for isotopy classes of embeddings of closed curves in 3-space. His method [7] [8] can be applied to various types of maps and the definition of the corresponding invariants is based, in each case, on the study of the topology of a discriminant subset determined by the subspace of maps under consideration.

(*) Work partially supported by CAPES, BEX0294/99-3 and CNPq, 142650/97-8.

(**) Work partially supported by DGCYT grant no.BFM 2000-1110

In [2], for instance, Arnol'd defined the three first order basic invariants for stable maps of the circle to the plane. On the other hand, Goryunov [5] studied stable maps of surfaces into 3-space, obtaining all first order local invariants for such maps. More recently, Ohmoto [6], applying ideas of Vassiliev and Goryunov, has given a general definition of first order local invariants for isotopy classes of stable smooth maps (between surfaces).

Here we consider the problem of defining global invariants of stable maps from a compact surface to the plane. We introduce invariants defined via the topological classification of the singular set of the map in question (which consists of a number of disjoint closed regular curves in the surface) and its image (which may contain cusps). These invariants depend on the isotopy classes of the singular set in the surface and its image in the plane. This leads to the definition of certain weighted graphs associated to collections of disjoint closed curves in a compact surface. These graphs provide global isotopy invariants of stable maps from a surface to the plane. Here we study some aspects of these invariants, in particular the question of which weighted graphs may be realized by stable maps. The number of edges of a given graph coincides with the number of connected components of the singular set of the corresponding map. This is, clearly, an isotopy invariant of the map. In section 5 we express this number in terms of certain (codimension one) transitions of the singular set.

2 STABLE MAPS FROM SURFACES TO THE PLANE

Let M be a compact, orientable surface with empty boundary. We say that two smooth maps f and g from M to the plane are *equivalent* (written $f \sim g$) if there are diffeomorphisms, $l : M \longrightarrow M$ and $k : \mathbb{R}^2 \longrightarrow \mathbb{R}^2$, such that $kf = gl$

A smooth map f is said to be *stable* if all maps sufficiently close to f (in the Whitney C^∞ topology) are equivalent to f.

Denote by Σf the singular set of f. We refer to its image $f(\Sigma f)$ as the *branch set* (of f). It is well known that if f is stable then its singularities are folds and (finitely many) cusps (see [4]). The singular set of f consists of (finitely many) disjoint embedded closed curves in M. On the other hand, the branch set consists of a finite number of immersed plane curves with a finite number of cusps and a finite number of transverse intersections and self-intersections (disjoint from the set of cusps).

An *isotopy* is a smooth 1-parameter family F_t of stable maps, $0 \leq t \leq 1$. The stable maps F_0 and F_1 are said to be *isotopic*. Isotopy classes correspond to path components of $\mathcal{E}(M, \mathbb{R}^2)$, the subspace of $C^\infty(M, \mathbb{R}^2)$ consisting of stable maps. Isotopy implies equivalence ([6]) but clearly not vice-versa.

If f and g are isotopic then Σf and Σg are isotopic subsets of M (i.e. there is a diffeomorphism of M which is isotopic to the identity and which takes Σf into Σg) and similarly for $f(\Sigma f)$ and $g(\Sigma g)$. Thus any isotopy invariant of the singular set or of the branch set of f will automatically be an isotopy invariant of f. In Section 4 we associate a weighted graph to Σf whose isomorphism class is an isotopy invariant of f. As for the branch set, one may, following Ohmoto, attempt to define isotopy invariants using Vassiliev-type invariants. The first global invariant of f that comes to mind is μ, the number of curves of Σf. In Section 6 we describe μ in terms of local transitions of $f(\Sigma f)$.

3   GRAPHS ASSOCIATED TO COLLECTIONS OF CLOSED CURVES ON SURFACES

Since the singular set Σf is an isotopy invariant of stable maps f, the number of connected components of Σf as well as the topological type of its complement in M are isotopy invariants. This information will be encoded in a weighted graph from which the pair $M, \Sigma f$ may be reconstructed (up to equivalence).

The complement in M of a collection $\mathcal{C}$ of disjoint embedded curves is the disjoint union of connected regions. The frontier of each region consists of a number of (boundary) curves. The weighted graph $\mathcal{G}(M, \mathcal{C})$ is defined as follows: There is a vertex for each connected component of $M - \mathcal{C}$ and an edge for each curve in $\mathcal{C}$. The two (or one) vertices of an edge correspond to the two (or one) regions whose boundaries contain the curve in question. Loops can occur. For example, the graph corresponding to a surface with a single (non separating curve) has one vertex and one edge. The *weight* $g(v)$ of a vertex v is the genus of the corresponding region. We recall that the *genus* of a surface with boundary is defined to be the genus of the closed surface obtained by adding a disk to each boundary component. For example, the genus of any planar region is zero. Figure (1) shows an example of such a weighted graph.

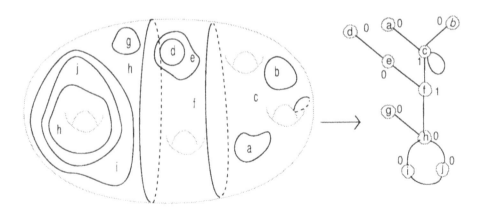

Figure 1: Weighted graph.

Two sets of curves $\mathcal{C}$ and $\mathcal{C}'$ in a surface M are said to be *isotopic* if there is an isotopy of M which takes $\mathcal{C}$ into $\mathcal{C}'$ This implies that there is a diffeomorphism of M into itself taking $\mathcal{C}$ into $\mathcal{C}'$. If $\mathcal{C}$ and $\mathcal{C}'$ are sets of curves in M and M' respectively then we say that the pairs $M, \mathcal{C}$ and $M', \mathcal{C}'$ are *equivalent* if there is a diffeomorphism between M and M' taking $\mathcal{C}$ to $\mathcal{C}'$. Clearly, equivalent pairs give rise to isomorphic weighted graphs (isomorphism of graphs preserving weights). In fact this defines a bijection between equivalence classes of pairs and isomorphism classes of weighted graphs (3.1).

THEOREM 3.1 *Every weighted graph is isomorphic to the graph of a pair $M, \mathcal{C}$ and this pair is unique up to equivalence. The genus $g(M)$ is given by*

$$g(M) = \alpha - \nu + 1 + \Sigma g(v),$$

where α is the number of edges and ν the number of vertices. and Σ represents the sum of the weights of all the vertices of the graph.

Proof: Just embed the graph in 3-space and let M_0 be the boundary of a suitable neighbourhood of the graph, with curves defined in the obvious way, one for each edge. M is then the connected sum of M_0 with surfaces of appropriate genus: for each vertex v we form the connected sum of the region of M_0 corresponding to v with a closed surface of genus $g(v)$. The weighted graph of the pair $M, \mathcal{C}$ is just the original graph. Conversely, a given pair $M, \mathcal{C}$ may be viewed as the connected sum of a pair $W, \mathcal{C}$ with surfaces of genus $g(v)$, one for each vertex. The two pairs have the same graph, but in the latter case all the weights are zero. The surface W, being orientable, is diffeomorphic to the boundary of an appropriate neighbourhood of the graph embedded in 3-space. Therefore, by uniqueness of connected sum, the pair $M, \mathcal{C}$ is equivalent to the pair obtained by reconstructing the graph as above. From the above description of the pair as a connected sum, the genus of M is the sum of $\sum_v g(v)$ and the genus of M_0. But the latter is equal to $\alpha - \nu + 1$, whence the formula for $g(M)$.

An immediate consequence of this is

COROLLARY 3.2 *For any pair $M, \mathcal{C}$ the surface M is a sphere if and only if the associated weighted graph is a tree all of whose weights are 0.*

We observe that the case of the sphere is special in the sense that two sets of curves in a sphere are isotopic if and only if they are equivalent. The reason is that any pair $M, \mathcal{C}$ can be taken into itself by an orientation reversing diffeomorphism so that if the diffeomorphism of the sphere taking one set of curves to another set of curves happens to be orientation-reversing then it can be replaced by an orientation preserving diffeomorphism which is then isotopic to the identity, so that the two sets of curves are isotopic in the sphere. For surfaces of higher genus, however, isotopy and equivalence are different conditions.

4   GRAPHS ASSOCIATED TO STABLE MAPS FROM A SURFACE TO THE PLANE

Any stable map f gives rise to a weighted graph $\mathcal{G}(M, \Sigma f)$. Clearly equivalent stable maps have isomorphic weighted graphs.

A natural question is now "which sets of curves may be realized as singular sets of stable maps?" In the case of the sphere, any nonempty set of curves can be realized if one allows cusps (if cusps are not allowed the realization question seems to be much harder). Recall that for M to be a sphere the graph must be a tree with all weights

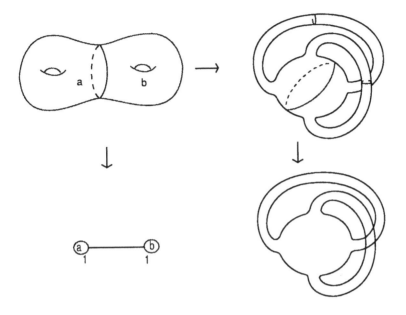

Figure 2: Orthogonal projection of a bitorus.

equal to 0. A tree with one edge is clearly realizable by orthogonal projection of a (round) sphere into the plane. Next consider a tree T with more than two edges. Choose an extreme edge (i.e. one which meets the rest of the tree in only one of its vertices, v say). By induction, the tree minus this edge can be realized. Inside the region corresponding to the vertex v we create a pair of cusps giving rise to a fold-curve (containing the two cusps) bounding a disk which is mapped to a "lips"-shaped region in the plane. This realises the tree T.

Consequently we have the following

THEOREM 4.1 *Any tree all of whose weights are zero may be realized as the graph of a stable map of the sphere into the plane.*

As for the question of realizing graphs by simple maps (i.e. maps which have no cusps) we offer the following examples. First, it is clear that if T is a tree realized by a simple stable map of the sphere into the plane then the tree obtained by adding a two-edge tree to T along a vertex v of T may also be similarly realized. Modify f inside a small disk contained in the region corresponding to v by introducing two parallel fold curves (without cusps). This has the effect of adding on a two-edge tree at v. Clearly, any tree all of whose weights are 0 and which can be obtained from a one-edge tree by successively adding on two-edge trees as above can be realized without cusps.

A second class of examples can be constructed by orthogonal projection of suitably embedded surfaces in 3-space. A simple example with weights 1 and 1 is given by a map of a surface of even genus with exactly one simple fold curve, as shown in figure 2.

5   VASSILIEV TYPE INVARIANTS

$C^\infty(M, \mathbb{R}^2)$ is a stratified set. The codimension zero strata are the connected components of $\mathcal{E}(M, \mathbb{R}^2)$. The complement of $\mathcal{E}(M, \mathbb{R}^2)$ (the *discriminant*) consists of strata of codimensions greater than zero.

First order Vassiliev invariants are defined as follows. Each codimension one stratum Y receives a "transition index" ξ_Y. Let F_t be a generic path in $C^\infty(M, \mathbb{R}^2)$, i.e. one whose end points $F_0 = f$ and $F_1 = g$ lie in $\mathcal{E}(M, \mathbb{R}^2)$ and which crosses the discriminant transversely in a (finite) number of points in the codimension one strata. Choose a base point f in $\mathcal{E}(M, \mathbb{R}^2)$. Since $C^\infty(M, \mathbb{R}^2)$ is path connected (being contractible), any other g in $\mathcal{E}(M, \mathbb{R}^2)$ may be reached from f by a generic paths C. Define $\Delta_C \xi$ to be the sum of the transition indices, one for each time the path crosses a 1-stratum Y (multiplied by ± 1 depending on which direction the path crosses Y).The increment $\Delta_C \xi$ depends in general, on the path C, but depends only on the end point g if and only if the transition indices ξ_Y satisfy a compatibility condition around the codimension two strata (the 1-cocycle condition).

The local transitions in the branch set have been described by E. Chíncaro [3]. These are illustrated in figure 3. The arrows signify the normal direction in which the number of pre-images increases (by two).

Using this classification Ohmoto proves

THEOREM 5.1 *([6]) For C^∞ stable mappings $M \longrightarrow \mathbb{R}^2$, the abelian group consisting of all local first order Vassiliev-type invariants is freely generated modulo constant functions by the following three elements:*

$$\Delta I_D = 2(\Delta T^0 + \Delta T^1 + \Delta T^2) + 2(\Delta C^1 + \Delta C^2) + \Delta S,$$

$$\Delta I_C = 2(\Delta B + \Delta L + \Delta S),$$

$$\Delta I_F = 2(\Delta T^0 + \Delta T^2) + \Delta C^1 + \Delta C^2.$$

We observe that, except for L and S, all the transitions shown in figure 3 involve (locally) more than one branch. We shall distinguish here between the cases in which such branches correspond (via f) to one, two or three connected components of Σf. The beaks transition is peculiar in that it alters not only the total number of cusps, but also the total number, I_μ, of connected components of Σf. Accordingly, we denote by B_+ the beaks transition corresponding to the birth of (a pair of) cusps where the number of components of Σf increases, and by B_- the beaks transition where the number of components decreases. Figure (4) shows a B_- followed by a B_+.

The remaining transitions, $(T^0, T^1, T^2, C^1, C^2, Q^0, Q^1)$ acquire a lower index denoting the number (1, 2 or 3) of connected components of Σf appearing in the transition. Thus

$$\Delta B = \Delta B_+ + \Delta B_-,$$

$$\Delta T^1 = \Delta T_1^1 + \Delta T_2^1,$$

$$\Delta T^0 = \Delta T_1^0 + \Delta T_2^0,$$

$$\Delta T^2 = \Delta T_1^1 + \Delta T_2^2,$$

$$\Delta C^1 = \Delta C_1^1 + \Delta C_2^1,$$

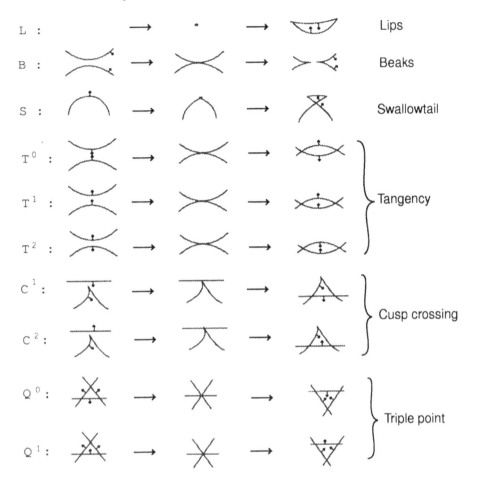

Figure 3: Local transitions of codimension one.

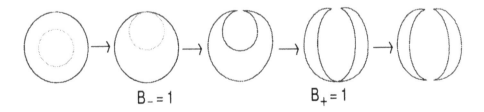

Figure 4: Beaks.

$$\Delta C^2 = \Delta C_1^2 + \Delta C_2^2,$$
$$\Delta Q^0 = \Delta Q_1^0 + \Delta Q_2^0 + \Delta Q_3^0,$$
$$\Delta Q^1 = \Delta Q_1^1 + \Delta Q_2^1 + \Delta Q_3^1.$$

The main result of this section is the following expression for ΔI_μ in terms of transitions.

THEOREM 5.2 *The increment of I_μ along any path tranversal to the strata of codimension one in D is given by*

$$\Delta I_\mu = \Delta B_+ - \Delta B_- + \Delta L.$$

Proof: It follows from the classification of transitions (see figure 1) that the only transitions altering the number of connected components of Σf are B and L. Now, a positive transition of type B_+ increases this number by one, whereas a positive transition of type B_- decreases it by one. On the other hand, a positive transition through the L stratum increases I_μ by one.   □

REMARKS:
   1) The invariants I_D, I_C and I_μ satisfy the relation

$$I_\mu + \frac{I_C}{2} + I_D = (1 + g(M)) \mod 2.$$

This follows from the fact that the only transitions which affect I_μ, I_C and I_D are S, L and B. It is easily seen that each of these transitions leaves the sum $I_\mu + \frac{I_C}{2} + I_D$ unchanged mod 2. The value of $I_\mu + \frac{I_C}{2} + I_D$ for a projection of a surface (embedded in $\mathbb{R}^3$) to the plane is $1 + g(M)$ whence the relation.

   2) For simple maps, I_C is clearly zero. Furthermore I_D is even for simple maps. For, if f is a simple map, the surface M is divided into two regions M_+ and M_- whose interiors are immersed (by f) into the plane with opposite orientations. M_+ and M_- meet in Σf which is the boundary of both. Let us write $\mathcal{R}$ for M_+ say (so that $\partial \mathcal{R} = \Sigma f$). The boundary $\mathcal{R}$ is immersed into the plane with transverse self-intersections. Consider the plane as the boundary $z = 0$ of the half-space $z \geq 0$. From the above immersion of $\mathcal{R}$ we may construct an immersion of $\mathcal{R}$ into the half space which is equal to the original immersion on $\partial \mathcal{R}$ and meets the plane $z = 0$ transversely in $\partial \mathcal{R}$. We may further modify the immersion of $\mathcal{R}$ (in the interior of $\mathcal{R}$) so that it intersects itself transversely. The self-intersection will be a compact 1-manifold (i.e. a disjoint union of circles and intervals, meeting the plane $z = 0$ transversely in its boundary which is exactly the set of self intersections of $\partial \mathcal{R}$. Hence $\partial \mathcal{R}$ consists of an even number of points as required.

REFERENCES

[1] V.I. Arnol'd. Plane curves, their invariants, perestroikas and classifications. Advances in Soviet Math. AMS, Providence, RI, 21:33–91, 1994.

[2] V.I. Arnol'd. Topological Invariants of Plane Curves and Caustics. University Lecture Series. AMS, Providence, RI, 5, 1994.

[3] E.A. Chíncaro. Bifurcations of Whitney Maps. ICEx, UFMG, 1978

[4] M. Golubitsky, V Guillemin. Stable Mappings and Their Singularities. Springer Verlag, Berlin and New York 1973.

[5] V. Goryunov. Local invariants of mappings of surfaces into three-space. The Arnold-Gelfand mathematical seminars, 223–255, Birkhäuser Boston, Boston, MA, 1997.

[6] T. Ohmoto. Vassiliev type invariants of order one of generic mappings from a surface to the plane. Topology of real singularities and related topics 55–68 (Japanese) Kyoto, 1997.

[7] V.A. Vassiliev. Cohomology of knot spaces. Advances in Soviet Math. AMS, Providence, RI, 1:23–69, 1990.

[8] V.A. Vassiliev. Complements of discriminants of smooth maps. Topology and applications, AMS, Providence, RI, 1992.

Cubics in R and C

RADMILA BULAJICH   Departamento de Matemáticas, Facultad de Ciencias, Universidad Autónoma del Edo. de Morelos 62210 Cuernavaca, Morelos, México
E-mail: bulajich@servm.fc.uaem.mx

LEON KUSHNER   Departamento de Matemáticas, Facultad de Ciencias, Universidad Nacional Autónoma de México 04510 México, D.F., Mexico
E-mail: kushner@servidor.unam.mx

SANTIAGO LÓPEZ DE MEDRANO   Instituto de Matemáticas Universidad Nacional Autónoma de México, 04510 México, D.F., Mexico
E-mail: santiago@matem.unam.mx

1   INTRODUCTION

In this article we study the stratification of the space of complex cubic forms in two variables under the classification by linear equivalence. The corresponding description for the space of real cubics was carried out by E.C. Zeeman [3] in a geometric way. We mainly study the topological characterization of the strata consisting of degenerate cubics and give a geometrical description of them inside $\mathbb{C}^4$. To present our method we start by applying it to the simpler and known case of real cubics and then proceed to the complex case.

The stratification of the real cubics plays an essential role in the theory of singularities of functions, as well as in many of its applications. The name "umbilic bracelet" given by E.C. Zeeman derives from the fact that it gives a geometric description of the relation between the umbilic catastrophes. (See [1],[2],[3]).

2   CUBICS IN $\mathbb{R}$

We consider the space of cubics in two variables over $\mathbb{R}$, $L^3(\mathbb{R}^2, \mathbb{R}) = \{\alpha x^3 + \beta x^2 y + \gamma x y^2 + \delta y^3\}$ (which we will identify with $\mathbb{R}^4$ by taking the coefficients as variables) and the group of linear $\mathbb{R}$-isomorphisms acting on it on the right. There are exactly five orbits, the orbits of: 0, x^3, $x^2 y$, $x(x^2 \pm y^2)$, which we will denote by 0, O_3, O_2 and $O_\pm$, respectively. Let $O_{3,2} = O_3 \cup O_2$.

The orbit O_3 is the set of cubics of the form $(ax+by)^3$ with $a, b \in \mathbb{R}$ and $a^2+b^2 \neq 0$, and it can be simply parametrized by the pair (a, b). Therefore it is diffeomorphic to $S^1 \times \mathbb{R}$.

The orbit O_2 is the set of cubics of the form $(ax + by)^2(cx + dy)$ with $a, b, c, d \in \mathbb{R}$ and $ac - bd \neq 0$. Developing the product $(ax + by)^2(cx + dy)$ one sees that O_2 is the image of the map

$$h_2 : \{(a, b, c, d) \in \mathbb{R}^4 : ad - bc \neq 0\} \to \mathbb{R}^4$$

$$h_2(a, b, c, d) = (a^2 c, 2abc + a^2 d, 2abd + b^2 c, b^2 d).$$

One computes easily that this mapping has rank 3 everywhere, so it can not be one-to-one. If two points (a, b, c, d) and (a', b', c', d') have the same image under h, then they give two cubics with the same zeroes, and so $(a, b) = t(a', b')$ and $(c, d) = s(c', d')$, $s, t \neq 0$. From the equality of the coefficients we have $s = t^{-2}$. So the fibers of h_2 are parametrized by

$$t \to \left(at, bt, ct^{-2}, dt^{-2}\right), \qquad \text{for} \quad t \neq 0.$$

In each fiber there are exactly two points closest to the origin. They correspond to the values

$$t = \pm 2^{\frac{1}{6}} \left(\frac{c^2 + d^2}{a^2 + b^2}\right)^{\frac{1}{6}},$$

and satisfy the equation $a^2 + b^2 = 2(c^2 + d^2)$. So h_2 restricted to the set $X_2 = \{(a, b, c, d) \in \mathbb{R}^4 : ad - bc \neq 0, \ a^2 + b^2 = 2(c^2 + d^2)\}$ is two-to-one. To get a one-to-one map we consider the quotient of X_2 by the map $(a, b, c, d) \to (-a, -b, c, d)$, which we denote by $\widetilde{X}_2$. Since h_2 has rank 3 everywhere, the induced map

$$\widetilde{h}_2 : \widetilde{X}_2 \to O_2$$

is a diffeomorphism.

To describe the union $O_{3,2}$ of these two orbits, we extend the parametrization h_2 to a parametrization $h_{3,2}$ defined in the set $\{(a, b, c, d) \in \mathbb{R}^4 : a^2+b^2 \neq 0, \ c^2+d^2 \neq 0\}$ by the same formula and again restrict to the set $X_{3,2} = \{(a, b, c, d) \in \mathbb{R}^4 : a^2 + b^2 = 2(c^2 + d^2) \neq 0\}$. When we pass to the quotient $\widetilde{X}_{3,2}$ by the antipodal map we obtain a homeomorphism

$$\widetilde{h}_{3,2} : \ \widetilde{X}_{3,2} \to O_{3,2}.$$

The restriction of $h_{3,2}$ to the subspace X_3 where $ad - bc = 0$ gives a diffeomorphism $\widetilde{h}_3 : \widetilde{X}_3 \to O_3$ which differs from the previous parametrization by a constant factor.

The map $\tilde{h}_{3,2}$ is not a diffeomorphism, because $h_{3,2}$ has rank less than 3 precisely at the points where $ad - bc = 0$. Now take a point p in this set and a vector v transversal to it at p which is in the kernel of $Dh_{3,2}$ and consider the straight line $p+tv$. Then $h_{3,2}(p+tv)$ is a simple cusp (equivalent to $t \to (t^2, t^3, 0, 0)$). Therefore, at the points of X_3 , the pair $(\mathbb{R}^4, O_{3,2})$ is locally diffeomorphic to the pair $(\mathbb{R}^2, \sigma) \times \mathbb{R}^2$, where σ is the curve $x_1^2 + x_2^3 = 0$.

To describe more precisely the topology of $O_{3,2}$ we consider the intersection $\overline{X}_{3,2} = X_{3,2} \cap S^3$. Since $X_{3,2}$ is conical we have that $X_{3,2}$ is homeomorphic to $\overline{X}_{3,2} \times \mathbb{R}$. The set $\overline{X}_{3,2}$ is given by $a^2 + b^2 = \frac{2}{3}$ and $c^2 + d^2 = \frac{1}{3}$, so it is homeomorphic to the torus $S^1 \times S^1$. The quotient is obtained by identifying antipodal points in the first factor S^1 of the product, so it is again homeomorphic to the torus. It follows that $\widehat{X}_{3,2}$, and therefore also $O_{3,2}$, is homeomorphic to $S^1 \times S^1 \times \mathbb{R}$. If we remove from $\overline{X}_{3,2}$ the points where $ad - bc = 0$ we obtain two copies of $S^1 \times \mathbb{R}$. It follows that O_2 is diffeomorphic to $S^1 \times \mathbb{R}^2$.

The beautiful geometry of these orbits inside $\mathbb{R}^4$ is well-known [3]. We give here our own description of this geometry:

Consider the isometry $T : \mathbb{R}^4 \to \mathbb{R}^4$ given by the matrix

$$\frac{1}{\sqrt{2}} \begin{pmatrix} 1 & 0 & -1 & 0 \\ 1 & 0 & 1 & 0 \\ 0 & 1 & 0 & 1 \\ 0 & 1 & 0 & -1 \end{pmatrix}.$$

Consider also the orthogonal projection π from $\mathbb{R}^4$ to the subspace of points of the form $(\alpha, 0, 0, \delta)$. Then we have:

The restriction of π to $T(O_{2,3})$ is a fiber bundle over $\mathbb{R}^2 \backslash \{0\}$ with fiber the deltoid and structural group $\mathbb{Z}_3$. In fact, every fiber is the image of the standard deltoid under an affine conformal transformation, and all the fibers over the points of a fixed circle centered at the origin are isometric.

This assertion follows from the fact that the set $T(O_{2,3})$ can be parametrized by the mapping

$\Phi(\rho, \theta, \varphi) = (\rho \cos(\theta), \rho(\cos(\theta - 2\varphi) + 2\cos(\varphi) - \cos(\theta))/2, \rho(\sin(\theta - 2\varphi) + 2\sin(\varphi) + \sin(\theta))/2, \rho \sin(\theta))$.

Therefore the fiber over each point $(\rho \cos(\theta), \rho \sin(\theta))$ with $\rho \neq 0$ is a deltoid which can be shown to be the image of the standard one $(\cos(-2\varphi) + 2\cos(\varphi), \sin(-2\varphi) + 2\sin(\varphi))$ under an affine, conformal mapping of $\mathbb{R}^2$.

In the usual description, the space of cubics is subject to a linear transformation that is not an isometry. Our description preserves the euclidean geometry of the orbits in the original coordinates, but their intersections with the unit sphere are not as regular.

3   CUBICS IN $\mathbb{C}$.

Our aim now is to describe topologically the stratification of the space of complex cubic forms in two variables. We consider the space of cubics in two variables over $\mathbb{C}$, $L^3(\mathbb{C}^2, \mathbb{C}) = \{\alpha z^3 + \beta z^2 w + \gamma z w^2 + \delta w^3\}$, which we will identify with $\mathbb{C}^4$, and the group of linear $\mathbb{C}$-isomorphisms acting on the right. There are exactly four orbits, the orbits of: 0, z^3, z^2w, $z(z^2 + w^2)$, which we will denote by 0, $\mathcal{O}_3$, $\mathcal{O}_2$ and $\mathcal{O}_1$, respectively. Let $\mathcal{O}_{3,2} = \mathcal{O}_3 \cup \mathcal{O}_2$.

The orbit $\mathcal{O}_3$ is the set of cubics of the form $(az+bw)^3$ with $a, b \in \mathbb{C}$ and $a\bar{a}+b\bar{b} \neq 0$. The pair (a, b) does not in this case give a good parametrization, because for ρ a cubic root of unity, (a, b) and $(\rho a, \rho b)$ give the same cubic form. Nevertheless, once we divide by this action of $\mathbb{Z}_3$ the parametrization becomes bijective. If we consider only the points (a, b) in the unit sphere S^3, the quotient is the lens space L_3. Therefore $\mathcal{O}_3$ is diffeomorphic to $L_3 \times \mathbb{R}$.

We now describe $\mathcal{O}_{3,2}$ using the map $h_{3,2} : \{(a, b, c, d) \in \mathbb{C}^4 : a\bar{a}+b\bar{b} \neq 0, c\bar{c}+d\bar{d} \neq 0\} \to \mathbb{C}^4$ with the same formula as in the real case:

$$h_{3,2}(a, b, c, d) = (a^2c, 2\,a\,b\,c + a^2d, 2\,a\,b\,d + b^2c, b^2d).$$

By the same argument as before, the fibers of this map are parametrized by

$$\tau \to \left(a\tau, b\tau, c\tau^{-2}, d\tau^{-2}\right)$$

where now $\tau \in \mathbb{C}\backslash\{0\}$. In each fiber, the set of points closest to the origin is a circle that corresponds to the values of τ such that

$$\tau\bar{\tau} = 2^{\frac{1}{3}} \left(\frac{c\bar{c} + d\bar{d}}{a\bar{a} + b\bar{b}}\right)^{\frac{1}{3}}$$

and they satisfy the equation $a\bar{a} + b\bar{b} = 2\left(c\bar{c} + d\bar{d}\right)$. So the restriction of $h_{3,2}$ to the subset $Z_{3,2} = \{(a, b, c, d) \in \mathbb{C}^4 : a\bar{a} + b\bar{b} = 2\left(c\bar{c} + d\bar{d}\right) \neq 0\}$ has compact fibers homeomorphic to S^1. To get a one-to-one map we need to identify each circle to a point, or, equivalently to take the quotient of $Z_{3,2}$ by the action of S^1 given by $g_u(a, b, c, d) = \left(au, bu, cu^{-2}, du^{-2}\right)$. Denoting this quotient by $\widetilde{Z}_{3,2}$, the induced map

$$\widetilde{h}_{3,2} : \widetilde{Z}_{3,2} \to \mathcal{O}_{3,2}$$

is a homeomorphism. When we restrict $h_{3,2}$ to $Z_3 = Z_{3,2} \cap \{(a, b, c, d) : ad - bc = 0\}$ and pass to the corresponding quotient we obtain a diffeomorphism

$$\widetilde{h}_3 : \widetilde{Z}_3 \to \mathcal{O}_3$$

which differs from the previous parametrization by a constant factor. The restriction of $h_{3,2}$ to the complementary $Z_2 = Z_{3,2} \cap \{(a, b, c, d) : ad - bc \neq 0\}$ again induces a diffeomorphism

$$\widetilde{h}_2 : \widetilde{Z}_2 \to \mathcal{O}_2.$$

The map $\widetilde{h}_{3,2}$ is not a diffeomorphism, because $h_{3,2}$ has rank less than 3 precisely at the points where $ad - bc = 0$. Again, a complex line transverse to this subset whose defining vector is in the kernel of $Dh_{3,2}$ is sent by the map as a simple cusp (equivalent to $\tau \to (\tau^2, \tau^3, 0, 0)$). Therefore, the pair $(\mathbb{C}^4, \mathcal{O}_{3,2})$ is locally diffeomorphic at the points of $\mathcal{O}_3$ to the pair $(\mathbb{C}^2, \sigma) \times \mathbb{C}^2$, where σ is the curve $z_1^2 + z_2^3 = 0$, which is well known to be a cone on the trefoil knot. Therefore the pair is not locally flat topologically.

To describe more precisely the topology of $\mathcal{O}_{3,2}$ we intersect the conical set $Z_{3,2}$ with the unit sphere S^7, thus obtaining a compact manifold $\overline{Z}_{3,2}$ such that $Z_{3,2} = \overline{Z}_{3,2} \times \mathbb{R}$. This intersection is given by $a\bar{a} + b\bar{b} = \frac{2}{3}$ and $c\bar{c} + d\bar{d} = \frac{1}{3}$, so it is diffeomorphic to the product $S^3 \times S^3$. We claim that the quotient of $\overline{Z}_{3,2}$ by the action g_u is diffeomorphic to the product $S^2 \times S^3$. This will follow from the case $k = -2$ of the following more general lemma:

LEMMA 3.1 *Let $g_{u,k}$ be the action of S^1 on $S^3 \times S^3$ given by $g_{u,k}(p,q) = (up, u^k q)$. Then for any integer k the action $g_{u,k}$ is smoothly conjugate to the action $g_{u,0}$ which is the Hopf action on the first factor. Therefore, for any integer k, the quotient of $S^3 \times S^3$ by the action $g_{u,k}$ is diffeomorphic to $S^2 \times S^3$.*

Proof: Define a map $F_k : S^3 \to S^3$ by

$$
F_k(z,w) = \begin{cases} \frac{(z^k, w^k)}{\|(z^k, w^k)\|} & \text{for} \quad k \geq 0, \\[2mm] \frac{(\bar{z}^k, \bar{w}^k)}{\|(z^k, w^k)\|} & \text{for} \quad k < 0. \end{cases}
$$

Considering now S^3 as the group of unit quaternions, F_k satisfies $F_k(up) = u^k F_k(p)$ for $u \in S^1$ and $p \in S^3$. Define the map $\Phi_k : S^3 \times S^3 \to S^3 \times S^3$ by

$$
\Phi_k(p,q) = (p, F_k(p)^{-1} q).
$$

Then Φ_k is a diffeomorphism which satisfies $\Phi_k(g_{u,k}(p,q)) = \Phi_k(up, u^k q) = (up, F_k(p)^{-1} q) = g_{u,0}(\Phi_k(p,q))$ and is therefore a conjugation between $g_{u,k}$ and $g_{u,0}$. $\square$

To describe the orbit $\mathcal{O}_2$ itself we take again the intersection $\overline{Z}_2 = Z_2 \cap S^7$, thus obtaining an open manifold such that $Z_2 = \overline{Z}_2 \times \mathbb{R}$. The difference $\overline{Z}_{3,2} \backslash \overline{Z}_2$ corresponds to the set of points of the form $(p, vp) \in S^3 \times S^3$, where $v \in S^1$. Now the diffeomorphism $(p,q) \to (p, qp^{-1})$ sends this difference to $S^3 \times S^1$ and under the induced action of S^1 any point in the complement $S^3 \times (S^3 \backslash S^1)$ is equivalent to a unique point of the form $(p, w + rj)$ where w is complex and $r > 0$. It follows that the quotient of $\overline{Z}_2$ by the action of g_u is diffeomorphic to the product of S^3 with an open disk and that $\overline{Z}_2$ is diffeomorphic to $S^3 \times \mathbb{R}^3$. Thus we have shown:

THEOREM 3.2 *The topological type of the orbits of the degenerate complex cubics is as follows: (i) $\mathcal{O}_{3,2}$ is homeomorphic to the product $S^2 \times S^3 \times \mathbb{R}$, (ii) $\mathcal{O}_3$ is diffeomorphic to $L_3 \times \mathbb{R}$, (iii) $\mathcal{O}_2$ is diffeomorphic to $S^3 \times \mathbb{R}^3$, (iv) $\mathcal{O}_{3,2}$ is smooth at the points of $\mathcal{O}_2$ while at the points of $\mathcal{O}_3$ is not locally flat, being transversally diffeomorphic to the cone on the trefoil knot.*

The geometry of these orbits inside $\mathbb{C}^4$ cannot be as nice and simple as in the real case: The set $\mathcal{O}_{2,3}$ does not project as nicely to a complex plane of $\mathbb{C}^4$, since it necessarily intersects the orthogonal subspace. Moreover, the fibers cannot be compact. Actually, not much seems to be gained by applying first a rotation (as we did in the real case), so it is simpler to project directly:

Consider the closure of $\mathcal{O}_{2,3}$ in $\mathbb{C}^4$, that is, $\bar{\mathcal{O}}_{2,3} = \mathcal{O}_{2,3} \cup \{0\}$ and the projection of $\mathbb{C}^4$ onto the subspace of points of the form $(\alpha, 0, 0, \delta)$. Let π be its restriction to $\bar{\mathcal{O}}_{2,3}$. Let $\varphi : \mathbb{C}^* \longrightarrow \mathbb{C}^2$ be given by

$$\varphi(\sigma) = (\frac{2\sigma^3 + 1}{\sigma^2}, \frac{\sigma^3 + 2}{\sigma}).$$

Its image is the *complex deltoid* Δ, the real one being given by taking σ in the unit circle. Then we have:

THEOREM 3.3 *The fibers of the projection* $\pi : \bar{\mathcal{O}}_{2,3} \longrightarrow \mathbb{C}^2$ *are as follows: (i) The generic fiber* $\pi^{-1}(\alpha, \delta)$ *for* $\alpha\delta \neq 0$ *is the image of the complex deltoid* Δ *by an element in the group of* 2×2 *diagonal complex matrices. Moreover,* π *restricted to the set* $\bar{\mathcal{O}}_{2,3} \cap \{\alpha\delta \neq 0\}$ *is a fiber bundle with fiber* Δ *and structural group* $\mathbb{Z}_3$. *(ii) The exceptional fiber* $\pi^{-1}(\alpha, \delta)$ *for* $\alpha\delta = 0$, α, δ *not both zero, is the union of a parabola and a double line tangent to it. (iii) The exceptional fiber* $\pi^{-1}(0,0)$ *is the union of two double lines. (iv) The restriction of* π *to* $\mathcal{O}_3$ *is a 3-fold branched covering over* $\mathbb{C}^2 \backslash \{0\}$, *branched over the set* $\alpha\delta = 0$. *Restricted further to the unit sphere, it is a 3-fold branched covering of the lens space* L_3 *onto a 3-sphere.*

Proof: (i) The set $\bar{\mathcal{O}}_{2,3} \cap \{\alpha\delta \neq 0\}$ can be reparametrized by setting $a = 1, c = \alpha, b = \xi, d = \delta/\xi^2$ in $h_{3,2}$ to obtain the one-to-one parametrization

$$\Psi(\xi, \alpha, \delta) = (\alpha, \frac{2\alpha\xi^3 + \delta}{\xi^2}, \frac{\alpha\xi^3 + 2\delta}{\xi}, \delta).$$

Here $\pi^{-1}(\alpha, \delta)$ is the curve given by the second and third coordinates, which resembles the complex deltoid. To see that it is a linear image of Δ it is better to use a second parametrization, by setting $a = r, b = s\sigma, c = r, d = s/\sigma^2$ in $h_{3,2}$ to obtain

$$\Phi(\sigma, r, s) = (r^3, r^2 s \frac{2\sigma^3 + 1}{\sigma^2}, r s^2 \frac{\sigma^3 + 2}{\sigma}, s^3).$$

This is a nine-to-one parametrization of $\bar{\mathcal{O}}_{2,3} \cap \{\alpha\delta \neq 0\}$ which has the advantage that now it is clear that the fiber over a point (r^3, s^3) with $\{rs \neq 0\}$ is the image of Δ under the linear map with matrix

$$\begin{pmatrix} r^2 s & 0 \\ 0 & s^2 r \end{pmatrix}.$$

It follows also that π restricted to the set $\bar{\mathcal{O}}_{2,3} \cap \{\alpha\delta \neq 0\}$ is a locally trivial fiber bundle with fiber Δ. When we follow a loop in the base space that winds around the set $\alpha = 0$ the value of r is multiplied by a cubic root of unity ρ and the deltoid is rotated by the matrix

$$\begin{pmatrix} \rho^2 & 0 \\ 0 & \rho \end{pmatrix}.$$

The effect of a loop that winds around the set $\delta = 0$ is a rotation by the inverse of this matrix (which is of the same form) and so the structural group of the bundle is the group of order 3 formed by those matrices.

(ii) Here it is better to consider the equation of the set $\bar{\mathcal{O}}_{2,3}$ given by setting the discriminant of the cubic equal to zero:

$$-27\alpha^2\delta^2 + 18\alpha\beta\gamma\delta + \beta^2\gamma^2 - 4\beta^3\delta - 4\alpha\gamma^3 = 0.$$

When $\alpha = 0$ and δ has a non zero fixed value, this equation reduces to

$$\beta^2(\gamma^2 - 4\beta\delta) = 0,$$

which represents the union of a parabola and the double line $\beta = 0$ tangent to it.

The case $\alpha \neq 0$ and $\delta = 0$ is analogous.

(iii) When $\alpha = \delta = 0$ the equation reduces further to

$$\beta^2\gamma^2 = 0,$$

which represents the union of two double lines.

(iv) Observe that $\mathcal{O}_3$ is the image by Φ of the set where σ is a cubic root of unity, but it is enough to consider the restriction of Φ to the set $\sigma = 1$, which is three-to-one. The map π restricted to $\mathcal{O}_3$ is also three-to-one and a local diffeomorphism, except at the set $\alpha\delta = 0$ where it is one-to-one and is equivalent transversely to the map $r \to r^3$. When restricted to the unit sphere these maps are respectively: the covering $S^3 \to L_3$ and a three-to-one covering $L_3 \to S^3$, branched over two circles. $\square$

REFERENCES

[1] V.I. Arnol'd, A.N. Varchenko, S.M. Goussein-Zadé. Singularités des Applications Différentiables, Editions Mir, Moscou, 1986.

[2] R. Thom. Structural Stability and Morphogenesis. An Outline of a General Theory of Models. W.A. Benjamin, Inc., 1975.

[3] E.C. Zeeman. The Umbilic Bracelet and the Double Cusp Catastrophe. In: Catastrophe Theory, Selected Papers 1972-77, Addison-Wesley, 1977, pp. 563–601.

Indices of Newton Non-Degenerate Vector Fields and a Conjecture of Loewner for Surfaces in $\mathbb{R}^4$

CARLOS GUTIERREZ Instituto de Ciências Matemáticas e de Computação, Universidade de São Paulo, Departamento de Matemática, Caixa Postal 668, 13560-970, São Carlos, SP, Brazil, E-mail: gutp@icmc.sc.usp.br

MARIA APARECIDA SOARES RUAS Instituto de Ciências Matemáticas e de Computação, Universidade de São Paulo, Departamento de Matemática, Caixa Postal 668, 13560-970, São Carlos, SP, Brazil, E-mail: maasruas@icmc.sc.usp.br

Abstract

We study the index of a vector field in $\mathbb{R}^2$, with isolated singularity, in terms of conditions on the Newton polyhedra associated to its coordinates. When the vector field is Newton non-degenerate, we show that its index is determined by the principal part of the Newton polyhedra. As a consequence we can prove that, under very mild conditions, the index of an isolated inflection point of a locally convex surface generically embedded in $\mathbb{R}^4$ is the same as the index of an umbilic point of a surface immersed in $\mathbb{R}^3$.

Research of the first author was partially supported by PRONEX/FINEP/MCT - grant # 76.97.1080.00 and research of the second author was partially supported by CNPq-grant #300066/88-0 and FAPESP, grant # 97/10735-3.

1   INTRODUCTION

This article has been motivated by the classical (local) Loewner's Conjecture which states that every umbilic of a smooth surface immersed in $\mathbb{R}^3$ must have index less than or equal to one. It was shown in [3] that the index of an isolated inflection point (in the sense of Little [8]) of a locally convex surface generically embedded in $\mathbb{R}^4$ is $\pm 1/2$.

Our first result, Theorem 1, was inspired in the result of M. Brunella and M. Miari [2], in which they proved that a Newton non-degenerate vector field in the plane possessing characteristic orbit, is topologically equivalent to its principal part. Under weaker nondegeneracy conditions given in terms of the Newton Polyhedra associated to the coordinates of the vector field, we prove that its index is determined by its principal part.

We believe that our conditions are better suited for the case of indices of vector fields. Also, as it should be, proofs are much shorter.

In our second result we show that, under very mild conditions given in terms of Newton Polyhedra, the index of an isolated inflection point is the same as the index of an umbilic point of a surface immersed in $\mathbb{R}^3$. It may be possible that the Loewner Conjecture can be extended to the case of isolated inflection points of locally convex surfaces in $\mathbb{R}^4$.

The authors wish to thank the referee and J. Llibre for their comments, which have been taken into account in this paper.

2   INDEX OF A VECTOR FIELD AT A SINGU-LARITY

Let $\mathcal{E}_n$ be the set of smooth germs $g : (\mathbb{R}^n, 0) \to \mathbb{R}$. To $g \in \mathcal{E}_n$ associate its formal Taylor series expansion at 0: $\hat{g}(x) = \sum a_k x^k$. Define supp $g = \{k \in \mathbb{Z}^n : a_k \neq 0\}$. Given $\Delta \subset \mathbb{Z}^n$, we define $g|_\Delta = \sum_{k \in \Delta} a_k x^k$.

The Newton Polyhedron of $g \in \mathcal{E}_n$, denoted by $\Gamma_+(g)$, is the convex hull in $\mathbb{R}_+^n$ of the set $\cup\{k + v : k \in \text{supp } g, v \in \mathbb{R}_+^n\}$. The Newton Diagram of g is the union $\Gamma(g)$ of all compact faces and vertices of $\Gamma_+(g)$. Observe that if $\hat{g} \neq 0$ and $\Gamma_+(g)$ has no compact faces, $\Gamma(g)$ will be the only vertex of $\Gamma_+(g)$; also, if $\hat{g} = 0$, then $\Gamma_+(g) = \emptyset$.

Let $X = (f, g) : (\mathbb{R}^2, 0) \to (\mathbb{R}^2, 0)$ be the germ of a vector field in the plane. We say that X is Newton non-degenerate if

(1) $\Gamma(f) \cup \Gamma(g)$ meets both $\{0\} \times \mathbb{R}^+$ and $\mathbb{R}^+ \times \{0\}$,

(2) for all pair of parallel faces $\Delta_1 \subset \Gamma(f)$ and $\Delta_2 \subset \Gamma(g)$, the equations $f|_{\Delta_1} = 0$ and $g|_{\Delta_2} = 0$ have no common solutions in $(\mathbb{R} \setminus 0)^2$,

(3) when $\hat{g} = 0$ (resp. when $\hat{f} = 0$), for every compact face Δ of $\Gamma_+(f)$ (resp. of $\Gamma_+(g)$) the equation $f|_\Delta = 0$ (resp. $g|_\Delta = 0$) has no solution in $(\mathbb{R} \setminus 0)^2$.

We say that X is Brunella-Miari non-degenerate, or shortly B.M. non-degenerate, if X is Newton non-degenerate and satisfies the following: let Δ be a compact face of $\Gamma_+(f)$ (resp. of $\Gamma_+(g)$) and let L be the only straight line containing Δ; if $L \cap \Gamma_+(g) = \emptyset$ (resp. $L \cap \Gamma_+(f) = \emptyset$), then the equation $f|\Delta = 0$ (resp. $g|\Delta = 0$) has no solution in $(\mathbb{R} \setminus 0)^2$.

This definition is equivalent to Brunella-Miari's non-degeneracy condition of [2, page 351] together with their assumption that $(0,0)$ is an isolated singularity of X (see [2, Theorem A]).

The vector field $Y(x,y) = (f(x,y), g(x,y)) = (x-y, x^2)$ is Newton non-degenerate, but it is B.M. degenerate. In fact, $\Gamma(f) = \Delta$ is the compact segment connecting $(1,0)$ with $(0,1)$) and $\Gamma(g) = (2,0)$. Let L denote the straight line containing Δ. Then $L \cap \Gamma_+(g) = \emptyset$ but the equation $f|\Delta = x - y = 0$ has solutions in $(\mathbb{R} \setminus 0)^2$.

We now state our main result.

THEOREM 2.1 *Let $X = (f,g) : (\mathbb{R}^2, 0) \to (\mathbb{R}^2, 0)$ be the germ of a smooth vector field in the plane. Let $X_\Gamma = \left(f \,|_{\Gamma(f)}, g \,|_{\Gamma(g)} \right)$ be the principal part of the vector field X. If X is Newton non-degenerate, then, there exists a neighborhood U of $(0,0)$ such that, $\forall (x,y) \in U \setminus \{(0,0)\}$ and $\forall s \in [0,1]$,*

$$0 < |sX(x,y) + (1-s)X_\Gamma(x,y)|$$
$$:= |s\,f(x,y) + (1-s)\,f|_{\Gamma(f)}(x,y)| + |s\,g(x,y) + (1-s)\,g|_{\Gamma(g)}(x,y)|$$

In particular $(0,0)$ is an isolated singularity of both X and X_Γ and

$$Index\ (X, 0) = Index\ (X_\Gamma, 0)$$

Proof: We shall only consider the case in which $\hat{f} \neq 0$ and $\hat{g} \neq 0$. The proof in the other case is similar. Let $S_+^1 = \{(a,b) \in \mathbb{R}^2 : a^2 + b^2 = 1,\ a > 0,\ b > 0\}$. Given $(a,b) \in S_+^1$ and $h \in \{f, f|_{\Gamma(f)}, g, g|_{\Gamma(g)}\}$, denote by

$$\Delta(a,b) = \Delta_h(a,b) = \{(x,y) \in \Gamma_+(h)\ :\ ax + by = \inf\{au + bv\ /\ (u,v) \in \Gamma_+(h)\}\}$$

As $\Gamma_+(h)$ is a convex subset of $\mathbb{R}_+^2$, we obtain:

(1a) $\Delta(a,b)$ is either a vertex or a compact face of $\Gamma(h)$;

(1b) if $(m_0, n_0), (m_1, n_1) \in \Delta(a,b)$, then
$(m_0, n_0) \cdot (a,b) = (m_1, n_1) \cdot (a,b)$, where "$\cdot$" denotes the usual inner product of $\mathbb{R}^2$.

(1c) if $(m_0, n_0) \in \Delta(a,b)$ and $\Gamma_+(h - h|_{\Delta(a,b)})$ is not empty, then, there exists $\epsilon(a,b,h) > 0$ such that

$$(m_0, n_0) \cdot (a,b) + \epsilon(a,b,h) = \inf\{(m,n) \cdot (a,b)/(m,n) \in \Gamma_+(h - h|_{\Delta(a,b)})\}.$$

This proof is organized as follows: We shall find, for each $(a,b) \in S_+^1 \cup \{(0,1),(1,0)\}$, a small curved sector of $\mathbb{R}^2$ having its vertex at the origin and such that X restricted to this sector satisfies the theorem. By local compactness, we will be able to cover a neighborhood of $(0,0) \in \mathbb{R}^2$ with finitely many of these sectors. In this way X restricted to this neighborhood of $(0,0)$ will satisfy the theorem. We shall accomplish this by studying, when $h \in \{f, g\}$, the following cases:

(i) $\Delta(a,b)$ is a compact face of $\Gamma(h)$ non-reduced to a point (and so, $(a,b) \in S_+^1$);

(ii) $(a,b) \in S_+^1$ and $\Delta(a,b)$ is a vertex of $\Gamma(h)$; and

(iii) $\Delta(1,0)$ and $\Delta(0,1)$ are one-point-sets (vertices of $\Gamma(h)$).

To simplify matters, we shall continue by assuming that

(2) $\Gamma_+(h - h|_{\Delta(a,b)}) \neq \emptyset$

Suppose first that $\Delta(a,b)$ is a segment and let (M,n) and (m,N) be its endpoints. We shall assume $0 \leq m < M$ and $0 \leq n < N$. Let $\epsilon = \epsilon(u,v,h)$ be defined in $B_\delta(a,b) = \{(u,v) \in S_+^1 : |(u-a, v-b)| < \delta\}$ by

$$\epsilon(u,v,h) = \inf\{(\mu,\nu)\cdot(u,v) \ : \ (\mu,\nu) \in \Gamma_+(h - h|_{\Delta(a,b)})\}$$
$$- \sup\{(\mu,\nu)\cdot(u,v) \ : \ (\mu,\nu) \in \Delta(a,b)\}.$$

If $\delta > 0$ is small enough, then $\epsilon(u,v,h)$ depends continuously on (u,v) and

(3)

$$\epsilon(u,v,h) \in (\frac{3}{4}\,\epsilon(a,b,h), \frac{5}{4}\,\epsilon(a,b,h));$$

This implies that, for all $(u,v) \in B_\delta(a,b)$,

(4)

$$(3/4)\,\epsilon(a,b,h) + u\cdot M + v\cdot n < \epsilon(u,v,h) + u\cdot M + v\cdot n$$
$$\leq \epsilon(u,v,h) + \sup\{(\mu,\nu)\cdot(u,v) \ : \ (\mu,\nu) \in \Delta(a,b)\}$$
$$= \inf\{(\mu,\nu)\cdot(u,v) : (\mu,\nu) \in \Gamma_+(h - h|_{\Delta(a,b)})\}$$

Therefore, if $\delta > 0$ and $\sigma > 0$ are small enough, there is a continuous function $R(t,u,v,h)$ defined for all $(t,u,v) \in [0,\sigma) \times B_\delta(a,b)$, such that:

(5)

$$h(t^u, t^v) = a_{m,N}\, t^{u\cdot m + v\cdot N} + \cdots + a_{M,n}\, t^{u\cdot M + v\cdot n}$$
$$+ t^{u\cdot M + v\cdot n}\, t^{(1/2)\epsilon(a,b,h)} R(t,u,v,h)$$

Let $\rho = \rho(u)$ be defined in $\{u \leq a : (u,v) \in B_\delta(a,b)\}$ by the equation

(6) $u = a\,\rho(u)$

Also, let $k = k(t,u,v)$ be defined in $(0,\sigma) \times \{(u,v) \in B_\delta(a,b) : u \leq a\}$ by the equation

(7)

$$t^v = k(t,u,v)\, t^{b\,\rho(u)}$$

As $\rho(u) \leq 1$ and $v \geq b$, we obtain that $0 < k(t,u,v) \leq 1$. Therefore, using (1b)-(1c) and (5)-(7), we obtain that, for all $(t,u,v) \in (0,\sigma) \times \{(u,v) \in B_\delta(a,b) : u \leq a\}$,

(8)

$$s\,h(t^u, t^v) + (1-s)\,h|_{\Gamma(h)}(t^u, t^v) =$$
$$= (s\,h((t^\rho)^a, k\,(t^\rho)^b) + (1-s)\,h|_{\Gamma(h)}((t^\rho)^a, k\,(t^\rho)^b))$$
$$= (k^n\,(t^\rho)^{a\cdot M + b\cdot n})\,(a_{m,N}\,k^{N-n} + \cdots + a_{M,n} + s\,t^{(1/2)\epsilon(a,b,h)}R(t,u,v,h)$$
$$+ (1-s)\,t^{(1/2)\epsilon(a,b,h|_{\Gamma(h)})}R(t,u,v,h|_{\Gamma(h)}))$$

This last expression shows that, under conditions above (in particular when $t > 0$ is small),

$$|s\,h(t^u, t^v) + (1-s)\,h|_{\Gamma(h)}(t^u, t^v)|$$

can only be zero nearby the real roots of the polynomial, in the variable K,

$$a_{m,N}\,K^{N-n} + \cdots + a_{M,n}$$

Suppose now that $\Delta(a,b) = \{a_{m,N}\} = \{a_{M,n}\}$ is a one-point-set. Then, proceeding as above we shall obtain the following relation which corresponds to item (8) above under the same assumptions:

(8')

$$s\,h(t^u, t^v) + (1-s)\,h|_{\Gamma(h)}(t^u, t^v) =$$
$$= (s\,h((t^\rho)^a, k\,(t^\rho)^b) + (1-s)\,h|_{\Gamma(h)}((t^\rho)^a, k\,(t^\rho)^b))$$
$$= (k^n\,(t^\rho)^{a\cdot M + b\cdot n})\,(a_{M,n} + s\,t^{(1/2)\epsilon(a,b,h)}R(t,u,v,h)$$
$$+ (1-s)\,t^{(1/2)\epsilon(a,b,h|_{\Gamma(h)})}R(t,u,v,h|_{\Gamma(h)}))$$

This expression shows us that, if $\delta > 0$ and $\sigma > 0$ are small enough, then for all $(t,u,v) \in (0,\sigma) \times \{(u,v) \in B_\delta(a,b) : u \le a\}$,

$$|s\,h(t^u, t^v) + (1-s)\,h|_{\Gamma(h)}(t^u, t^v)| > 0$$

Therefore, using the fact that X is Newton non-degenerate and by extending the arguments above to $\{(u,v) \in B_\delta(a,b) : v \le b\}$, we obtain,

(9) If $(a,b) \in S^1_+$, then there are $\delta > 0$ and $\sigma > 0$ such that, $\forall (u,v) \in B_\delta(a,b)$ and $\forall\, 0 < t < \sigma$,

$$|s\,f(t^u, t^v) + (1-s)\,f|_{\Gamma(f)}(t^u, t^v)| + |s\,g(t^u, t^v) + (1-s)\,g|_{\Gamma(g)}(t^u, t^v)| > 0.$$

Now, if $\Delta(1,0) = (0,N)$, then it may be seen that there exists $\rho > 0$ such that for all $0 \le |y| < \rho$ and $-1 \le r \le 1$,

$$h(r\,y^{N+1}, y) = a_{0,N}y^N + y^{N+1}R(y,r)$$

for some continuous real valued function $R = R(y,r)$. It follows that $|h|$ restricted to a set of the form $\{(x,y) \;:\; 0 < |y| < \rho, -\rho^{N+1} \le x \le \rho^{N+1}\}$ is positive.

Proceeding similarly when $\Delta(0,1) = (M,0)$ and using the fact that X is Newton non-degenerate, we may conclude that there exists $\rho > 0$ such that, for all $s \in [0,1]$, the restriction of $|s\,h + (1-s)\,h|_{\Gamma(h)}$ to the set $\Sigma(\rho)$, which is the union of

$$\{(x,y) \ : \ 0 < |y| < \rho, \ -\rho^{N+1} \le x \le \rho^{N+1}\}$$

and

$$\{(x,y) \ : \ 0 < |x| < \rho, \ -\rho^{M+1} \le y \le \rho^{M+1}\},$$

is positive.

Using the fact that $\{(a,b) \in \mathbb{R}^2 : a^2 + b^2 = 1, \ a \ge 0, \ b \ge 0\}$ is compact and assuming that $\rho > 0$ is small enough, we may cover a neighborhood of $(0,0)$ in $\{(x,y) \in \mathbb{R}^2 : x \ge 0, y \ge 0\}$ with the union of $(0,0) \cup \Sigma(\rho)$ with finitely many sectors of the form

$$\{(x,y) \in \mathbb{R}^2 : x = t^u, y = t^v, (u,v) \in B_\delta(a,b), 0 < t < \sigma\}$$

and for which (9) is true.

Similarly, we may cover a neighborhood of $(0,0)$ in $\mathbb{R}^2$ with the union of $(0,0) \cup \Sigma(\rho)$ with finitely many sectors of one of the forms

$$\{(x,y) \in \mathbb{R}^2 : x = t^u \quad , \quad y = t^v \quad , \quad (u,v) \in B_\delta(a,b), \quad 0 < t < \sigma\}$$
$$\{(x,y) \in \mathbb{R}^2 : x = -t^u \quad , \quad y = t^v \quad , \quad (u,v) \in B_\delta(a,b), \quad 0 < t < \sigma\}$$
$$\{(x,y) \in \mathbb{R}^2 : x = -t^u \quad , \quad y = -t^v \quad , \quad (u,v) \in B_\delta(a,b), \quad 0 < t < \sigma\}$$
$$\{(x,y) \in \mathbb{R}^2 : x = t^u \quad , \quad y = -t^v \quad , \quad (u,v) \in B_\delta(a,b), \quad 0 < t < \sigma\}$$

and for which the statement that corresponds to (9) is true. The theorem follows from this.   □

3   LOCALLY STRICTLY CONVEX SURFACES OF $\mathbb{R}^4$

The asymptotic line fields associated to an embedding f of a surface M in $\mathbb{R}^4$ were studied in [3]. A generically embedded surface $f(M)$ always has an open region at which there are defined two line fields V_1 and V_2 of asymptotic directions. These two fields eventually collapse onto a unique one over a curve in M. Their integral lines are called asymptotic lines and their singular points are the inflection points of the embedding as defined by Little in [8]. The geometry of a generically embedded surface in the neighborhood of an inflection point was studied in [9]. The differential equation of the asymptotic lines ([3]) of M is the following binary equation:

$$\begin{vmatrix} f_{1x} & f_{2x} & f_{3x} & f_{4x} & 0 \\ f_{1y} & f_{2y} & f_{3y} & f_{4y} & 0 \\ f_{1xx} & f_{2xx} & f_{3xx} & f_{4xx} & dy^2 \\ f_{1xy} & f_{2xy} & f_{3xy} & f_{4xy} & -dxdy \\ f_{1yy} & f_{2yy} & f_{3yy} & f_{4yy} & dx^2 \end{vmatrix} = 0; \qquad (3.1)$$

where (f_1, f_2, f_3, f_4) are the coordinate functions of the embedding.

In this section we study singular points of this equation when M is locally strictly convexly embedded in $\mathbb{R}^4$.

DEFINITION 3.1 *A hyperplane H in $\mathbb{R}^4$ is a nonsingular support hyperplane to M at the point p if it is tangent to M at p, $M \cap H = \{p\}$ and p is a non-degenerate critical point of the linear projection $\pi : M \to L_H$, where L_H is the line in $\mathbb{R}^4$, passing through p, orthogonal to H.*

DEFINITION 3.2 *The embedding $f : M \to \mathbb{R}^4$ is locally strictly convex if every point admits a nonsingular support hyperplane.*

When the embedding $f : M \to \mathbb{R}^4$ is locally strictly convex, the pair of transversal foliations on M induced by the asymptotic lines are -away from a discrete set of singularities- globally defined ([3]); their singular points are isolated and coincide with the inflection points of imaginary type.

We shall study the indices of these inflection points with respect to either of the asymptotic foliations. The index does not depend on the specific foliation $\mathcal{A}_i$ because $\mathcal{A}_1$ and $\mathcal{A}_2$ are transversal to each other in the complement of the inflection points.

This section is motivated by a famous conjecture due to Loewner which states that -with respect to either of the foliations induced by the principal lines of curvature of an embedded surface in $\mathbb{R}^3$ - there are no umbilics of index bigger than one. This Loewner Conjecture has been asserted to be true for analytic surfaces by several authors among which H. Hamburger [6], G. Bol [1], T. Klotz [7] and C. J. Titus [10]. In the following we shall show that under very mild conditions, the index of an isolated inflection point - with respect to $\mathcal{A}_1$ - is the same as the index of an umbilic point of a surface immersed in $\mathbb{R}^3$.

The following lemma gives a useful normal form for M in a neighborhood of an isolated inflection point of a strictly convexly embedded surface.

LEMMA 3.3 *Let M be locally strictly convexly embedded in $\mathbb{R}^4$ and $p \in M$ be an isolated inflection point. Then, up to a rigid motion of $\mathbb{R}^4$, it can be assumed that $p = (0,0,0,0)$ and that M, around p, admits a parametrization of the form*

$$\alpha : (x,y) \to (x, y, \frac{1}{2}(x^2 + y^2) + F(x,y), G(x,y)),$$

where $J^2 F(0,0) = J^2 G(0,0) = 0$,

Proof: We can assume $p = (0,0,0,0)$ and that the embedding is in Monge's form

$$\alpha : (x,y) \to (x, y, f_1(x,y), f_2(x,y)),$$

where $J^1 f_1(0,0) = J^1 f_2(0,0) = 0$. Since p is an inflection point of the embedding, the second fundamental form has rank 1 at p. Hence, the 2-jets of f_1 and f_2 are linearly dependent at zero. By assumption the embedding is locally strictly convex, hence by a rigid motion in $\mathbb{R}^4$, we can assume that M has the desired normal form.   □

For the parametrization α of Lemma 3.1, the differential equation (3.1) takes the form

$$\begin{vmatrix} dy^2 & -dxdy & dx^2 \\ 1 + F_{xx} & F_{xy} & 1 + F_{yy} \\ G_{xx} & G_{xy} & G_{yy} \end{vmatrix} = 0; \tag{3.2}$$

This equation may be rewritten as:

$$(G_{xy} + \Delta_{12}(F,G))dx^2 \ + \ \begin{aligned} & (G_{yy} - G_{xx} + \Delta_{13}(F,G)dxdy \\ & + (-G_{xy} - \Delta_{23}(F,G))dy^2 = 0, \end{aligned} \tag{3.3}$$

where

$$\begin{aligned} \Delta_{12}(F,G) &= G_{xy}F_{xx} - G_{xx}F_{xy}, \\ \Delta_{13}(F,G) &= F_{xx}G_{yy} - G_{xx}F_{yy}, \\ \Delta_{23}(F,G) &= G_{xy}F_{yy} - G_{yy}F_{xy}. \end{aligned}$$

THEOREM 3.4 *Let* $F, G : \mathbb{R}^2, 0 \to (\mathbb{R}, 0)$ *be germs of smooth functions such that* $J^2 F(0) = J^2 G(0) = 0$. *Let* X, X_1, X_2 *be germs of vector fields in* $\mathbb{R}^2, 0$ *defined by*

$$\begin{aligned} X(x,y) &= (G_{xx} - G_{yy}, \ 2G_{xy}), \\ X_1(x,y) &= (G_{xx} - G_{yy} - \Delta_{13}(F,G), \ 2G_{xy} + 2\Delta_{12}(F,G)), \\ X_2(x,y) &= (G_{xx} - G_{yy} - \Delta_{13}(F,G), \ 2G_{xy} + 2\Delta_{23}(F,G)); \end{aligned}$$

Suppose that X *is Newton non-degenerate and that* $X_\Gamma = (X_1)_\Gamma = (X_2)_\Gamma$. *Then*

(i) *the index at 0, of (either of the two foliations induced by) equation (3.3), is the same as the index at 0 of (either of the two foliations induced by) equation*

$$G_{xy}dx^2 + (G_{yy} - G_{xx})dxdy - G_{xy}dy^2 = 0. \tag{3.4}$$

This common index is half of the index of X *at 0.*

(ii) *there exists a smooth surface in* $\mathbb{R}^3$ *such that the differential equation of its principal lines of curvature -in a suitable coordinate system- is precisely equation (3.4).*

Proof: By the same argument as that of the Proof of Theorem 2.1, there exists a neighborhood V of $0 \in \mathbb{R}^2$ such that, for all $s \in [0,1]$, and -when restricted to V- the vector fields

$$\begin{aligned} X &= & (G_{xx} - G_{yy}, \ 2G_{xy}), \\ X_1^{(s)} &= (G_{xx} - G_{yy} - \Delta_{13}(sF,G), \ 2G_{xy} + 2\Delta_{12}(sF,G)), \\ X_2^{(s)} &= (G_{xx} - G_{yy} - \Delta_{13}(sF,G), \ 2G_{xy} + 2\Delta_{23}(sF,G)), \end{aligned} \tag{3.5}$$

have 0 as the only singularity. Therefore, for all $s \in [0,1]$ and for every point of $V \setminus \{0\}$,

$$(G_{xx} - G_{yy} + \Delta_{13}(sF,G))^2 + (2G_{xy} + \Delta_{12}(sF,G))(2G_{xy} + \Delta_{23}(sF,G)) > 0.$$

This and Homotopy Theory [5] imply that, given $s \in [0,1]$, the index at 0 of equation

$$(G_{xy} + \Delta_{12}(sF,G))dx^2 \ + \ \begin{aligned} & (G_{yy} - G_{xx} + \Delta_{13}(sF,G)dxdy \\ & + (-G_{xy} - \Delta_{23}(sF,G))dy^2 = 0, \end{aligned} \tag{3.6}$$

does not depend on s. Taking $s = 0, 1$, the first statement of item (i) follows. It is not difficult to check the second statement of (i) (see, for instance, [15]). The proof of (ii) can be found in [4].   □

In the following corollary we show that the above result always holds when the germ G is quasi-homogeneous and X is Newton non-degenerate. We recall that the germ $G : \mathbb{R}^2, 0 \to \mathbb{R}, 0$ is quasi-homogeneous with weights $w = (w_1, w_2) \in \mathbb{N}^2$ and quasi-degree $d \in \mathbb{N}$, if $G(\lambda^{w_1} x, \lambda^{w_2} y) = \lambda^d G(x, y)$, for all $\lambda > 0$.

COROLLARY 3.5 *Let G, $j^2 G(0,0) = 0$, be a quasi-homogeneous polynomial defining the vector field $X = (G_{xx} - G_{yy}, 2G_{xy})$ which is Newton non-degenerate. Then, the index of equation (3.3) does not depend on $F : \mathbb{R}^2, 0 \to \mathbb{R}, 0$, for any F such that $j^2 F(0,0) = 0$, and it is half of the index of the vector field X at 0.*

Proof: Let us assume that G is quasi-homogeneous with weights (α, β), $\alpha \le \beta$, and quasi-degree d. Then, while the function $2G_{xy}$ is clearly quasi-homogeneous of type $(\alpha, \beta; d - \alpha - \beta)$, this is not always true for the function $G_{yy} - G_{xx}$. However, a direct analysis of the possibilities shows that in any case, the remainders $\Delta_{12}(F, G)$, $\Delta_{13}(F, G)$, $\Delta_{23}(F, G)$ satisfy the hypothesis of Theorem 3.2.   □

REFERENCES

[1] G. Bol. Über Nabelpunkte auf einer Eifläche. Math Z. 49:389–410 1943/1944.

[2] M. Brunella, M. Miari. Topological equivalence of a plane vector field with its principal part defined through Newton Polyhedra. Jour. Diff. Eq. 85:338–366, 1990.

[3] R.A. Garcia, D.K.H. Mochida, M.C. Romero Fuster, M.A.S. Ruas. Inflection points and topology of surfaces in 4−space. Trans. Amer. Math. Soc. 352:3029–3043, 2000.

[4] C. Gutierrez, F. Mercury, F. Sánchez-Bringas. On a Carathéodorys Conjecture: analyticity versus smoothness. Experimental Mathematics, 5:133–38, 1996.

[5] V. Guillemin, A. Pollack. Differential topology. Englewood Cliffs, N. J., Prentice-Hall, 1974

[6] H. Hamburguer. Beweis einer Carathéodoryschen Vermütung. Ann. of Math. 41:63–68, 1940. II, III, Acta Math. 73:174–332, 1941.

[7] T. Klotz. On Bol's proof of Carathéodory's Conjecture. Comm. Pure Appl. Math. 12:277–311, 1959.

[8] J.A. Little. On singularities of submanifolds of higher dimensional euclidean spaces. Ann. Mat. Pura et Appl. 83 4A:261–336, 1969.

[9] D.K.H. Mochida, M.C. Romero Fuster, M.A.S. Ruas. The geometry of surfaces in 4-space from a contact viewpoint. Geometriae Dedicata. 54:323–332, 1995.

[10] C.J. Titus. A proof of a conjecture of Loewner and of the conjecture of Carathéodory on umbilic points. Acta Math. 131:43–77, 1973.

Generic Singularities of H-Directions *

LUIS FERNANDO O. MELLO   Instituto de Ciências Escola Federal de Engenharia de Itajubá 37.500-000, Itajubá, MG, Brazil E-mail:lfmelo@cpd.efei.br

Dedicated to Professor Maurício M. Peixoto on the occasion of his 80th birthday

Abstract

On immersed surfaces in R^4 we can consider the *H-lines*, which are characterized by the condition that their tangent lines are carried over the directions of the mean curvature vector field. We call the singularities of these lines *H-singularities*. In this paper we analyze the H-singularities and we show that their generic local configurations are analogous to Darbouxian umbilic points on surfaces immersed in R^3.

1   INTRODUCTION

The local configurations of singularities of line fields on surfaces in R^4 are known in only very few cases: the local configurations of the axiumbilic points, which are the singularities of the axial curvature direction fields, studied by Gutierrez, Guadalupe, Tribuzy and Guíñez in [4] and by Garcia and Sotomayor in [3], and the local configurations of inflection points, which are the singularities of the asymptotic direction fields, studied by Garcia, Mochida, Fuster and Ruas in [5].

For the study of the theory of surfaces in R^4 see the papers of Wong [8], Little [6] and Asperti [1].

In this paper we study the H-singularities on oriented smooth surface immersed in R^4. They consist of the singularities of the H-lines, which are characterized by the condition that tangent lines are carried over the directions of the mean curvature vector field.

This paper is organized as follows:

---

* This work is supported in part by CAPES-Brazil.

In section 2 we give the precise definitions of H-lines and their singularities, the H-singularities. We analyze the differential equation of the H-lines and we show that it fits in the class of quadratic or binary differential equations.

In section 3 we characterize the H-singularities which are locally structurally stable. The characterization is expressed in a Monge chart. These singularities are analogous to Darbouxian umbilic points on surfaces in R^3 and are called H-singularities of S-type.

2   DIFFERENTIAL EQUATION FOR H-LINES

Let $\alpha : M^2 \to R^4$ be a C^r, $r \geq 4$, immersion of an oriented smooth surface into R^4, which is endowed with the euclidean inner product $\langle \cdot, \cdot \rangle$ and oriented by a once for all fixed orientation. Denote respectively by TM and NM the tangent and the normal bundles of α and by T_pM and N_pM the respective fibers, i.e., the tangent and the normal planes at $p \in M$. Assume that (u,v) is a positive chart of M and that $\{\alpha_u, \alpha_v, N_1, N_2\}$ form a positive frame of R^4 where $\{N_1, N_2\}$ is a frame of vector fields orthonormal to α. In such a chart (u,v) the first fundamental form of α, I_α, is given by

$$I = I_\alpha = \langle d\alpha, d\alpha \rangle = Edu^2 + 2Fdudv + Gdv^2,$$

where $E = \langle \alpha_u, \alpha_u \rangle$, $F = \langle \alpha_u, \alpha_v \rangle$ and $G = \langle \alpha_v, \alpha_v \rangle$.

The second fundamental form of α, II_α, is defined in terms of the NM-valued quadratic form

$$II = II_\alpha = \langle d^2\alpha, N_1 \rangle N_1 + \langle d^2\alpha, N_2 \rangle N_2 = II_{1,\alpha}N_1 + II_{2,\alpha}N_2,$$

where

$$II_i = II_{i,\alpha} = e_idu^2 + 2f_idudv + g_idv^2,$$

and $e_i = \langle \alpha_{uu}, N_i \rangle$, $f_i = \langle \alpha_{uv}, N_i \rangle$, and $g_i = \langle \alpha_{vv}, N_i \rangle$, for $i = 1,2$.

We have the following functions associated to α (for more details see [6]):

1. The *mean curvature vector* of α

$$H = H_\alpha = H_1N_1 + H_2N_2, \qquad (2.1)$$

where

$$H_i = H_{i,\alpha} = \frac{Eg_i - 2Ff_i + Ge_i}{2(EG - F^2)},$$

for $i = 1,2$;

2. The *normal curvature* of α

$$k_N = k_{N,\alpha} = \frac{(g_2 - e_2)f_1 + (e_1 - g_1)f_2}{EG - F^2}; \qquad (2.2)$$

3. The *resultant* Δ of $II_{1,\alpha}$ and $II_{2,\alpha}$

$$\Delta = \Delta_\alpha = \frac{1}{4} \begin{vmatrix} e_1 & 2f_1 & g_1 & 0 \\ e_2 & 2f_2 & g_2 & 0 \\ 0 & e_1 & 2f_1 & g_1 \\ 0 & e_2 & 2f_2 & g_2 \end{vmatrix}; \qquad (2.3)$$

4. The *normal curvature vector* of α defined by $\eta : TM \to NM$, where $\eta(p,v) = \frac{II(p,v)}{I(p,v)}$.

We have the following results:

- The image of the unitary circle S^1 of T_pM by $\eta(p) : T_pM \to N_pM$ describes an ellipse in N_pM called *ellipse of curvature* of α at p and denoted by $\varepsilon_\alpha(p)$;

- This ellipse may degenerate into a line segment, a circle or a point;

- The center of the ellipse of curvature is the mean curvature vector H. Therefore the ellipse of curvature $\varepsilon_\alpha(p)$ can be expressed as $(\eta - H) + H$;

- The area of $\varepsilon_\alpha(p)$ is given by $\frac{\pi}{2}|k_N(p)|$;

- The map $\eta(p)$ restricted to S^1 being quadratic, is a double covering of the ellipse of curvature $\varepsilon_\alpha(p)$. Thus every point of the ellipse corresponds to two diametrically opposed points of the unitary tangent circle;

- The ellipse of curvature is invariant by rotations in the tangent and normal planes.

A point $p \in M$ is called a *minimal point* of α if $H(p) = 0$ and it is called an *inflection point* of α if $\Delta(p) = k_N(p) = 0$. It follows that p is an inflection point if and only if

$$e_1 f_2 - e_2 f_1 = e_1 g_2 - e_2 g_1 = f_1 g_2 - f_2 g_1 = 0.$$

Thus $p \in M$ is an inflection point if and only if the ellipse of curvature $\varepsilon_\alpha(p)$ is a radial line segment.

Considering the above results we have the following remark: from any well-defined continous choice of points on the ellipse of curvature, continous *tangent direction fields* may be constructed on M. If the choice is not well-defined for special points of M we say that they are *singular points* of the direction field. Consider some examples of this remark.

Asymptotic direction fields. Suppose that the origin of N_pM does not belong to the region bounded by the ellipse of curvature $\varepsilon_\alpha(p), \forall p \in M$. The two points on $\varepsilon_\alpha(p)$ at which the lines generated by the normal curvature vectors are tangents to $\varepsilon_\alpha(p)$ induce a pair of directions in T_pM, which in general is not orthogonal. Making this construction for all $p \in M$ we define two tangent direction fields on M, called *asymptotic direction fields* (see figure 1). The singularities of these fields are the points where the ellipse of curvature becomes a radial line segment, i.e., the *inflection points*. Garcia, Mochida, Fuster and Ruas studied these direction fields and their singularities in [5].

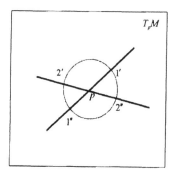
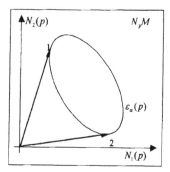

Figure 1 Asymptotic direction fields.

Axial curvature direction fields. The two maxima and two minima of the norm of $(\eta - H)(p)$ determine four points on the ellipse of curvature $\varepsilon_\alpha(p)$ which are their vertices. This construction induces eight points on the unitary circle $S^1 \subset T_p M$ which define four directions in the tangent plane. Making this construction for all $p \in M$ we define a *field of 2-crossings* on M, called *axial curvature direction fields* (see figure 2). This construction fails at the *axiumbilic points*, where the ellipse of curvature becomes a circle or a point. Gutierrez, Guadalupe, Tribuzi and Guíñez in [4] and Garcia and Sotomayor in [3] studied this case.

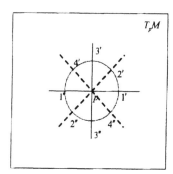
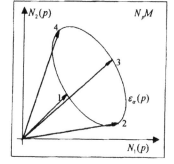

Figure 2 Axial curvature direction fields.

H-direction fields. The line spanned by the mean curvature vector $H(p)$ will meet the ellipse of curvature $\varepsilon_\alpha(p)$ in two points. This construction induces two orthogonal directions on $T_p M$. Making this construction for all $p \in M$ we define two direction fields on M, called *H-direction fields* (see figure 3). The singularities of these fields, called *H-singularities*, are the points where $H = 0$ (minimal points) or at which the ellipse of curvature becomes a radial line segment (inflection points). The set of H-singularities will be denoted by $S(\alpha)$.

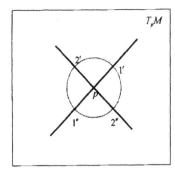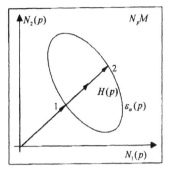

Figure 3 *H*-direction fields.

An *H-line* is a regular curve $\varphi : (a, b) \to M$ such that at each of its points the tangent line is an *H*-direction and contains any regular curve with this property with intersects it.

The differential equation of the *H*-lines is given by

$$\eta \wedge H = 0, \tag{2.4}$$

which is a quadratic differential equation of the form

$$A(u, v)du^2 + 2B(u, v)dudv + C(u, v)dv^2 = 0, \tag{2.5}$$

where

$$A = A(u, v) = (e_1 g_2 - e_2 g_1)E + 2(e_2 f_1 - e_1 f_2)F, \tag{2.6}$$

$$B = B(u, v) = (f_1 g_2 - f_2 g_1)E + (e_2 f_1 - e_1 f_2)G, \tag{2.7}$$

$$C = C(u, v) = 2(f_1 g_2 - f_2 g_1)F + (e_2 g_1 - e_1 g_2)G. \tag{2.8}$$

The *H*-singularities are determined by $A = B = C = 0$ in (2.5). But is immediate that the equation $EC = 2FB - GA$ holds.

We have established the following .

PROPOSITION 2.1 Let $\alpha : M^2 \to R^4$ be a C^r, $r \geq 4$, immersion of a smooth and oriented surface in R^4. With the above notations we have:

1. The differential equation of the *H*-lines is given by (2.5);

2. The *H*-singularities of α are given by $A = B = 0$, where A and B are defined in (2.6) and (2.7) respectively.

In isothermic coordinates where $E = G = \lambda^2$ and $F = 0$, equation (2.5) has the form

$$A_1(u,v)du^2 + 2B_1(u,v)dudv - A_1(u,v)dv^2 = 0, \tag{2.9}$$

where

$$A_1 = e_1 g_2 - e_2 g_1, \tag{2.10}$$

and

$$B_1 = f_1(e_2 + g_2) - f_2(e_1 + g_1). \tag{2.11}$$

LEMMA 2.2 $A(p) = B(p) = 0$ if and only if p is either a minimal point an inflection point of M.

Proof: It is enough to prove the lemma in the isothermic case. We make

$$0 = A_1 f_1 + B_1 g_1 = (e_1 + g_1)(f_1 g_2 - f_2 g_1)$$

and

$$0 = A_1 f_2 + B_1 g_2 = (e_2 + g_2)(f_1 g_2 - f_2 g_1).$$

If $f_1 g_2 - f_2 g_1 \neq 0$ then $e_1 + g_1 = e_2 + g_2 = 0$ and this implies that $H(p) = 0$, i.e., p is a minimal point. If $f_1 g_2 - f_2 g_1 = 0$ then $e_1 f_2 - e_2 f_1 = e_1 g_2 - e_2 g_1 = f_1 g_2 - f_2 g_1 = 0$ and this implies that p is an inflection point. The reciprocal is immediate.

From the lemma 2.2 we have $S(\alpha) = S_1(\alpha) \cup S_2(\alpha)$, where $S_1(\alpha)$ is the set of minimal points of M and $S_2(\alpha)$ is the set of inflection points of M.

3 H-SINGULARITIES OF S-TYPE

We take the surface M in a Monge chart, i.e., the surface M is the graph of the map $\alpha(u,v) = (u, v, S(u,v), R(u,v))$, where S and R are C^r, $r \geq 4$, functions defined on a neighborhood $U \subset R^2$ of $(0,0)$ with the conditions

$$S(0,0) = R(0,0) = S_u(0,0) = R_u(0,0) = S_v(0,0) = R_v(0,0) = 0.$$

For each point $\alpha(u,v) \in M$ the tangent plane $T_{\alpha(u,v)}M$ is generated by

$$\{\alpha_u(u,v) = (1, 0, S_u(u,v), R_u(u,v)), \alpha_v(u,v) = (0, 1, S_v(u,v), R_v(u,v))\}$$

and the normal plane $N_{\alpha(u,v)}M$ is generated by $\{\bar{N}_1, \bar{N}_2\}$, where

$$\bar{N}_1 = (-S_u, -S_v, 1, 0)$$

and

$$\bar{N}_2 = (-R_u(1+S_v{}^2)+S_u S_v R_v, -R_v(1+S_u{}^2)+S_u S_v R_v, -S_u R_u - S_v R_v, 1+S_u{}^2+S_v{}^2).$$

Therefore
$$E = \langle \alpha_u, \alpha_u \rangle, F = \langle \alpha_u, \alpha_v \rangle, G = \langle \alpha_v, \alpha_v \rangle,$$
$$e_i = \langle \alpha_{uu}, N_i \rangle, f_i = \langle \alpha_{uv}, N_i \rangle, g_i = \langle \alpha_{vv}, N_i \rangle,$$

where
$$N_i = \frac{\bar{N}_i}{|\bar{N}_i|},$$

for $i = 1, 2$.

Write the Taylor expansion of the functions S and R near $(0,0)$

$$S(u,v) = \frac{s_{20}}{2}u^2 + s_{11}uv + \frac{s_{02}}{2}v^2 + \frac{a}{6}u^3 + \frac{d}{2}u^2v + \frac{b}{2}uv^2 + \frac{c}{6}v^3 + O(4), \qquad (3.1)$$

$$R(u,v) = \frac{r_{20}}{2}u^2 + r_{11}uv + \frac{r_{02}}{2}v^2 + \frac{\bar{a}}{6}u^3 + \frac{\bar{d}}{2}u^2v + \frac{\bar{b}}{2}uv^2 + \frac{\bar{c}}{6}v^3 + O(4). \qquad (3.2)$$

Thus the coefficients of the first and the second fundamental forms in the Monge chart are given by

$$E = 1 + O(2), F = O(2), G = 1 + O(2),$$

$$e_1 = s_{20} + au + dv + O(2), f_1 = s_{11} + du + bv + O(2), g_1 = s_{02} + bu + cv + O(2),$$

$$e_2 = r_{20} + \bar{a}u + \bar{d}v + O(2), f_2 = r_{11} + \bar{d}u + \bar{b}v + O(2), g_2 = r_{02} + \bar{b}u + \bar{c}v + O(2).$$

We define

$$J = s_{02}r_{20} - s_{20}r_{02}, K = s_{02}\bar{a} + r_{20}b - s_{20}\bar{b} - r_{02}a, L = s_{02}\bar{d} + r_{20}c - s_{20}\bar{c} - r_{02}d,$$

$$M = s_{11}(r_{20}+r_{02}) - r_{11}(s_{20}+s_{02}), N = s_{11}(\bar{a}+\bar{b}) - r_{11}(a+b) + (r_{20}+r_{02})d - (s_{20}+s_{02})\bar{d},$$

$$P = s_{11}(\bar{c}+\bar{d}) - r_{11}(c+d) + (r_{20}+r_{02})b - (s_{20}+s_{02})\bar{b}.$$

The differential equation of the *H*-lines (2.5) in the Monge chart has the form

$$C(u,v)dv^2 + 2B(u,v)dudv + A(u,v)du^2 = 0, \qquad (3.3)$$

where
$$C(u,v) = J + Ku + Lv + Q_1(u,v),$$
$$B(u,v) = M + Nu + Pv + Q_2(u,v),$$
$$A(u,v) = -J - Ku - Lv + Q_3(u,v),$$

Q_1, Q_2 and Q_3 are of order $O(2)$.

At the point $(0,0)$ we have

$$\alpha_u(0,0) = (1,0,0,0), \alpha_v(0,0) = (0,1,0,0), N_1(0,0) = (0,0,1,0), N_2(0,0) = (0,0,0,1),$$

$$E(0,0) = G(0,0) = 1, F(0,0) = 0,$$

$$e_1(0,0) = s_{20}, f_1(0,0) = s_{11}, g_1(0,0) = s_{02},$$

$$e_2(0,0) = r_{20}, f_2(0,0) = r_{11}, g_2(0,0) = r_{02}.$$

From (3.3), the condition for $(0,0)$ to be an H-singularity is that

$$J = s_{02}r_{20} - s_{20}r_{02} = 0$$

and

$$M = s_{11}(r_{20} + r_{02}) - r_{11}(s_{20} + s_{02}) = 0.$$

We put $p = \frac{dv}{du}$ and from (3.3) we consider the function

$$T(u,v,p) = (J+Ku+Lv+Q_1)p^2 + 2(M+Nu+Pv+Q_2)p - (J+Ku+Lv+Q_3). \quad (3.4)$$

We define the surface $W = T^{-1}(0)$. If $(0,0)$ is an H-singularity then $(0,0,p) \in W$, for all p. Furthermore the surface W is smooth in a neighborhood of the axis p if and only if $(0,0)$ is a transversal H-singularity, according to [7] and [2].

Suppose that $(0,0)$ is a transversal H-singularity. In this case we consider the following vector field on W

$$X(u,v,p) = (T_p(u,v,p), pT_p(u,v,p), -(T_u(u,v,p) + pT_v(u,v,p))). \quad (3.5)$$

The vector field X has generically either one or three singularities on axis p, which are either saddles or nodes. Furthermore we have only one of the following three possibilities according to [7] and [2]:

1. The vector field X has one singularity of saddle type. In this case $(0,0)$ is a transversal H-singularity of the type S_1 (figure 4);

2. The vector field X has three singularities, one node and two saddles. In this case $(0,0)$ is a transversal H-singularity of the type S_2 (figure 4);

3. The vector field X has three singularities of saddle type. In this case $(0,0)$ is a transversal H-singularity of the type S_3 (figure 4).

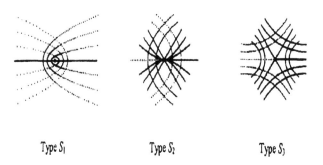

Type S_1 Type S_2 Type S_3

Figure 4 H-singularities of S-type.

REMARK 3.1 The index $i = 1,2,3$ of S_i denotes the number of *separatrices* of the H-singularity. These are H-lines which approach the H-singularity and which separate regions of different patterns of approach to it. Figure 4 shows the local *H*-configuration of the three different types of *H*-singularity denoted by S_1, S_2 and S_3 and called *H-singularity of S-type*. The index is $\frac{1}{2}$ in S_1 and S_2, and $-\frac{1}{2}$ in S_3.

Let $I^k = I^k(M^2, R^4)$ be the space of C^k immersions of M into R^4, where M is a smooth oriented surface.

THEOREM 3.2 Let $I_c^k \subset I^k$ be the space of C^k immersions of M into R^4, where M is a compact smooth oriented surface, endowed with the C^k topology. Then the subset T^k of immersions for which all *H*-singularities are of the *S*-type is open and dense in I_c^k, for all $k \geq 4$.

Proof: The proof of this theorem follows the same lines as the Darbouxian case treated in detail in [7].

3.1   The case of minimal points

Suppose that $p \in M$ satisfies $H(p) = 0$ and $k_N(p) \neq 0$. By a rotation in the normal plane it is possible to write $s_{20} = s_{02} = 0$. By a rotation in the uv-plane it is possible to write $b = -a$. Thus (3.1) and (3.2) can be written as

$$S(u,v) = Auv + \frac{a}{6}u^3 + \frac{d}{2}u^2v - \frac{a}{2}uv^2 + \frac{c}{6}v^3 + O(4), \qquad (3.1.1)$$

$$R(u,v) = \frac{B}{2}(u^2 - v^2) + Cuv + \frac{\bar{a}}{6}u^3 + \frac{\bar{d}}{2}u^2v + \frac{\bar{b}}{2}uv^2 + \frac{\bar{c}}{6}v^3 + O(4). \qquad (3.1.2)$$

This implies that
$$J = 0, K = 0, L = B(c+d), M = 0,$$
$$N = A(\bar{a} + \bar{b})$$

and
$$P = A(\bar{c} + \bar{d}) - C(c+d).$$

We define
$$W_1 = B(c+d),$$
$$W_2 = A(\bar{a} + \bar{b})$$

and
$$W_3 = C(c+d) - A(\bar{c} + \bar{d}).$$

LEMMA 3.1.1 With the above construction $p \in M$ is a transversal *H*-singularity if and only if
$$W_1 W_2 \neq 0. \qquad (3.1.3)$$

Proof: The condition of transversality is given by $NL \neq 0$, which is equivalent to (3.1.3).

THEOREM 3.1.2 Consider a transversal *H*-singularity for which $H(p) = 0$ and $k_N(p) \neq 0$. Then we have:

1. If

$$\frac{W_2}{W_1} > \frac{1}{2}\left[\left(\frac{W_3}{W_1}\right)^2 + 1\right],$$ (3.1.4)

so the H-singularity is of type S_1;

2. If

$$\frac{1}{2}\left[\left(\frac{W_3}{W_1}\right)^2 + 1\right] > \frac{W_2}{W_1} > 0, 2W_2 \neq W_1,$$ (3.1.5)

so the H-singularity is of type S_2;

3. If

$$\frac{W_2}{W_1} < 0,$$ (3.1.6)

so the H-singularity is of type S_3.

Proof: The singularities of the vector field X (3.5) on the axis p are given by

$$\varphi(p) = T_u(0,0,p) + pT_v(0,0,p) = 0.$$

But

$$0 = \varphi(p) = p\left[B(c+d)p^2 - 2\left(C(c+d) - A(\bar{c}+\bar{d})\right)p + 2A(\bar{a}+\bar{b}) - B(c+d)\right].$$

Therefore the vector field X has:

1. One singularity which is a saddle if and only if (3.1.4) holds;

2. Three singularities, namely one node and two saddles, if and only if (3.1.5) holds;

3. Three singularities which are all saddles if and only if (3.1.6) holds.

The local configurations of the H-singularities of S-type illustrated in Figure 4 are obtained by projection on the uv-plane of the local configurations of the above singularities, according to [7].

3.2   The case of inflection points

Suppose that $p \in M$ is an inflection point and that $H(p) \neq 0$. By a rotation in the normal plane it is possible to write $r_{20} = r_{02} = r_{11} = 0$. By a rotation in the uv-plane it is possible to write $\bar{b} = \bar{a}$. Thus (3.1) and (3.2) can be written as

$$S(u,v) = \frac{A}{2}(u^2 + v^2) + Buv + O(3),$$ (3.2.1)

$$R(u,v) = \frac{\bar{a}}{6}u^3 + \frac{\bar{d}}{2}u^2v + \frac{\bar{a}}{2}uv^2 + \frac{\bar{c}}{6}v^3 + O(4).$$ (3.2.2)

This implies that

$$J = 0, K = 0, L = A(\bar{d} - \bar{c}), M = 0,$$

$$N = 2(\bar{a}B - A\bar{d})$$

and

$$P = B(\bar{c} + \bar{d}) - 2A\bar{a}.$$

We define

$$\bar{W}_1 = A(\bar{d} - \bar{c}),$$

$$\bar{W}_2 = \bar{a}B - A\bar{d}$$

and

$$\bar{W}_3 = B(\bar{c} + \bar{d}) - 2A\bar{a}.$$

LEMMA 3.2.1 With the above construction $p \in M$ is a transversal *H*-singularity if and only if

$$\bar{W}_1 \bar{W}_2 \neq 0. \tag{3.2.3}$$

Proof: The condition of transversality is given by $NL \neq 0$, which is equivalent to (3.2.3).

THEOREM 3.2.2 Consider a transversal *H*-singularity which is an inflection point and $H(p) \neq 0$. Then we have:

1. If

$$\frac{\bar{W}_2}{\bar{W}_1} > \frac{1}{4}\left[\left(\frac{\bar{W}_3}{\bar{W}_1}\right)^2 + 1\right], \tag{3.2.4}$$

   then the *H*-singularity is of type S_1;

2. If

$$\frac{1}{4}\left[\left(\frac{\bar{W}_3}{\bar{W}_1}\right)^2 + 1\right] > \frac{\bar{W}_2}{\bar{W}_1} > 0, 4\bar{W}_2 \neq \bar{W}_1, \tag{3.2.5}$$

   then the *H*-singularity is of type S_2;

3. If

$$\frac{\bar{W}_2}{\bar{W}_1} < 0, \tag{3.2.6}$$

   then the *H*-singularity is of type S_3.

Proof: The proof follows the same lines as that of theorem 3.1.2.

Acknowledgment: The results here presented are part of my doctoral dissertation at Universidade de São Paulo (IME). I wish to thank my advisor, Professor Jorge Sotomayor, for helpful suggestions.

REFERENCES

[1] A.C. Asperti. Immersions of surfaces into 4-dimensional spaces with nonzero normal curvature. Ann. Mat. Pura Appl. 125:313–329, 1980.

[2] J.W. Bruce, F. Tari. On binary differential equations. Nonlinearity 8:255–271, 1995.

[3] R. Garcia, J. Sotomayor. Lines of axial curvature on surfaces immersed in R^4. Diff. Geom. and its Applications 12:253–269, 2000.

[4] C. Gutierrez, I. Guadalupe, R. Tribuzy, V. Guíñez. Lines of curvature on surfaces immersed in R^4. Bol. Soc. Bras. Mat. 28:233–251, 1997.

[5] R. Garcia, D. Mochida, M.R. Fuster, M.A.S. Ruas. Inflection points and topology of surfaces in 4-space. Trans. Am. Math. Soc. 352:3029–3043, 2000.

[6] J.A. Little. On singularities of submanifolds of a higher dimensional Euclidean space. Ann. Mat. Pura Appl. 83:261–335, 1969.

[7] C. Gutierrez, J. Sotomayor. Lines of curvature and umbilical points. IMPA, Rio de Janeiro, 1991.

[8] W.C. Wong. A new curvature theory for surfaces in Euclidean 4-space. Comm. Math. Helv. 26:152–170, 1952.

Vertices of Curves on Constant Curvature Manifolds

CLAUDIA C. PANSONATO Depto Matemática - CCNE - UFSM 97119-900 Santa Maria-RS Brazil E-mail: pansonat@ccne.ufsm.br

SUELI I. R. COSTA Instituto de Matemática - Unicamp 13083-970 Campinas-SP Brazil E-mail: sueli@ime.unicamp.br

Abstract

A vertex of a curve on a Riemannian manifold of any dimension can be defined as a critical point of its first curvature at which the second curvature also vanishes. We prove here that under conformal parametrization there is a correspondence between Euclidean and Riemannian vertices if and only if the manifold has constant curvature. A four-vertex theorem for spherical curves is then obtained as a corollary. Special results and a characterization in terms of singularities of the evolute surface for the three dimensional case are also presented.

1 INTRODUCTION

This paper is an extension of some concepts and results concerning higher order singularities and geometrical contact of curves to constant curvature manifolds. It is an intrisic Riemannian approach of curvatures, contact and vertices via conformal parametrizations.

The connection between higher order singularities of height and distance functions and geometrical contact of Euclidean curves have been studied in different ways by several authors since long ago. We may quote, for instance, works like [2], [3], [13], [1], [7] and [11].

A vertex of a plane curve can be viewed as a critical point of its curvature and also as a contact point of order at least three (a 4-point contact in [4] notation) with the osculating circle. This characterization can be naturally extended to higher dimensional Riemannian manifold in terms of a higher order singularity of the geodesic curvature functions. On the other hand, it allows, for constant curvature manifolds, a conformal approach like the one done in [5].

Previous works more related to what is done here are [6] and [16]. This last one is linked to [2] and concerning vertices of a curve, defined as higher contact with the osculating hypersphere, in the hyperbolic n-dimensional space (Euclidean approach). We may say that the approach here besides being Riemannian is "dual" to that one in the sense that we deal with singularities linked to higher order contact with osculating circles. We point out that, for a curve on a three-dimensional manifold, both types of vertices determine singularities on the evolute surface.

In section 2 we deduce general expressions for Riemannian first and second curvatures of a curve on a n-dimensional manifold and specialize them for the constant curvature ones. Riemannian contact is discussed and it is stated that the vanishing of the first and the second curvature corresponds to contact of order two and three, respectively, with a totally geodesic submanifold (Proposition 2). The main result in section 3 is proved in Theorem 1 which states that under a conformal parametrization vertices of curves correspond to Euclidean vertices if and only if the manifold has constant curvature. This extends a result presented in [6] for two-dimensional manifolds. A Four-Vertex Theorem for spherical curves is then obtained as a corollary (Theorem 2). Section 4 is devoted to curves on three-dimensional manifolds. Particular results (Theorems 3 and 4), also linked to the second curvature changes of sign, are deduced and for spherical curves vertices are characterized as cuspidal edges of the evolute surface.

2   HIGHER ORDER CONTACT BETWEEN CURVES AND TOTALLY GEODESIC SUBMANIFOLDS

We consider here regular C^4- curves, $\alpha : I \to M$ on a n-dimensional Riemannian manifold M, $n \geq 3$. We denote by k_{g_1} and k_{g_2} the Riemannian first and second curvature functions of α.

For a conformal parametrization $\psi : U \to M$, where $U \subset \mathbb{R}^n$ is an open set, we denote by $g_{ij} = \delta_{ij} G^2$ the metric coefficients. We start getting expressions for the Riemannian curvatures of α in terms of the Euclidean curvatures of its conformal pre-image $\beta = \psi^{-1} \circ \alpha$ which are to be used in the following sections. Considering constant curvature manifolds we discuss relations between Riemannian contact and the vanishing of the first and the second curvatures.

2.1 General Expressions for Riemannian Curvatures

In the expressions to be deduced we follow the notations established in [14]: s is the Riemannian arc length of α, s_e the Euclidean arc length of β; k_1, k_2 are the curvatures of β; $\langle,\rangle$ is the inner product in M and $\langle,\rangle_e$ is the Euclidean inner product; $\mathbf{V}_i$, $i = 1, \cdots, n$ are the Frenet frame vectors of $\alpha \subset M$ in the tangent space and $\mathbf{v}_i$, $i = 1, \cdots, n$ is the Frenet frame of $\beta \subset \mathbb{R}^n$.

We start determining an expression for the first Riemannian curvature of α in terms of the metric G and the Euclidean first curvature of its pre-image. Since $k_{g_1}(s) = \left| \dfrac{D'\mathbf{V}_1}{ds} \right|$, where $\dfrac{D'\mathbf{V}_1}{ds}$ is the covariant derivative of $\mathbf{V}_1 = \dfrac{d\alpha}{ds}$, we must develop an appropriate expression for the covariant derivative. If $x_i(s)$, $i = 1, \cdots, n$ are the coordinates of $\beta(s)$,

$$\frac{D'\mathbf{V}_1}{ds} = \sum_{k=1}^{n} \left\{ \frac{d^2 x_k}{ds^2} + \sum_{ij} \Gamma_{ij}^k \frac{dx_j}{ds} \frac{dx_i}{ds} \right\} \frac{\partial}{\partial x_k}.$$

Calculating the Christoffel symbols Γ_{ij}^k of the metric $g_{ij} = \delta_{ij} G^2$, we get $\Gamma_{ij}^k = 0$ if the three indices are different, and

$$\Gamma_{ij}^i = \frac{G_{x_j}}{G}, \quad \Gamma_{ii}^j = -\frac{G_{x_j}}{G}, \quad \Gamma_{ii}^i = \frac{G_{x_i}}{G}.$$

Then, we have

$$\frac{D'\mathbf{V}_1}{ds} = \sum_{k=1}^{n} \left\{ \frac{d^2 x_k}{ds^2} + \frac{G_{x_k}}{G} \left[\left(\frac{dx_k}{ds} \right)^2 - \sum_{i \neq k} \left(\frac{dx_i}{ds} \right)^2 \right] \right. $$
$$\left. + 2 \frac{dx_k}{ds} \left[\sum_{i \neq k} \frac{G_{x_i}}{G} \frac{dx_i}{ds} \right] \right\} \frac{\partial}{\partial x_k}.$$

Let θ_i, $i = 1, \cdots n$, be the angles between α and the i-th coordinate curve. Then:

$$\cos\theta_i = G \frac{dx_i}{ds} \quad \text{and} \quad -\sin\theta_i \frac{d\theta_i}{ds} = \left[\sum_{j=1}^{n} G_{x_j} \frac{dx_j}{ds} \right] \frac{dx_i}{ds} + G \frac{d^2 x_i}{ds^2}.$$

Replacing these values in the above expression and taking $\mathbf{t}_k = \dfrac{1}{G} \dfrac{\partial}{\partial x_k}$, lead us to

$$\frac{D'\mathbf{V}_1}{ds} = \sum_{k=1}^{n} \left\{ -\sin\theta_k \frac{d\theta_k}{ds} - \cos\theta_k \langle \nabla \left(\frac{1}{G} \right), \mathbf{v}_1 \rangle_e + \frac{\partial}{\partial x_k} \left(\frac{1}{G} \right) \right\} \mathbf{t}_k, \qquad (2.1)$$

where $\nabla \left(\dfrac{1}{G} \right)$ is the gradient and $\mathbf{v}_1 = \dfrac{d\beta}{ds_e} = (\cos\theta_1, \cdots, \cos\theta_n)$.

We can also express the Euclidean curvature k_1 and vector $\mathbf{v}_2$ in terms of θ_i.

Since $\dfrac{d^2\beta}{ds_e^2} = G \left(-\sin\theta_1 \dfrac{d\theta_1}{ds}, \cdots, -\sin\theta_n \dfrac{d\theta_n}{ds} \right)$,

$$k_1 = G \sqrt{\sum_{i=1}^{n} \left(\sin \theta_i \frac{d\theta_i}{ds} \right)^2} \qquad (2.2)$$

and

$$\mathbf{v}_2 = \frac{G}{k_1} \left(-\sin \theta_1 \frac{d\theta_1}{ds}, \cdots, -\sin \theta_n \frac{d\theta_n}{ds} \right). \quad (\text{if } k_1 \neq 0) \qquad (2.3)$$

So from (2.1) and (2.2) we get an extension of *Liouville Formula* [15] for n-dimensional manifolds:

$$k_{g_1} = \left(\frac{k_1^2}{G^2} - 2 \sum_{k=1}^{n} \frac{\partial}{\partial x_k} \left(\frac{1}{G} \right) \sin \theta_k \frac{d\theta_k}{ds} - \langle \nabla \left(\frac{1}{G} \right), \mathbf{v}_1 \rangle_e^2 + \left| \nabla \left(\frac{1}{G} \right) \right|_e^2 \right)^{1/2}. \quad (2.4)$$

If $k_1 \neq 0$, an alternative expression for k_{g_1} can be given:

$$k_{g_1} = \left(\frac{k_1^2}{G^2} + \frac{2k_1}{G} \langle \nabla \left(\frac{1}{G} \right), \mathbf{v}_2 \rangle_e - \langle \nabla \left(\frac{1}{G} \right), \mathbf{v}_1 \rangle_e^2 + \left| \nabla \left(\frac{1}{G} \right) \right|_e^2 \right)^{1/2}. \quad (2.5)$$

We will not deduce here an explicit expression for the second Riemannian curvature. This is done (in the next Proposition 1) just for constant curvature manifolds. If $k_{g_1} \neq 0$, $\mathbf{V}_2 = \frac{1}{k_{g_1}} \frac{D' \mathbf{V}_1}{ds}$ and k_{g_2} is defined by

$$\begin{aligned} |k_{g_2}| &= \left| \frac{D' \mathbf{V}_2}{ds} + k_{g_1} \mathbf{V}_1 \right| \\ &= \left| -\frac{1}{k_{g_1}^2} \frac{dk_{g_1}}{ds} \frac{D' \mathbf{V}_1}{ds} + \frac{1}{k_{g_1}} \frac{D'^2 \mathbf{V}_1}{ds^2} + k_{g_1} \mathbf{V}_1 \right|. \end{aligned} \qquad (2.6)$$

Calculating the covariant derivative,

$$\begin{aligned} \frac{D'^2 \mathbf{V}_1}{ds^2} = \sum_{k=1}^{n} \Big\{ &- \cos \theta_k \left(\frac{d\theta_k}{ds} \right)^2 - \sin \theta_k \frac{d^2 \theta_k}{ds^2} \\ &+ \cos \theta_k \left(-k_{g_1}^2 + \frac{k_1^2}{G^2} \right) + \sin \theta_k \frac{d\theta_k}{ds} \langle \nabla \left(\frac{1}{G} \right), \mathbf{v}_1 \rangle_e \\ &- \cos \theta_k \langle \frac{d}{ds} \nabla \left(\frac{1}{G} \right), \mathbf{v}_1 \rangle_e + \frac{1}{G} \sum_{i=1}^{n} \frac{\partial^2}{\partial x_k \partial x_i} \left(\frac{1}{G} \right) \cos \theta_i \Big\} \mathbf{t}_k. \end{aligned} \qquad (2.7)$$

The two first terms of the last summation can be developed as follows:

• If $k_1 \neq 0$,

$$\begin{aligned} \cos \theta_k \left(\frac{d\theta_k}{ds} \right)^2 + \sin \theta_k \frac{d^2 \theta_k}{ds^2} = &-\frac{k_1}{G^2} \left(\frac{d\mathbf{v}_2}{ds_e} \right)_k + \\ &+ \sin \theta_k \frac{d\theta_k}{ds} \left(\langle \nabla \left(\frac{1}{G} \right), \mathbf{v}_1 \rangle_e + \frac{1}{Gk_1} \frac{dk_1}{ds_e} \right), \end{aligned} \qquad (2.8)$$

where $\left(\frac{d\mathbf{v}_2}{ds_e} \right)_k$ is the k-th coordinate of $\frac{d\mathbf{v}_2}{ds_e}$.

- If $k_1 = 0$,

$$\cos\theta_k \left(\frac{d\theta_k}{ds}\right)^2 + \sin\theta_k \frac{d^2\theta_k}{ds^2} = -\frac{1}{G^2}\left(\frac{d^3\beta}{ds_e^3}\right)_k, \tag{2.9}$$

where $\left(\frac{d^3\beta}{ds_e^3}\right)_k$ is the k-th coordinate of $\frac{d^3\beta}{ds_e^3}$.

Developing the part of $\frac{D^{2\prime}\mathbf{V}_1}{ds^2}$ which involves $\frac{d}{ds}\nabla\left(\frac{1}{G}\right)$, we get

$$
\begin{aligned}
\frac{D'^2\mathbf{V}_1}{ds^2} = \sum_{k=1}^{n} \Bigg\{ & -\cos\theta_k\left(\frac{d\theta_k}{ds}\right)^2 - \sin\theta_k\frac{d^2\theta_k}{ds^2} \\
& + \cos\theta_k\left(-k_{g_1}^2 + \frac{k_1^2}{G^2}\right) + \sin\theta_k\frac{d\theta_k}{ds}\left\langle\nabla\left(\frac{1}{G}\right),\mathbf{v}_1\right\rangle_e \\
& + \frac{1}{G}\Bigg\{\sum_{i\neq k}\left[\cos\theta_k\cos^2\theta_i\left(\frac{\partial^2}{\partial x_k^2}\left(\frac{1}{G}\right) - \frac{\partial^2}{\partial x_i^2}\left(\frac{1}{G}\right)\right)\right. \\
& + \frac{\partial^2}{\partial x_i \partial x_k}\left(\frac{1}{G}\right)\cos\theta_i\left(1 - 2\cos^2\theta_k\right)\Bigg] \\
& + \cos\theta_k \sum_{\substack{i,j\neq k \\ i\neq j}}\frac{\partial^2}{\partial x_i\partial x_j}\left(\frac{1}{G}\right)\cos\theta_i\cos\theta_j \Bigg\}\Bigg\}\mathbf{t}_k.
\end{aligned}
\tag{2.10}
$$

2.2 Constant Curvature Manifolds

For a manifold with constant curvature K it is always possible to find a conformal parametrization [14] such that the metric is given by

$$g_{ij} = \frac{\delta_{ij}}{(1 + \frac{K}{4}\sum_{i=1}^{n} x_i^2)^2}. \tag{2.11}$$

But this is equivalent to say (up to translation) that $g_{ij} = \delta_{ij}G^2$ satisfies

$$
\begin{cases}
\dfrac{\partial^2}{\partial x_i \partial x_j}\left(\dfrac{1}{G}\right) \equiv 0, \quad i\neq j \\[2mm]
\dfrac{\partial^2}{\partial x_i^2}\left(\dfrac{1}{G}\right) - \dfrac{\partial^2}{\partial x_j^2}\left(\dfrac{1}{G}\right) \equiv 0
\end{cases}
\tag{2.12}
$$

Then we have

PROPOSITION 2.1. *Let $\alpha : I \to M$ be a regular C^4- curve on a n-dimensional constant curvature manifold M and $k_{g_1}(s) \neq 0$. Under the above conformal parametrization, its second curvature, k_{g_2}, is given by*

1. *For $k_1 \neq 0$,*

$$
\begin{aligned}
|k_{g_2}| = \frac{1}{k_{g_1}G^2}\Bigg[& k_1^2\left|\frac{d\mathbf{v}_2}{ds_e}\right|_e^2 + \left(\frac{dk_1}{ds_e}\right)^2 - k_1^4 - \frac{1}{k_{g_1}^2}\left(k_1\left\langle\nabla\left(\frac{1}{G}\right),\frac{d\mathbf{v}_2}{ds_e}\right\rangle_e\right. \\
& + \frac{dk_1}{ds_e}\left(\frac{k_1}{G} + \left\langle\nabla\left(\frac{1}{G}\right),\mathbf{v}_2\right\rangle_e\right) + k_1^2\left\langle\nabla\left(\frac{1}{G}\right),\mathbf{v}_1\right\rangle_e\right)^2\Bigg]^{1/2}
\end{aligned}
\tag{2.13}
$$

2. For $k_1 = 0$,

$$|k_{g_2}| = \frac{1}{k_{g_1} G^2} \left[-\frac{1}{k_{g_1}^2} \left\langle \nabla \left(\frac{1}{G} \right), \frac{d^3\beta}{ds_e^3} \right\rangle_e^2 + \left| \frac{d^3\beta}{ds_e^3} \right|_e^2 \right]^{1/2}. \qquad (2.14)$$

Proof: We observe first that $\left\langle \dfrac{D'\mathbf{V}_1}{ds}, \mathbf{V}_1 \right\rangle = 0 \Rightarrow \left\langle \dfrac{D'^2\mathbf{V}_1}{ds^2}, \mathbf{V}_1 \right\rangle = -k_{g_1}^2$. Differentiating $k_{g_1}^2 = \left\langle \dfrac{D'\mathbf{V}_1}{ds}, \dfrac{D'\mathbf{V}_1}{ds} \right\rangle$ with respect to s, we also have,

$$\frac{dk_{g_1}}{ds} = \frac{1}{k_{g_1}} \left\langle \frac{D'\mathbf{V}_1}{ds}, \frac{D'^2\mathbf{V}_1}{ds^2} \right\rangle. \qquad (2.15)$$

Then, by (2.6), we get

$$|k_{g_2}| = \frac{1}{k_{g_1}} \left(\left| \frac{D'^2\mathbf{V}_1}{ds^2} \right|^2 - \left(\frac{dk_{g_1}}{ds} \right)^2 - k_{g_1}^4 \right)^{1/2}.$$

On the other hand, it follows from (2.10) and (2.12) that if M has constant curvature, $\dfrac{D'^2\mathbf{V}_1}{ds^2}$ is given by

$$\frac{D'^2\mathbf{V}_1}{ds^2} = \sum_{k=1}^{n} \left\{ -\cos\theta_k \left(\frac{d\theta_k}{ds} \right)^2 - \sin\theta_k \frac{d^2\theta_k}{ds^2} \right.$$
$$\left. + \cos\theta_k \left(-k_{g_1}^2 + \frac{k_1^2}{G^2} \right) + \sin\theta_k \frac{d\theta_k}{ds} \left\langle \nabla \left(\frac{1}{G} \right), \mathbf{v}_1 \right\rangle_e \right\} \mathbf{t}_k.$$

If $k_1 \neq 0$, we replace (2.8) in $\dfrac{D'^2\mathbf{V}_1}{ds^2}$ and get

$$\frac{D'^2\mathbf{V}_1}{ds^2} = \sum_{k=1}^{n} \left\{ \frac{k_1}{G^2} \left(\frac{d\mathbf{v}_2}{ds_e} \right)_k - \frac{1}{Gk_1} \frac{dk_1}{ds_e} \sin\theta_k \frac{d\theta_k}{ds} + \cos\theta_k \left(-k_{g_1}^2 + \frac{k_1^2}{G^2} \right) \right\} \mathbf{t}_k. \qquad (2.16)$$

Then, from (2.1), (2.15) and (2.16),

$$\frac{dk_{g_1}}{ds} = \frac{1}{k_{g_1} G^2} \left[k_1 \left\langle \nabla \left(\frac{1}{G} \right), \frac{d\mathbf{v}_2}{ds_e} \right\rangle_e + \frac{dk_1}{ds_e} \left(\frac{k_1}{G} + \left\langle \nabla \left(\frac{1}{G} \right), \mathbf{v}_2 \right\rangle_e \right) + k_1^2 \left\langle \nabla \left(\frac{1}{G} \right), \mathbf{v}_1 \right\rangle_e \right].$$
$$(2.17)$$

Summing up we finalize the proof of part 1. The proof of part 2 is done analogously utilizing (2.9) in $\dfrac{D'^2\mathbf{V}_1}{ds^2}$ and noting that $\left\langle \mathbf{v}_1, \dfrac{d^3\beta}{ds_e^3} \right\rangle = 0$ when $k_1 = 0$. $\qquad \square$

We now will establish relations between Riemannian contact and the vanishing of curvatures. The following definition extends the classical contact definition between curves in the Euclidean space $\mathbb{R}^n$.

DEFINITION 2.2. *Let* $\alpha : I \to M$ *and* $\gamma : I \to M$ *be regular* C^{m+1}- *curves on a constant curvature n-manifold parametrized by their arc lengths s and t, respectively. We say* α *and* γ *have* **Riemannian contact** *of order m (exactly) at a common point P if*

$$\alpha(s_0) \quad = \quad \gamma(t_0) = P$$

$$\left.\frac{d\alpha}{ds}\right|_{s=s_0} \quad = \quad \left.\frac{d\gamma}{dt}\right|_{t=t_0}$$

$$\left.\frac{D'\alpha'}{ds}\right|_{s=s_0} \quad = \quad \left.\frac{D'\gamma'}{dt}\right|_{t=t_0}$$

$$\vdots$$

$$\left.\frac{D'^{m-1}\alpha'}{ds^{m-1}}\right|_{s=s_0} \quad = \quad \left.\frac{D'^{m-1}\gamma'}{dt^{m-1}}\right|_{t=t_0}$$

$$\text{and}$$

$$\left.\frac{D'^{m}\alpha'}{ds^{m}}\right|_{s=s_0} \quad \neq \quad \left.\frac{D'^{m}\gamma'}{dt^{m}}\right|_{t=t_0}$$

REMARK: This geometric definition extends the usual contact notion in $\mathbb{R}^n$, which can also be formulated as $\lim\limits_{s \to s_0} \dfrac{d(\alpha(s), \gamma(s))}{s^k} = 0$ if and only if $k \leq n$, where d is the Euclidean distance. However, we may point out that, like in the Euclidean case, if two regular curves have same derivatives with respect to a common parameter, the same condition must be satisfied for their arc length derivatives. This means that the previous definition can be re-written as: *two curves have contact of order m (exactly) at a point* $P = \alpha(t_0) = \gamma(t_0)$ *if and only if for some parametrization we have*

$$\alpha(t_0) = \gamma(t_0), \cdots, \left.\frac{D'^{m-1}\alpha}{dt^{m-1}}\right|_{t=t_0} = \left.\frac{D'^{m-1}\gamma}{dt^{m-1}}\right|_{t=t_0} \text{ but } \left.\frac{D'^{m}\alpha}{dt^{m}}\right|_{t=t_0} \neq \left.\frac{D'^{m}\gamma}{dt^{m}}\right|_{t=t_0}.$$

For constant curvature manifolds the contact notion is preserved under conformal parametrization, namely

PROPOSITION 2.3. *Under the same conditions of definition 2.1,* α *and* γ *have Riemannian contact of order m (exactly) at a common point P if and only if their conformal pre-images have same order of contact at* $\psi^{-1}(P)$.

Proof: If $\mathbf{u}(s) = (u_1, \cdots, u_n)$ are the coordinates of a vector field defined on α, then using the metric expression (2.11) we get

$$\frac{D'\mathbf{u}}{ds} \quad = \quad \sum_{i=1}^{n} \left\{ \frac{du_k}{ds} - \frac{KG}{2} \left[u_k x_k \frac{dx_k}{ds} + \left(\sum_{i \neq k} x_i \frac{dx_i}{ds} \right) u_k \right.\right.$$

$$\left.\left. + \left(\sum_{i \neq k} x_i u_i \right) \frac{dx_k}{ds} - \left(\sum_{i \neq k} \frac{dx_i}{ds} u_i \right) x_k \right] \right\} \frac{\partial}{\partial x_k}.$$

From this and reminding that $\dfrac{ds_e}{ds} = \dfrac{1}{G}$, we can get through a recursive process that equality between covariant derivatives (with respect to s) is equivalent to equality between common derivatives (with respect to s_e). $\qquad\qquad \square$

Proposition 2 allows us do not distinguish between Riemannian and Euclidean contact. So, from now one, we just say contact when referred to constant curvature manifold.

DEFINITION 2.4. *A curve* $\alpha : I \to M$ *has contact of order m (exactly) with a submanifold $S \subset M$ at a point P if there exists at least one curve γ on S which has a contact of order m with α at P and there does not exist a curve on S which has a contact of order greater than m with α at P.*

The next result establishes the relation between the vanishing of the first and the second Riemannian curvatures and the contact with totally geodesic submanifolds, ie, submanifolds $N \subset M$ such that every geodesic in N with the induced metric is also a geodesic in M.

PROPOSITION 2.5. *Let $\alpha : I \to M$ be a regular C^4-curve on a n-dimensional constant curvature manifold and $P = \alpha(s_0)$. Then:*

  1. *$k_{g_1}(s_0) = 0$ if and only if α has contact of order 2 (at least) with a geodesic of M at P.*

     And, if $k_{g_1}(s_0) \neq 0$ at P,

  2. *$k_{g_2}(s_0) = 0$ if and only if α has contact of order 3 (at least) with a totally geodesic 2-submanifold of M.*

Proof: The proof is quite long and can be sketched as follows (see [10] for the extended version). Since M has constant curvature, its universal covering is isometric to $\mathbb{R}^n$, $\mathbb{S}^n$ or $\mathbb{H}^n$. We start showing that when $M = \mathbb{S}^n$ or $M = \mathbb{H}^n$, with the metric given in (2.11), the totally geodesic j-submanifolds of M are image either of the j-planes through the origin in $\mathbb{R}^n$ or j-spheres contained in $(j + 1)$-planes through the origin in $\mathbb{R}^n$ whose center C and ray r satisfy the condition

$$|C|_e^2 = r^2 - \frac{4}{K}. \tag{2.18}$$

The proof of the part 1 is then divided in two cases: $k_1 \neq 0$ and $k_1 = 0$. For the first case, we get from (2.5) and (2.11) that

$$k_{g_1} = \left[\left(\frac{k_1}{G} + \frac{K}{2} \langle \beta, \mathbf{v}_2 \rangle_e \right)^2 + \frac{K^2}{4} \left(|\beta|_e^2 - \langle \beta, \mathbf{v}_2 \rangle_e^2 - \langle \beta, \mathbf{v}_1 \rangle_e^2 \right) \right]^{1/2},$$

where $\beta = \psi^{-1}(\alpha)$.

So, $k_{g_1} = 0 \Leftrightarrow |\beta|_e^2 - \langle \beta, \mathbf{v}_1 \rangle_e^2 - \langle \beta, \mathbf{v}_2 \rangle_e^2 = 0$ and $\langle \beta, \mathbf{v}_2 \rangle_e^2 = \dfrac{-2k_1}{GK}$.

The first condition is equivalent to say that the osculating (Euclidean) plane must pass through the origin. The second condition implies that the center C of the Euclidean osculating circle $(C = \beta + \dfrac{1}{k_1} \mathbf{v}_2)$ satisfies

$$|C|_e^2 = -\frac{4}{K} + \frac{1}{k_1^2}.$$

And this implies, by (2.18) and the first condition, that the image of the osculating circle is a geodesic of M. Since the osculating circle has contact of order 2 (at least) with the curve, this completes the proof of part 1 for $k_1 \neq 0$.

If $k_1 = 0$, we can derive from (2.4) and (2.11) that

$$k_{g_1} = \frac{K}{2} \left[|\beta|_e^2 - \langle \beta, \mathbf{v}_1 \rangle_e^2 \right]^{1/2}.$$

Then, $k_{g_1} = 0 \Leftrightarrow \beta = \lambda \mathbf{v}_1$, $\lambda \in \mathbb{R}$, and this means that the image of the tangent line is a geodesic of M. Besides, from $k_1 = 0$ we have that the curve has contact of order 2 (at least) with the tangent line.

The proof of part 2 follows similar steps but now considering three different cases ($k_1 \neq 0$ and $k_2 \neq 0$; $k_1 \neq 0$ and $k_2 = 0$; $k_1 = 0$). We utilize (2.11) and the expressions for k_{g_2} given in Proposition 1. In each case we show that $k_{g_2} = 0$ is equivalent to either the osculating plane must pass through the origin or the center C and the ray r of osculating sphere must satisfy (2.18).   □

3   CORRELATING RIEMANNIAN TO EUCLIDEAN VERTICES

In this section we obtain an extension to the n-dimensional case of a result presented in [6] for a curve on a constant curvature manifold: its Riemannian vertices are correlated to the vertices of its conformal pre-image. For doing so we adapted a definition of vertex of an Euclidean curve from [5], which restricted to 2-dimensional manifolds or plane curves, is the standard one.

DEFINITION 3.1. *Let $\alpha : I \to M$ be a regular C^4- curve on a n-dimensional Riemannian manifold with never vanishing first curvature. A vertex is a point where* $k_{g_2} = \dfrac{dk_{g_1}}{ds} = 0.$

Developing the calculations presented in **2.1**, we get the following

LEMMA 3.2. *Under a conformal parametrization, Euclidean vertices of any curve in $\mathbb{R}^n$ correspond to Riemannian vertices of the curve on the manifold if and only if*

$$\begin{cases} \dfrac{\partial^2}{\partial x_i \partial x_j} \left(\dfrac{1}{G} \right) \equiv 0, \quad i \neq j \\[3mm] \dfrac{\partial^2}{\partial x_i^2} \left(\dfrac{1}{G} \right) - \dfrac{\partial^2}{\partial x_j^2} \left(\dfrac{1}{G} \right) \equiv 0 \end{cases} \tag{3.1}$$

Proof: Since the above equations are the same conditions of (2.12) and $k_1 \neq 0$ we have, using (2.17) in (2.13), that

$$k_{g_2}^2 = \frac{1}{k_{g_1}^2 G^4} \left[k_1^2 \left| \frac{d\mathbf{v}_2}{ds_e} \right|_e^2 + \left(\frac{dk_1}{ds_e} \right)^2 - k_1^4 - \frac{1}{k_{g_1}^2} \left(\frac{dk_{g_1}}{ds} \right)^2 \right].$$

Taking into account that $\dfrac{d\mathbf{v}_2}{ds_e} = -k_1 \mathbf{v}_1 + \mathbf{V}$, where $V \in [\mathbf{v}_1, \mathbf{v}_2]^\perp$, we have:

$$k_{g_2} = \frac{dk_{g_1}}{ds} = 0 \Leftrightarrow k_1 |\mathbf{V}|_e = \frac{dk_1}{ds_e} = 0 \Leftrightarrow k_2 = \frac{dk_1}{ds_e} = 0.$$

The reciprocal is not straightforward. We will show that if equations (3.1) are not satisfied, there are curves for which Riemannian vertices are not correspondent to the Euclidean ones.

Let us suppose, for instance, that $\dfrac{\partial^2}{\partial x_2 \partial x_3}\left(\dfrac{1}{G}\right) \neq 0$ at a point P. For any given unit vector $\mathbf{v}_1 = (\cos\theta_1, \cdots, \cos\theta_n)$ in $\mathbb{R}^n$ and any positive number r there is a circle through P with tangent vector $\mathbf{v}_1$ and radius r. We choose such a circle in order to get $k_{g_1} \neq 0$ at P.

Starting from directions $\mathbf{v}_1 = (\cos\theta, \sin\theta, 0, \cdots, 0)$ and $\mathbf{v}_2 = (-\sin\theta, \cos\theta, 0, \cdots, 0)$. For $\theta = \dfrac{\pi}{2}$, we have from (2.5) that

$$
\begin{aligned}
k_{g_1}^2 &= \frac{k_1^2}{G^2} - \frac{2k_1}{G}\frac{\partial}{\partial x_1}\left(\frac{1}{G}\right) - \left[\frac{\partial}{\partial x_2}\left(\frac{1}{G}\right)\right]^2 + \left|\nabla\left(\frac{1}{G}\right)\right|^2 \\
&= \left[\frac{k_1}{G} - \frac{\partial}{\partial x_1}\left(\frac{1}{G}\right)\right]^2 + \sum_{i=3}^{n}\left(\frac{\partial}{\partial x_i}\left(\frac{1}{G}\right)\right)^2.
\end{aligned}
$$

Varying the radius r $(r = \dfrac{1}{k_1})$ of the circles contained in the plane $x_1 x_2$, with tangent vector $\mathbf{v}_1$ (and $\theta = \dfrac{\pi}{2}$) we set $\dfrac{k_1}{G} - \dfrac{\partial}{\partial x_1}\left(\dfrac{1}{G}\right) \neq 0$ and then $k_{g_1} \neq 0$.

Since for circle $k_2 = \dfrac{dk_1}{ds_e} = 0$, from (2.8) we get

$$
\cos\theta_k \left(\frac{d\theta_k}{ds}\right)^2 + \sin\theta_k \frac{d^2\theta_k}{ds^2} = \frac{k_1^2}{G^2}\cos\theta_k + \sin\theta_k \frac{d\theta_k}{ds}\left\langle \nabla\left(\frac{1}{G}\right), \mathbf{v}_1 \right\rangle_e.
$$

Using this expression in (2.10) and $\mathbf{v}_1 = (0, 1, 0, \cdots, 0)$ results

$$
\frac{D'^2 \mathbf{V}_1}{ds^2} = -k_{g_1}^2 \, \mathbf{V}_1 + \frac{1}{G}\left\{\sum_{\substack{k=1 \\ k\neq 2}}^{n} \frac{\partial^2}{\partial x_2 \partial x_k}\left(\frac{1}{G}\right)\right\}\mathbf{t}_k,
$$

So, from (2.6) we get

$$
|k_{g_2}| = \left| -\frac{1}{k_{g_1}^2}\frac{dk_{g_1}}{ds}\frac{D'\mathbf{V}_1}{ds} + \frac{1}{Gk_{g_1}}\left\{\sum_{\substack{k=1 \\ k\neq 2}}^{n}\frac{\partial^2}{\partial x_2 \partial x_k}\left(\frac{1}{G}\right)\right\}\mathbf{t}_k \right|,
$$

and if $\dfrac{dk_{g_1}}{ds} = 0$, we can establish

$$
|k_{g_2}| = \frac{1}{Gk_{g_1}}\sqrt{\sum_{\substack{k=1 \\ k\neq 2}}^{n}\left[\frac{\partial^2}{\partial x_2 \partial x_k}\left(\frac{1}{G}\right)\right]^2} \neq 0,
$$

since $\dfrac{\partial^2}{\partial x_2 \partial x_3}\left(\dfrac{1}{G}\right) \neq 0$. This means there is no correspondence between Euclidean vertices $\left(k_2 = \dfrac{dk_1}{ds_e} = 0\right)$ and the Riemannian ones $\left(k_{g_2} = \dfrac{dk_{g_1}}{ds} = 0\right)$.

Suppose now that $\left(\dfrac{\partial^2}{\partial x_1^2}\left(\dfrac{1}{G}\right)-\dfrac{\partial^2}{\partial x_2^2}\left(\dfrac{1}{G}\right)\right)\neq 0$ at a point P. Like in the previous case we consider circles trough P in the $x_1 x_2$-plane and tangent to $\mathbf{v}_1$ for $\theta=\dfrac{\pi}{4}$. Trough an analogous reasoning we can deduce that if $\dfrac{dk_{g_1}}{ds}=0$, then

$$
|k_{g_2}| = \frac{\sqrt{2}}{2Gk_{g_1}}\left\{\frac{1}{2}\left(\frac{\partial^2}{\partial x_1^2}\left(\frac{1}{G}\right)-\frac{\partial^2}{\partial x_2^2}\left(\frac{1}{G}\right)\right)^2\right.
$$
$$
\left.+\sum_{k=3}^{n}\left[\frac{\partial^2}{\partial x_1 \partial x_k}\left(\frac{1}{G}\right)+\frac{\partial^2}{\partial x_2 \partial x_k}\left(\frac{1}{G}\right)\right]^2\right\}^{1/2},
$$

what means that $k_{g_2}\neq 0$ due our initial hypothesis. $\qquad\square$

We now point out that the function $\dfrac{1}{G}$ satisfies the condition (1) if and only if

$$
\frac{1}{G}=A\sum_{i=1}^{n}x_i^2+\sum_{i=1}^{n}B_i x_i+C,
$$

where A, B_i and C are constant. Besides, this means that the manifold has constant curvature

$$
K=4AC-\sum_{i=1}^{n}B_i^2.
$$

As a direct consequence of the previous lemma and this remark, we get the following:

THEOREM 3.3. *Under a conformal parametrization of a n dimensional manifold M, vertices of C^4-curves on M correspond to Euclidean vertices of their pre-images if and only if M has constant curvature.*

In the next section we show for the 3-dimensional case that this correspondence can be stated even in a stronger way.

REMARK: A. Kneser, in his classical four-vertex theorem article [8], already pointed out the correspondence between vertices of a plane curve and the vanishing torsion points of its stereographic image. We remark that at those points of spherical curves, we have also the vanishing of the curvature derivative. *Hence, simple closed spherical curves have at least four Euclidean vertices in our sense.*

We can establish an analogous property for curves on Riemannian manifolds. Instead of spheres we will consider curves $\alpha:I\to M$ on a totally umbilic 2-submanifold of M. If M has constant curvature, we may call such curves *spherical.* ¿From the previous theorem we can extend the four-vertex theorem:

THEOREM 3.4. *(Four-Vertex Theorem) Let M be a n dimensional complete constant curvature manifold and $\alpha:I\to M$ a closed simple null homotopic spherical curve. Then α has at least four vertices.*

Proof: We consider first $n = 3$. α is a spherical curve homotopic to zero what means it can be lifted, via the universal covering map, $\pi : \tilde{M} \to M$, to a spherical curve $\tilde{\alpha}$ on $\tilde{M}$. Since M has constant curvature, its universal covering is isometric to $\mathbb{R}^3$, $\mathbb{S}^3$ or $\mathbb{H}^3$. The conformal pre-image of the totally umbilic 2-submanifold of $\mathbb{S}^3$ or $\mathbb{H}^3$ is contained in an Euclidean sphere or in a plane. This implies, by the previous remark and Theorem 3.3, that $\tilde{\alpha}$ must have at least four vertices. The local isometry between M and $\tilde{M}$ completes the proof for $n = 3$. The extension for general n, is done by considering the identification of a totally geodesic 3-dimensional submanifold of the covering space which contains the umbilical 2-submanifold with $\mathbb{R}^3$, $\mathbb{S}^3$ or $\mathbb{H}^3$. $\qquad\square$

As we will see in the next section, in the 3-dimensional case, Theorem 3.4 can be specialized by asserting at those vertices the second curvature also changes its sign.

We point out that some authors ([12], [7]) also define a vertex of a curve in $\mathbb{R}^3$ as a *flattening point*, that is, just a vanishing torsion point. In [12] it is shown that if a closed space curve is convex (that is, contained in the boundary of its convex hull) it has at least four vanishing torsion points. Next we make use of Theorem 3.3 to exhibit a family of convex curves in $\mathbb{R}^3$ having no vertex in our sense.

We consider the family of curves on $\mathbb{S}^3$ given by

$$\alpha_\theta(t) = (\cos\theta\cos t, \cos\theta\sin t, \sin\theta\cos(nt), \sin\theta\sin(nt)), \quad \theta \in \left(0, \frac{\pi}{2}\right)$$

where $n \in \mathbb{Z}$. Under the stereographic projection, $\beta_\theta = \pi(\alpha_\theta)$ are curves on a standard torus in $\mathbb{R}^3$. If $\left|1 - n^2\right|_e \left|\sin\theta\right|_e < 1$, we can assert that those curves are convex because their orthogonal projection on $\mathbb{R}^2$ are convex plane curves with nowhere vanishing curvature. Since α_θ curves have constant nonvanishing curvatures, k_{g_1} and k_{g_2}, it follows from Theorem 3.3 that β_θ curves have no vertex.

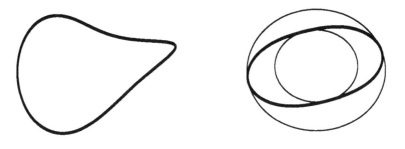

Figure 1:   A convex curve with no vertex and its projection in $x_1 x_2$-plane

4 THREE-DIMENSIONAL CASE

Vertices of a curve $\beta : I \to \mathbb{R}^n$ are also points where it has a contact with its osculating circle of order at least three. An vertex is called *ordinary* if this contact is of order exactly three. In the 3-dimensional case, $n = 3$, this means $k_2 = \dfrac{dk_1}{ds_e} = 0$ and $\dfrac{d^2 k_1}{ds_e^2} \neq 0$ or $\dfrac{dk_2}{ds_e} \neq 0$.

Considering the space curve $\gamma = \beta + \dfrac{1}{k_1} \mathbf{v}_2$ given by the centers of the osculating circles of β, we have $\dfrac{d\gamma}{ds_e} = -\dfrac{1}{k_1^2} \dfrac{dk_1}{ds_e} \mathbf{v}_2 + \dfrac{k_2}{k_1} \mathbf{v}_3$. We can check that the vertices of β are singular points and the ordinary vertices are cusps of γ.

For constant curvature 3-manifolds we can state a correspondence between ordinary vertices.

THEOREM 4.1. *For a C^4-curve on a constant curvature 3-manifold ordinary vertices are correspondent to Euclidean ordinary vertices of its conformal pre-image.*

Proof: If $k_1 \neq 0$ and $n = 3$, we may replace the expressions $\dfrac{d\mathbf{v}_2}{ds_e} = -k_1 \mathbf{v}_1 + k_2 \mathbf{v}_3$ and

$$k_{g_1}^2 = \left(\frac{k_1}{G} + \left\langle \nabla \left(\frac{1}{G} \right), \mathbf{v}_2 \right\rangle_e \right)^2 + \left\langle \nabla \left(\frac{1}{G} \right), \mathbf{v}_3 \right\rangle_e^2 \qquad (4.1)$$

in (2.13) and get

$$k_{g_2} = \frac{1}{G^2 k_{g_1}^2} \left[k_2 \left(\frac{k_1^2}{G} + k_1 \left\langle \nabla \left(\frac{1}{G} \right), \mathbf{v}_2 \right\rangle_e \right) - \frac{dk_1}{ds_e} \left\langle \nabla \left(\frac{1}{G} \right), \mathbf{v}_3 \right\rangle_e \right]. \qquad (4.2)$$

From (2.17) and replacing $\dfrac{d\mathbf{v}_2}{ds_e} = -k_1 \mathbf{v}_1 + k_2 \mathbf{v}_3$, we have

$$\frac{dk_{g_1}}{ds} = \frac{1}{G^2 k_{g_1}} \left[\frac{dk_1}{ds_e} \left(\frac{k_1}{G} + \left\langle \nabla \left(\frac{1}{G} \right), \mathbf{v}_2 \right\rangle_e \right) + k_1 k_2 \left\langle \nabla \left(\frac{1}{G} \right), \mathbf{v}_3 \right\rangle_e \right]$$

Differentiating (4.2) and the above expression with respect to s at a Euclidean (and then Riemannian) vertex point $P = \alpha(s_0)$ we get

$$\frac{dk_{g_2}}{ds} \bigg|_{s=s_0} = \frac{1}{k_{g_1}^2 G^3} \left[\frac{dk_2}{ds_e} \left(\frac{k_1^2}{G} + k_1 \left\langle \nabla \left(\frac{1}{G} \right), \mathbf{v}_2 \right\rangle_e \right) - \frac{d^2 k_1}{ds_e^2} \left\langle \nabla \left(\frac{1}{G} \right), \mathbf{v}_3 \right\rangle_e \right]$$

$$(4.3)$$

and

$$\frac{d^2 k_{g_1}}{ds^2} \bigg|_{s=s_0} = \frac{1}{G^3 k_{g_1}} \left[\frac{d^2 k_1}{ds_e^2} \left(\frac{k_1}{G} + \left\langle \nabla \left(\frac{1}{G} \right), \mathbf{v}_2 \right\rangle_e \right) + \frac{dk_2}{ds_e} k_1 \left\langle \nabla \left(\frac{1}{G} \right), \mathbf{v}_3 \right\rangle_e \right]$$

$$(4.4)$$

From (4.1), (4.3) and (4.4):

$$\left(k_{g_1}\frac{dk_{g_2}}{ds}\right)^2 + \left(\frac{d^2k_{g_1}}{ds^2}\right)^2 = \frac{1}{G^6}\left(\left(\frac{dk_2}{ds_e}\right)^2 k_1^2 + \left(\frac{d^2k_1}{ds_e^2}\right)^2\right),$$

what concludes the proof.   □

The second curvature $k_{g_2} = \left\langle \dfrac{D'\mathbf{V}_2}{ds}, \mathbf{V}_3 \right\rangle_e$ of a curve on a 3-dimensional manifold has a sign. Since a convex curve in $\mathbb{R}^3$ has at least four changes of its torsion sign [7] (at those points it stays at one side of its osculating plane) we may ask if a similar result concerning k_{g_2} changes of sign still holds for a spherical curve on constant curvature manifold M^3.

Under the same reasoning done in the proof of Theorem 2 we consider α a local lifting to the universal covering of a spherical curve on M^3 and its conformal pre-image $\beta = \psi^{-1}(\alpha)$, a spherical curve in $\mathbb{R}^3$. Hence $k_1 \neq 0$ and

$$\beta - C = -\frac{1}{k_1}\mathbf{v}_2 - r\frac{k_g}{k_1}\mathbf{v}_3,$$

where C and r are the center and the radius of the sphere and k_g is the geodesic curvature of $\beta \subset \mathbb{R}^3$, $|k_g| = \dfrac{\sqrt{k_1^2 r^2 - 1}}{r}$ [9]. For a spherical curve we also have

$$\frac{dk_1}{ds_e} = -rk_2k_1k_g. \tag{4.5}$$

From these and from (2.11) and (4.2), we get

$$k_{g_2} = \frac{k_2 k_1^2}{k_{g_1}^2 G^2}\left[1 + \frac{K}{4}\left(|C|_e^2 - r^2\right)\right],$$

what implies k_{g_2} and k_2 have the same sign.

We finally point out that changes of sign of k_2 of a spherical curve in $\mathbb{R}^3$ correspond to cuspidal edges in the evolute surface of the curve, that is, the surface defined as a envelope of the family of the curve normal planes [4]. Considering the standard parametrization of the evolute surface of a space curve $X(s_e, \lambda) = \beta + \dfrac{1}{k_1}\mathbf{v}_2 + \lambda\mathbf{v}_3$, we see that the singularities are given by the vertices and the centers of the osculating spheres. Using (4.5) we can conclude that, for a spherical curve, the points at which the torsion changes its sign generate at least four cuspidal edges on the evolute surface. For $\lambda = -r\dfrac{k_g}{k_1}$ and any s_e, the evolute surface pass through the center of the sphere — a degenerate singularity.

Summing up we can state the following

THEOREM 4.2. *Let M^3 be a constant curvature 3-manifold and $\alpha : I \to M$ a regular closed spherical null homotopic C^4-curve. Then α has at least four points at which the second curvature changes its sign. Those points correspond to cuspidal edges on the pre-image curve evolute surface.*

Figure 2 ilustrates the evolute surface of a spherical curve with just four (Riemannian/ Euclidean) vertices.

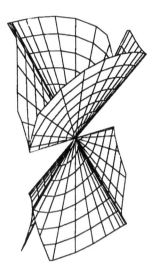

Figure 2: The evolute surface S of α, a anti-stereographic projection of an ellipse. S has four cuspidal edges and one degenerate singularity.

REFERENCES

[1] V. Arnol'd. On the number of flattening points on space curves. Amer. Math. Soc. Transl. 171 2:11–22, 1996.

[2] M. Barner. Über die Mindestanzahl stationärer Schmiegebenen bei geschlossenen streng-konvexen Raumkurven. Abh. Math. Sem. Univ. Hamburg 20:196–215, 1956.

[3] J.W. Bruce, P.J. Giblin. Generic isotopies of space curves. Glasgow Math. J. 29:41–63, 1987.

[4] ———. Curves and singularities 2nd ed. Cambridge University Press, Cambridge, 1992.

[5] G. Cairns, R. Sharpe, L. Webb. Conformal invariants for curves and surfaces in three dimensional space forms. Rocky Mountain Journal of Mathematics. 24 3:933–959, 1994.

[6] S.I.R. Costa, M. Firer. Four-or-more-vertex theorems for constant curvature manifolds. In: Real and Complex Singularities, Research in Mathematics Series, Ed. J. Bruce, F. Tari. Champman & Hall/CRC 412:164–172, 2000.

[7] S.I.R. Costa, M.C. Romero-Fuster. Nowhere vanishing torsion closed curves alawys hide twice. Geom. Dedicata. 66:1–17, 1997.

[8] A. Kneser. Bemerkunger über die Anzahi der Extrema der Krümmung auf gescholossenen Kurven und über verwandt Fragen in einer nicht-euklidischen Geometrie. Festschrift Heinrich Weber, Teubner 1912 pp. 170–180.

[9] R.S. Millman, G.D. Parker. Elements of differential geometry. Prentice-Hall Inc., Englewood Cliffs, New Jersey, 1977.

[10] C.C. Pansonato. Contact and vertices of curves on constant curvature manifolds. Ph.D. thesis, Instituto de Matemática-Unicamp, 2001.

[11] M.C. Romero-Fuster, E. Sanabria-Codesal. Generalized evolutes, vertices and conformal invariants of curves in $\mathbb{R}^{n+1}$. Indag. Mathem. N. S. 10 2:297–305, 1999.

[12] V.D. Sedykh. Four vertices of a convex space curve. Bull. London Math. Soc. 26:177–180, 1994.

[13] ———. Connection of Lagrangian singularities with Legendrian singularities under stereographic projection. Russian Acad. Sci. Sb. Math. 83 2:533–540, 1995.

[14] M. Spivak. A comprehensive introduction to differential geometry. Publish or Perish, Inc.. Houston, 1979.

[15] D.J. Struik. Lectures on classical differential geometry. Dover Publications Inc. New York, 1961.

[16] R. Uribe-Vargas. On the $(2k + 2)$-vertex and $(2k + 2)$-flattening theorems in higher dimensional Lobatchevskian space. C. R. Acad. Sci. Paris. 325:505–510, 1997.

Projections of Hypersurfaces in $\mathbf{R}^4$ to Planes

ANA CLAUDIA NABARRO* Instituto de Ciências Matemáticas e de Computação, Universidade de São Paulo, Departamento de Matemática, Caixa Postal 668, CEP 13560-970, São Carlos, SP, Brazil, E-mail: acnabarro@uem.br

Abstract

We study in this paper orthogonal projections of generic embedded hypersurfaces in $\mathbb{R}^4$ to 2-spaces. We list all the generic local singularities of such projections and establish geometric criteria for recognition of the singularities of codimension ≤ 1 and for their versal unfoldings. We also determine a stratification of the set of singular planes of projections.

1 INTRODUCTION

We study in this paper the $\mathcal{A}$-classification of the singularities of the orthogonal projections of a generic embedded hypersurface M in $\mathbb{R}^4$ to a 2-dimensional plane. The singularities of such maps measure, for instance, the contact of M with 2-dimensional planes.

The contact of smooth varieties with degenerate objects (lines, planes, circles, spheres, etc.) is measured locally by the $\mathcal{K}$-classification of the singularities of some map-germs [20], although in practice (right-left) $\mathcal{A}$-equivalence classes are sought as they yield a finer classification, a richer set of invariants, and more geometrical information. This approach allowed the discovery of beautiful geometric results of

* Current address: Universidade Estadual de Maringá, Departamento de Matemática, Av. Colombo 5790, Campus Universitário, CEP 87020-900, Maringá, PR, Brazil.

surfaces in $\mathbb{R}^3$ (for example, the study of singularities of the Gauss map [1], the classification of the umbilic points and the structure of the focal set at such points, the flat geometry of the focal set, the ridge and subparabolic curves, the changes on these set, etc.; see [2] for references). These results are being extended to surfaces in $\mathbb{R}^4$ in [5, 8, 17, 18, 24].

The work in this paper is part of our investigation in [22] of the flat geometry of generic embedded hypersurfaces in $\mathbb{R}^4$. In [23], we study the contact of M with hyperplanes and lines, and here we deal with the contact with planes.

There is a natural family of projections of $M \subset \mathbb{R}^4$ to planes, parametrised by the Grassmanian of 2-planes in $\mathbb{R}^4$, $G(2,4)$. Let a and b form an orthonormal basis of the plane of projection $u \in G(2,4)$ then the family of projections to 2-spaces is given by

$$
\begin{aligned}
\Pi: \quad M \times G(2,4) &\rightarrow \mathbb{R}^2 \\
(p,u) &\mapsto (<p,a>,<p,b>).
\end{aligned}
$$

Given $u \in G(2,4)$ the map Π_u measures the contact between M and the plane orthogonal to u, the kernel of Π_u. (Note that Π_u is of corank 1.) The $\mathcal{A}$-equivalence class of Π_u does not depend on the choice of orthonormal basis a, b, so in effect Π is a 4-parameter family [6]. Therefore the generic $\mathcal{A}$-equivalence classes of Π_u are those of $\mathcal{A}_e$-codimension ≤ 4. (In the presence of moduli, the generic cases are when the union of the orbits along the moduli parameters form a set of $\mathcal{A}_e$-codimension ≤ 4.) This fact is a direct consequence of Montaldi's Theorem in [21], and follows from the fact that the family Π defined on the ambient space is $\mathcal{A}$-versal. So the first task is to determine the normal forms of the singularities of corank 1 and codimension ≤ 4 of germs $\mathbb{R}^3, 0 \rightarrow \mathbb{R}^2, 0$. Those that are $\mathcal{A}$-equivalent to $(x, f(x,y) \pm z^2)$ are classified by Rieger and Ruas in [26]. We classify in section 2 those germs of $\mathcal{A}_e$-codimension ≤ 4 that are not covered in [26]. The main tools used are the complete transversal technique in [3, 7] and the software "Transversal" developed by Neil Kirk [14].

In section 3, we give geometric criteria for recognition of the codimension ≤ 1 singularities of germs $\mathbb{R}^3, 0 \rightarrow \mathbb{R}^2, 0$ and in section 4 criteria for determining when a deformation of a codimension 1 singularity is versal. Finally, we determine in section 5 a partition of the set of singular planes by the singularity type of Π_u at a fixed point on M.

2   CLASSIFICATION RESULTS

As highlighted in the introduction, the generic singularities of Π are those of $\mathcal{A}_e$-codimension (or codimension of the stratum) ≤ 4. Before giving a complete list of such singularities we need some notation and preliminaries. (Although our interest is in germs $\mathbb{R}^3, 0 \rightarrow \mathbb{R}^2, 0$ we set these notation in more general terms.)

Let $f : \mathbb{R}^n, 0 \rightarrow \mathbb{R}^p, 0$ be a smooth map-germ and denote the Lie group of right-left equivalences by $\mathcal{A} = Diff(\mathbb{R}^n, 0) \times Diff(\mathbb{R}^p, 0)$ which acts on the space of smooth germs $\mathcal{E}(n,p)$ as follows: $(h,k).f = h \circ f \circ k^{-1}$, where $(h,k) \in \mathcal{A}$. We say that f is $\mathcal{A}$-equivalent to g if there is $(h,k) \in \mathcal{A}$ such that $g = h \circ f \circ k^{-1}$. Let $J^k(n,p)$ denote the space of kth order Taylor polynomials without constant terms, and write $j^k f$ for the k-jet of f. A germ f is said to be k-determined if, for any g such that $j^k f = j^k g$, it follows that f is equivalent to g.

Denote by $\mathcal{E}(n)$ and $\mathcal{E}(p)$ the local rings of function-germs in source and target whose respective maximal ideals are $m(n)$ and $m(p)$. Let θ_f denote the $\mathcal{E}(n)$-module of vector fields over f, and set $\theta_n = \theta_{(1_{\mathbf{R}^n})}$ and $\theta_p = \theta_{(1_{\mathbf{R}^p})}$ where $1_{\mathbf{R}^n}$ and $1_{\mathbf{R}^p}$ denote respectively the identity mapping of $\mathbf{R}^n$ and $\mathbf{R}^p$. One can define a $\mathcal{E}(n)$-homomorphism

$$tf : \theta_n \rightarrow \theta_f$$
$$\phi \mapsto df \circ \phi$$

and a $\mathcal{E}(p)$-homomorphism

$$wf : \theta_p \rightarrow \theta_f$$
$$\psi \mapsto \psi \circ f.$$

The tangent space to the $\mathcal{A}$-orbit at f is then given by $L\mathcal{A} \cdot f = tf(m(n).\theta_n) + wf(m(p).\theta_p)$. The $\mathcal{A}$-codimension of f is given by $cod(\mathcal{A}, f) = dim_{\mathbb{R}} m(n).\theta_f / L\mathcal{A} \cdot f$ and the $\mathcal{A}_e$-codimension by

$$\mathcal{A}_e\text{-}codim(f) = dim_{\mathbb{R}}(\mathcal{E}(n,p)/L\mathcal{A}_e \cdot f),$$

where

$$L\mathcal{A}_e \cdot f = tf(\theta_n) + wf(\theta_p).$$

The space $L\mathcal{A}_e \cdot f$ is called the extended tangent space.

The more recent advances in determinacy and classification theory are given in the articles by Bruce, Kirk, du Plessis and Wall [3, 7]. We state below a corollary of one of the key results in [7] that can be used to list singularities of map germs $\mathbf{R}^n, 0 \rightarrow \mathbf{R}^p, 0$.

Let $\mathcal{G}$ be a subgroup of one of the Mather's group and define $\mathcal{G}_s$ to be the subgroup of $\mathcal{G}$ whose elements have the s-jet equal to the identity.

COROLLARY 2.1 *[7]* f *is* $k - \mathcal{G}_s$-*determined* $(\mathcal{G} = \mathcal{L}$ *or* $\mathcal{A}$, $s \geq 1)$ *if and only if*

$$m_n^{k+1}.\theta_f \subset L\mathcal{G}_s.f + m_n^{k+1}.f^*(m_p).\mathcal{E}_n + m_n^{2k+2}.\mathcal{E}(n,p).$$

Finite determinacy means that the classification problem can be reduced to the space of k-jets.

Let $H^{k+1}(n, p)$ be the space of germs of homogeneous polynomials of degree $k+1$ in $\mathcal{E}(n, p)$.

PROPOSITION 2.2 (Complete Transversal to jets [3]) *Let* $\mathcal{G}$ *be a Mather's group and denote the tangent space to the* $J^{k+1}\mathcal{G}_1$-*orbit of* $j^{k+1}f$ *by* $L(J^{k+1}\mathcal{G}_1).j^{k+1}f$. *Then given* $f \in m(n).\mathcal{E}(n,p)$ *and* $T \subset H^{k+1}(n,p)$ *such that*

$$H^{k+1}(n, p) \subset L(J^{k+1}\mathcal{G}_1).j^{k+1}f + T$$

any $(k+1)$-*jet* $j^{k+1}g$, *with* $j^k g = j^k f$, *is in the same* $\mathcal{G}_1$-*orbit of* $j^{k+1}f + t$, *for some* $t \in T$.

The techniques developed in [3, 7] provide a very efficient, wide-ranging classification scheme involving algebraic calculations which may be reduced to finite dimensional symbolic problems. However, the calculations can become very intensive and repetitive. In [14], Kirk produced a package, called "Transversal", which consists of a collection of procedures which run under the symbolic algebra system Maple and deals with problems in classification and unfolding theory for these equivalence relations. Recent applications of "Transversal" [4, 10, 11, 12, 13, 29] represent some of the most extensive classifications carried out to date. In this paper we use "Transversal" to classify germs $\mathbb{R}^3, 0 \to \mathbb{R}^2, 0$.

Germs $\mathbb{R}^3, 0 \to \mathbb{R}^2, 0$ of corank 1 that can be written in the form $(x, f(x, y) \pm z^2)$ in some coordinate system are classified by Rieger and Ruas in [26].

THEOREM 2.3 [26] *The germs* $\mathbb{R}^3, 0 \to \mathbb{R}^2, 0$ *of corank 1 and* $\mathcal{A}_e$*-codimension (or codimension of the stratum)* ≤ 4, *that are equivalent to one of the form* $(x, f(x, y) \pm z^2)$ *are given below, where all the classes are simple with the exception of* $8, 9, 10, 15, 18$ *and* 19.

Name	Normal form	$\mathcal{A}_e$-cod
1	(x, y)	0
2 (fold)	$(x, y^2 \pm z^2)$	0
3 (cusp)	$(x, xy + y^3 \pm z^2)$	0
4_k (lips/beaks when $k = 2$)	$(x, y^3 + (\pm 1)^{k-1} x^k y \pm z^2), 2 \leq k \leq 5$	$k - 1$
5 (swallowtail)	$(x, xy + y^4 \pm z^2)$	1
6	$(x, xy + y^5 \pm y^7 \pm z^2)$	2
7	$(x, xy + y^5 \pm z^2)$	3
8	$(x, xy + y^6 \pm y^8 + ay^9 \pm z^2)$	4(3†)
9	$(x, xy + y^6 + y^9 \pm z^2)$	4
10	$(x, xy + y^7 \pm y^9 + ay^{10} + by^{11} \pm z^2)$	6*(4†)
11_{2k+1}	$(x, xy^2 + y^4 + y^{2k+1} \pm z^2), 2 \leq k \leq 4$	k
12	$(x, xy^2 + y^5 + y^6 \pm z^2)$	3
13	$(x, xy^2 + y^5 \pm y^9 \pm z^2)$	4
15	$(x, xy^2 + y^6 + y^7 + ay^9 \pm z^2)$	5(4†)
16	$(x, x^2 y + y^4 \pm y^5 \pm z^2)$	3
17	$(x, x^2 y + y^4 \pm z^2)$	4
18	$(x, x^2 y + xy^3 + ay^5 + y^6 + by^7 \pm z^2)$	6*(4†)
19	$(x, x^3 y + ax^2 y^2 + y^4 + x^3 y^2 \pm z^2)$	5(4†)

    †   codimension of the stratum
    *   excluding exceptional values of the moduli.

We complete in this section the list in Theorem 2.3 and obtain the remaining germs of codimension ≤ 4. These germs are those that can not be written in the form $(x, f(x, y) \pm z^2)$.

THEOREM 2.4 *The* $\mathcal{A}$*-classes of the singularities* $\mathbb{R}^3, 0 \to \mathbb{R}^2, 0$ *of corank 1 and* $\mathcal{A}_e$*-codimension (or codimension of the stratum in the presence of moduli)* ≤ 4 *that are not listed in Theorem 2.3 are given in Table 1.*

Name	Normal form	$\mathcal{A}_e$-cod	δ
N_1	$(x, xy + y^3 + ay^2 z + z^3 \pm z^5)$ $a \neq 0$, $4a^3 \pm 27 \neq 0$	3 (2†)	5
N_2	$(x, xy + y^3 + ay^2 z + z^3)$, $a \neq 0$, $4a^3 + 27 \neq 0$	4 (3†)	5
N_3	$(x, xy + y^3 + z^3 \pm y^3 z)$	3	4
N_4	$(x, xy \pm y^2 z + z^3 \pm z^5)$	3	5
N_5	$(x, xy \pm y^3 + yz^2 + z^5)$	3	5
N_6	$(x, xy \pm y^2 z + z^3)$	4	5
N_7	$(x, xy + y^3 + z^3 \pm y^4 z)$	4	5
N_8	$(x, xy + z^3 \pm y^4 + y^3 z + ay^4 z + b(y^6 + \lambda y^5 z))$, $b \neq 0$	6 (4†)	6
N_9	$(x, xy + yz^2 \pm y^4 + z^5 + ay^6)$	5 (4†)	6
N_{10}	$(x, xy + y^2 z + yz^3 \pm z^4 + az^6)$	5 (4†)	6
N_{11}	$(x, xy \pm y^3 + yz^2 + z^7)$	4	7
N_{12}	$(x, xyz \pm y^2 z + z^3 + ax^2 y + bx^2 z + cyz^3 + z^4)$	7* (4†)	4

Table 1

where λ is a constant, and δ denotes the determinacy degree of the germs.

    †   codimension of the stratum
    * $4b - 1 \neq 0$ and $4b - 1 + 6ac \neq 0$.

Proof: We are interested in germs of corank 1, so we can write $F(x,y,z) = (x, g(x,y,z))$. Let $j^2 g = x\phi(x, y, z) + \psi(y, z)$. Then the germs that can not be written in the form $(x, f(x, y) \pm z^2)$ are those with $\psi \equiv 0$. Therefore to complete the list in Theorem 2.3, we need to consider the cases where $j^2 g = x\phi(x, y, z)$. It is not difficult to show that we can change coordinates so that $j^2 F$ is given by (x, xy) or $(x, 0)$. We now follow these germs and carry out the classification inductively on the jet level, using the complete transversal method [3] and the "Transversal" package [14]. We start by considering the 2-jet (x, xy).

(1) $j^2 F = (x, xy)$.
We have $F_x = (1, y)$, $F_y = (0, x)$, $F_z = (0, 0)$. Therefore,

$$H^3(3, 2) \subset T(J^3 \mathcal{A}_1).j^3 F + \mathbb{R}.\{y^2 z, y^3, yz^2, z^3\}$$

and by Proposition 2.2 any 3-jet with a 2-jet (x, xy) is equivalent to $(x, xy + ay^2 z + by^3 + cyz^2 + dz^3)$ for some $a, b, c, d \in \mathbb{R}$. We can now make linear changes of coordinates and obtain the following orbits in $J^3(3, 2)$:

$$(x, xy + y^3 + ay^2 z + z^3),$$
$$(x, xy + y^3 + z^3),$$
$$(x, xy \pm y^2 z + z^3),$$
$$(x, xy + z^3),$$
$$(x, xy \pm y^3 + yz^2),$$
$$(x, xy + yz^2),$$
$$(x, xy + y^2 z),$$
$$(x, xy + y^3),$$
$$(x, xy),$$

where a in the first orbit is a smooth modulus. Using "Transversal" we show that the last 2 cases lead to germs of $\mathcal{A}_e$- codimension ≥ 5 and therefore will not be followed.

- The 3-jet $(x, xy + y^3 + ay^2z + z^3)$:

  Using "Transversal" we show that the 4-transversal is empty and the 5-transversal is given by $\mathbb{R}.\{z^5\}$. The orbits in $J^5(3,2)$ are $(x, xy+y^3+ay^2z+z^3\pm z^5)$ if $4a^3\pm 27 \neq 0$ and $a \neq 0$, and $(x, xy + y^3 + ay^2z + z^3)$ if $4a^3 + 27 \neq 0$ and $a \neq 0$. Both germs are 5-determined, the first (N_1) has codimension 3 (the codimension of the stratum is 2) and the second (N_2) is of codimension 4 (the codimension of the stratum is 3). The exceptional case $4a^3 \pm 27 = 0$ will not be followed as the current version of "Transversal" does deal with constant such as $(27/4)^{2/3}$. We shall only consider the case $a = 0$.

- The 3-jet $(x, xy + y^3 + z^3)$:

  A 4-transversal is given by $\mathbb{R}.\{y^3z\}$. The orbits in $J^4(3,2)$ are $(x, xy+y^3+z^3\pm y^3z)$ and $(x, xy+y^3+z^3)$. The first case (N_3) is 5-determined and has codimension 3. For the second case, a 5-transversal is given by $\mathbb{R}.\{y^4z\}$ so in $J^5(3,2)$ we have two orbits, namely $(x, xy + y^3 + z^3 \pm y^4z)$ and $(x, xy + y^3 + z^3)$. The form N_7 is 6-determined and has codimension 4. The later yields germs of codimension ≥ 5.

- The 3-jet $(x, xy \pm y^2z + z^3)$:

  The 4-transversal is empty and a 5-transversal is given by $\mathbb{R}.\{z^5\}$. The orbits in $J^5(3,2)$ are $(x, xy\pm y^2z+z^3\pm z^5)$ and $(x, xy\pm y^2z+z^3)$. Both germs are 5-determined with the first (N_4) having codimension 3 and the second (N_6) is of codimension 4.

- The 3-jet $(x, xy + z^3)$:

  A 4-transversal is given by $\mathbb{R}.\{y^4, y^3z\}$ so the orbits in $J^4(3,2)$ are $(x, xy + z^3 + y^3z \pm y^4)$ and $(x, xy + z^3 + y^3z)$. A 5-transversal of the first germ is given by $\mathbb{R}.\{y^4z\}$ so any germ in $J^5(3,2)$ with a 4-jet $(x, xy + z^3 + y^3z \pm y^4)$ is equivalent to $(x, xy + z^3 + y^3z \pm y^4 + ay^4z)$ for some $a \in \mathbb{R}$. Here the parameter a is a smooth modulus. The orbits in $J^6(3,2)$ are given by $(x, xy + z^3 \pm y^4 + y^3z + ay^4z + b(y^6 + \lambda y^5z))$ where b is also a smooth modulus and λ is a constant. These germs (N_8) are 6-determined and have codimension 6 (the codimension of the stratum is 4) provided $b \neq 0$.

  The orbit in $J^5(3,2)$ whose 4-jet is equal to $(x, xy + z^3 + y^3z)$ are $(x, xy + z^3 + y^3z \pm y^5)$ and $(x, xy + z^3 + y^3z)$. These lead to germs of codimension greater or equal to 5.

- The 3-jet $(x, xy \pm y^3 + yz^2)$:

  The 4-transversal is empty and a 5-transversal is given by $\mathbb{R}.\{z^5\}$, so the orbits in $J^5(3,2)$ are $(x, xy \pm y^3 + yz^2 + z^5)$ and $(x, xy \pm y^3 + yz^2)$. The first germ (N_5) is 5-determined and has codimension 3. The second germ has an empty 6-transversal but a 7-transversal is given by $\mathbb{R}.\{z^7\}$. In $J^7(3,2)$, the orbit (N_{11}) $(x, xy \pm y^3 + yz^2 + z^7)$ is 7-determined and has codimension 4. The remaining orbit leads to germs of higher codimensions.

- The 3-jet $(x, xy + yz^2)$:

  A 4-transversal is given by $\mathbb{R}.\{y^4\}$ and a 5-transversal of the orbit $(x, xy+yz^2\pm y^4)$ is given by $\mathbb{R}.\{z^5\}$. The orbits in $J^5(3,2)$ are $(x, xy + yz^2 \pm y^4 + z^5)$ and $(x, xy + yz^2 \pm y^4)$. This second orbit generates germs of codimension ≥ 5. A 6-transversal of the first germ is given by $\mathbb{R}.\{y^6\}$, therefore the germs in $J^5(3,2)$ with a 5-jet as above are equivalent to $(x, xy + yz^2 \pm y^4 + z^5 + ay^6)$, for some $a \in \mathbb{R}$. In this case

a is a smooth modulus. These germs (N_9) are 6-determined and have codimension 5 (the codimension of the stratum is equal to 4).

- The 3-jet $(x, xy + y^2 z)$:

A 4-transversal is given by $\mathbb{R}.\{z^4, yz^3\}$ and the orbits in $J^4(3,2)$ are $(x, xy + y^2 z + yz^3 \pm z^4)$ and $(x, xy + y^2 z \pm z^4)$. The second orbit generates germs of higher codimension. A 5-transversal of the first orbit is empty and a 6-transversal is given by $\mathbb{R}.\{z^6\}$. Therefore any germ in $J^6(3,2)$ with a 5-jet as above is equivalent to $(x, xy + y^2 z + yz^3 \pm z^4 + az^6)$ for some $a \in \mathbb{R}$. Here a is a smooth modulus. These germs (N_{10}) are 6-determined and have codimension 5 (the codimension of the stratum is equal to 4).

(2) $j^2 F = (x, 0)$.

A 4-transversal is given by $\mathbb{R}.\{xyz, z^3, y^2 z, x^2 y, x^2 z\}$. There are various cases to consider here but the least degenerate case (we are seeking strata of codimension ≤ 4) is $(x, xyz \pm y^2 z + z^3 + ax^2 y + bx^2 z)$. Then a 5-transversal is given by $\mathbb{R}.\{yz^3, z^4\}$. The least degenerate case is given by $(x, xyz \pm y^2 z + z^3 + ax^2 y + bx^2 z + cyz^3 + z^4)$ (N_{12}) which is 4-determined and has codimension 7 (the codimension of the stratum is 4), provided $4b - 1 \neq 0$ and $4b - 1 + 6ac \neq 0$. Any other case is part of (or leads to) a stratum with $\mathcal{A}_e$-codimension greater or equal to 5. □

We also use "Transversal" to obtain normal forms for the versal unfoldings of the germs in Table 1.

PROPOSITION 2.5 *Versal unfoldings of the germs listed in Table 1 are given in Table 2.*

Name	Versal unfolding
N_1	$(x, xy + y^3 + ay^2 z + z^3 \pm z^5 + u_1 z + u_2 y^2)$
N_2	$(x, xy + y^3 + ay^2 z + z^3 + u_1 z + u_2 y^2 + u_3 y^5)$,
N_3	$(x, xy + y^3 + z^3 \pm y^3 z + u_1 z + u_2 yz + u_3 y^2 z)$
N_4	$(x, xy \pm y^2 z + z^3 \pm z^5 + u_1 z + u_2 y^2 + u_3 y^3)$
N_5	$(x, xy \pm y^3 + yz^2 + z^5 + u_1 z + u_2 y^2 + u_3 z^3)$
N_6	$(x, xy \pm y^2 z + z^3 + u_1 z + u_2 y^2 + u_3 y^3 + u_4 y^4 z)$
N_7	$(x, xy + y^3 + z^3 \pm y^4 z + u_1 z + u_2 yz + u_3 y^2 z + u_4 y^3 z)$
N_8	$(x, xy + z^3 \pm y^4 + y^3 z + ay^4 z + by^6 + cy^5 z + u_1 z + u_2 yz + u_3 y^2 + u_4 y^3)$
N_9	$(x, xy + yz^2 \pm y^4 + z^5 + ay^6 + u_1 z + u_2 y^2 + u_3 y^3 + u_4 z^3)$
N_{10}	$(x, xy + y^2 z + yz^3 \pm z^4 + az^6 + u_1 z + u_2 z^2 + u_3 yz^2 + u_4 z^3)$
N_{11}	$(x, xy \pm y^3 + yz^2 + z^7 + u_1 z + u_2 y^2 + u_3 y^3 + u_4 y^5)$
N_{12}	$(x, xyz \pm y^2 z + z^3 + ax^2 y + bx^2 z + cyz^3 + z^4 + u_1 y + u_2 z + u_3 xy + u_4 y^2)$

Table 2

REMARK 2.1 It is of interest to calculate more invariants associated to the singularities in Table 1. This is part of future investigation.

3   GEOMETRIC CRITERIA FOR RECOGNITION OF THE SINGULARITIES OF $\mathcal{A}_e$-codim ≤ 1

Geometric criteria for recognition of the singularities of $\mathcal{A}_e$-codim ≤ 1 of germs $\mathbb{R}^2, 0 \to \mathbb{R}^2, 0$ are given in [15] and [28]. We adapt these criteria to germs $F : \mathbb{R}^3, 0 \to \mathbb{R}^2, 0$. When F has corank 1, we can change coordinates and write $F(x, y, z) = (x, f(x, y, z))$. The differential of F at (x, y, z) is then given by

$$DF(x, y, z) = \begin{pmatrix} 1 & 0 & 0 \\ \frac{\partial f}{\partial x}(x, y, z) & \frac{\partial f}{\partial y}(x, y, z) & \frac{\partial f}{\partial z}(x, y, z) \end{pmatrix}$$

so that the critical set of F is given by

$$\Sigma = \{(x, y, z) : \frac{\partial f}{\partial y}(x, y, z) = \frac{\partial f}{\partial z}(x, y, z) = 0\}.$$

Let $G : \mathbb{R}^3, 0 \to \mathbb{R}^2, 0$ such that $G(x, y, z) = (f_y(x, y, z), f_z(x, y, z))$. $\Sigma = G^{-1}(0)$ is locally a smooth curve if G is regular, that is, if G has maximal rank at $(0, 0, 0)$.

It is easy to show that if Σ is smooth, then F can be written in some coordinate systems in the form $(x, f(x, y) \pm z^2)$. In this case $\Sigma = \{(x, y, 0); f_y(x, y) = 0\}$. Let $\phi : I, 0 \to \mathbb{R}^3, 0$ be a parametrization of Σ, where I is a neighbourhood of 0 in $\mathbb{R}$. The order of contact of Σ with the kernel of $DF(0, 0, 0)$ ($ker(DF(0, 0, 0))$) is the order of the vanishing of the derivatives of $F \circ \phi$ at 0. This order of contact is independent of the parametrization of Σ and is an $\mathcal{A}$-invariant of the map F.

We shall use this order of contact to recognize geometrically the fold, cusp and swallowtail singularities in the list of Theorem 2.3. When the critical set is smooth, we can set $F(x, y, z) = (x, f(x, y) \pm z^2)$ as the order of contact is invariant.

The following is well known:

PROPOSITION 3.1 *Let* $F : \mathbb{R}^3, 0 \to \mathbb{R}^2, 0$ *a singular germ with a smooth critical set* Σ *and let* ϕ *be a parametrization of* Σ. *Then*

(i) F *is a fold if and only if* $\frac{\partial}{\partial t}(F \circ \phi)(0) \neq 0$.

(ii) F *is a cusp if and only if* $\frac{\partial}{\partial t}(F \circ \phi)(0) = 0$ *and* $\frac{\partial^2}{\partial t^2}(F \circ \phi)(0) \neq 0$.

Proof: Let $F(x, y, z) = (x, f(x, y) \pm z^2)$ and ϕ a local parametrization of $\Sigma = \{(x, y, 0); f_y(x, y) = 0\}$.

So $\phi(t) = (\phi_1(t), 0)$ with $\phi_1(0) = 0$ and $\frac{\partial}{\partial t}\phi(t) = (-\frac{\partial^2 f}{\partial y^2}(\phi_1(t)), \frac{\partial^2 f}{\partial x \partial y}(\phi_1(t)), 0)$. Then

$$\begin{aligned} \frac{\partial}{\partial t}(F \circ \phi)(t) \quad &= DF(\phi(t)).\frac{\partial}{\partial t}\phi(t) \\ &= \begin{pmatrix} 1 & 0 & 0 \\ \frac{\partial f}{\partial x}(\phi_1(t)) & \frac{\partial f}{\partial y}(\phi_1(t)) & 0 \end{pmatrix} \begin{pmatrix} -\frac{\partial^2 f}{\partial y^2}(\phi_1(t)) \\ \frac{\partial^2 f}{\partial x \partial y}(\phi_1(t)) \\ 0 \end{pmatrix} \\ &= \begin{pmatrix} -\frac{\partial^2 f}{\partial y^2}(\phi_1(t)) \\ -\frac{\partial f}{\partial x}(\phi_1(t)).\frac{\partial^2 f}{\partial y^2}(\phi_1(t)) + \frac{\partial f}{\partial y}(\phi_1(t)).\frac{\partial^2 f}{\partial x \partial y}(\phi_1(t)) \end{pmatrix} \\ &= -\frac{\partial^2 f}{\partial y^2}(\phi_1(t)).\begin{pmatrix} 1 \\ \frac{\partial f}{\partial x}(\phi_1(t)) \end{pmatrix} \end{aligned} \quad (3.1)$$

and at the origin $\frac{\partial}{\partial t}(F \circ \phi)(0) = -\frac{\partial^2 f}{\partial y^2}(0,0) \cdot \left(\frac{1}{\frac{\partial f}{\partial x}(0,0)} \right)$. Therefore $\frac{\partial}{\partial t}(F \circ \phi)(0) \neq 0$

if and only if $\frac{\partial^2 f}{\partial y^2}(0,0) \neq 0$, if and only if F is a fold.

Differentiating (3.1) we get:

$$
\begin{aligned}
\frac{\partial^2}{\partial t^2}(F \circ \phi)(t) &= \frac{\partial}{\partial t}\left(-\frac{\partial^2 f}{\partial y^2}(\phi_1(t)) \cdot \left(\frac{1}{\frac{\partial f}{\partial x}(\phi_1(t))} \right) \right) \\
&= -\{ -\frac{\partial^3 f}{\partial x \partial y^2}(\phi_1(t)) \cdot \frac{\partial^2 f}{\partial y^2}(\phi_1(t)) + \\
&\quad + \frac{\partial^3 f}{\partial y^3}(\phi_1(t)) \cdot \frac{\partial^2 f}{\partial x \partial y}(\phi_1(t)) \} \left(\frac{1}{\frac{\partial f}{\partial x}(\phi_1(t))} \right) \\
&\quad - \frac{\partial^2 f}{\partial y^2}(\phi_1(t)) \cdot \left(\frac{1}{\frac{\partial}{\partial t}(\frac{\partial f}{\partial x}(\phi_1(t)))} \right)
\end{aligned}
\tag{3.2}
$$

If $\frac{\partial}{\partial t}(F \circ \phi)(0) = 0$, that is $\frac{\partial^2 f}{\partial y^2}(0,0) = 0$, then

$$
\frac{\partial^2}{\partial t^2}(F \circ \phi)(0) = -\frac{\partial^3 f}{\partial y^3}(0,0) \cdot \frac{\partial^2 f}{\partial x \partial y}(0,0) \left(\frac{1}{\frac{\partial f}{\partial x}(0,0)} \right).
$$

Hence $\frac{\partial}{\partial t}(F \circ \phi)(0) = 0$ and $\frac{\partial^2}{\partial t^2}(F \circ \phi)(0) \neq 0$ if and only if $\frac{\partial^2 f}{\partial y^2}(0,0) = 0$ and $\frac{\partial^2 f}{\partial x \partial y}(0,0) \neq 0$ and $\frac{\partial^3 f}{\partial y^3}(0,0) \neq 0$, if and only if F is a cusp. □

The conditions in Proposition 3.1 reflect the order of contact of Σ with the kernel of $DF(0,0,0)$.

COROLLARY 3.2 *A singular germ $F : \mathbb{R}^3, 0 \to \mathbb{R}^2, 0$ of corank 1 and with a smooth critical set is a fold if and only if Σ is transversal to the set $\ker(DF(0,0,0))$ at the origin. It is a cusp if and only if Σ and $\ker(DF(0,0,0))$ have 2-point contact at the origin.*

We observe that the previous two results are the statement that the fold and cusp orbits coincide with Boardman strata $\Sigma^{1,0}$ and $\Sigma^{1,1,0}$, respectively.

PROPOSITION 3.3 *Let $F : \mathbb{R}^3, 0 \to \mathbb{R}^2, 0$ be a singular germ with a smooth critical set Σ and let ϕ be a parametrization of Σ. Then F is a swallowtail if and only if $\frac{\partial}{\partial t}(F \circ \phi)(0) = \frac{\partial^2}{\partial t^2}(F \circ \phi)(0) = 0$ and $\frac{\partial^3}{\partial t^3}(F \circ \phi)(0) \neq 0$.*

Proof: From the proof of Proposition 3.1 we have $\frac{\partial}{\partial t}(F \circ \phi)(0) = \frac{\partial^2}{\partial t^2}(F \circ \phi)(0) = 0$ if and only if $\frac{\partial^2 f}{\partial y^2}(0,0) = \frac{\partial^3 f}{\partial y^3}(0,0) \cdot \frac{\partial^2 f}{\partial x \partial y}(0,0) = 0$.

Differentiating the expression (3.2) and setting $\frac{\partial^2 f}{\partial y^2}(0,0) = \frac{\partial^3 f}{\partial y^3}(0,0) \cdot \frac{\partial^2 f}{\partial x \partial y}(0,0) = 0$, yield

$$
\frac{\partial^3}{\partial t^3}(F \circ \phi)(0) = -\frac{\partial^4 f}{\partial y^4}(0,0) \cdot (\frac{\partial^2 f}{\partial x \partial y}(0,0))^2 \left(\frac{1}{\frac{\partial f}{\partial x}(0,0)} \right).
$$

Then, $\frac{\partial}{\partial t}(F \circ \phi)(0) = \frac{\partial^2}{\partial t^2}(F \circ \phi)(0) = 0$ and $\frac{\partial^3}{\partial t^3}(F \circ \phi)(0) \neq 0$ if and only if $\frac{\partial^2 f}{\partial y^2}(0,0) = \frac{\partial^3 f}{\partial y^3}(0,0) = 0$, $\frac{\partial^2 f}{\partial x \partial y}(0,0) \neq 0$, and $\frac{\partial^4 f}{\partial y^4}(0,0) \neq 0$, if and only if F is a swallowtail. □

Again the conditions in Proposition 3.3 express the order contact of the critical set Σ with the $\ker(DF(0,0,0))$.

COROLLARY 3.4 *A singular germ $F : \mathbb{R}^3, 0 \to \mathbb{R}^2, 0$ of corank 1 and with a smooth critical set is a swallowtail if and only if the critical set Σ has contact 3 with the kernel of $DF(0,0,0)$ at the origin.*

When F has a lips/beaks singularity, its critical set Σ is singular. Then we need additional algebraic conditions for recognizing these singularities.

PROPOSITION 3.5 *A singular germ that can be written in the form $F = (x, f(x,y) \pm z^2)$ is a lips/beaks if and only if $\frac{\partial^2 f}{\partial y^2}(0,0) = \frac{\partial^2 f}{\partial x \partial y}(0,0) = 0$, $\frac{\partial^3 f}{\partial y^3}(0,0) \neq 0$ and $(\frac{\partial^3 f}{\partial x \partial y^2}(0,0))^2 - \frac{\partial^3 f}{\partial x^2 \partial y}(0,0) \cdot \frac{\partial^3 f}{\partial y^3}(0,0) \neq 0$.*

Proof: The proof follows by a direct calculation. □

PROPOSITION 3.6 *The germ $\frac{\partial f}{\partial y} : \mathbb{R}^2, 0 \to \mathbb{R}, 0$ is a germ of a Morse function if and only if $\frac{\partial^2 f}{\partial y^2}(0,0) = \frac{\partial^2 f}{\partial x \partial y}(0,0) = 0$ and $(\frac{\partial^3 f}{\partial x \partial y^2})^2(0,0) - \frac{\partial^3 f}{\partial x^2 \partial y}(0,0) \cdot \frac{\partial^3 f}{\partial y^3}(0,0) \neq 0$.*

Proof: The proof follows immediately by considering the Taylor expression of $\frac{\partial f}{\partial y}$ at the origin,

$$\frac{\partial f}{\partial y}(x,y) = \quad \frac{\partial^2 f}{\partial x \partial y}(0,0).x + \frac{\partial^2 f}{\partial y^2}(0,0).y + $$
$$\frac{1}{2}\{\frac{\partial^3 f}{\partial x^2 \partial y}(0,0).x^2 + 2\frac{\partial^3 f}{\partial x \partial y^2}(0,0).xy + \frac{\partial^3 f}{\partial y^3}.y^2\} + O_3(x,y).$$

□

As $\mathcal{A}$-equivalence preserves the singularity of the critical set, we have the following result.

COROLLARY 3.7 *Let $F(x,y,z) = (x, f(x,y) \pm z^2))$. If $\frac{\partial^3 f}{\partial y^3}(0,0) \neq 0$, then F is a lips/beaks if and only if the critical set has a non-degenerate curve singularity.*

REMARK **3.1** The above geometric criteria do not extend to higher codimension singularities. The reason is that when Σ is smooth its order of contact k with the kernel of $DF(0,0,0)$ determines the A_{k-1} $\mathcal{K}$-class of the projection. When $k \geq 4$, there are several $\mathcal{A}$-orbits of the projection inside the A_{k-1}-orbit. One needs algebraic conditions to distinguish between these $\mathcal{A}$-orbits.

4 GEOMETRIC CRITERIA FOR RECOGNITION OF THE VERSAL UNFOLDINGS SINGULARITIES OF $\mathcal{A}_e$-codim $= 1$

Geometric criteria for determining when a deformation of a lips/beaks or swallowtail singularity of a germ $\mathbb{R}^2, 0 \to \mathbb{R}^2, 0$ is versal are given in [27]. We extend these results to germs $F : \mathbb{R}^3, 0 \to \mathbb{R}^2, 0$. We deal in more details with the lips/beaks case.

Let $F : \mathbb{R}^3 \times \mathbb{R}, 0 \to \mathbb{R}^2 \times \mathbb{R}, 0$ be an unfolding of a germ $F_0(x,y,z,0) = (x, f_0(x,y) \pm z^2)$ that has a lips/beaks singularity at the origin. Then the critical set of F_0 can be considered as a plane curve with a Morse singularity and the unfolding F induces an unfolding of this Morse singularity.

THEOREM 4.1 *An unfolding F of a lips/beaks singularity of F_0 is $\mathcal{A}_e$-versal if and only if induces a $\mathcal{K}$-versal unfolding of the Morse singularity of the critical set of F_0.*

Proof: Let $\tilde{F}$ be an unfolding of a lips/beaks and write

$$\tilde{F}(x, y, z, u) = (\tilde{F}_1(x, y, z, u), \tilde{F}_2(x, y, z, u), u).$$

CLAIM: $\tilde{F}$ is equivalent to an unfolding written in the form

$$F(x, y, z, u) = (x, F_2(x, y, z, u), u).$$

Proof of the claim: Without loss of generality we can assume that $\frac{\partial \tilde{F}_1}{\partial x} \neq 0$ at the origin. Then the local diffeomorphism

$$
\begin{aligned}
H: \quad &\mathbb{R}^3 \times \mathbb{R} && \to \mathbb{R}^3 \times \mathbb{R} \\
&(x, y, z, u) && \mapsto (\tilde{F}_1(x, y, z, u), y, z, u)
\end{aligned}
$$

makes the following diagram commute

$$
\begin{array}{ccc}
\mathbb{R}^3 \times \mathbb{R} & \xrightarrow{\ \ \tilde{F}\ \ } & \mathbb{R}^2 \times \mathbb{R} \\
(x, y, z, u) & & (\tilde{F}_1(x, y, z, u), \tilde{F}_2(x, y, z, u), u) \\
\Big\downarrow H & & \Big\downarrow id \\
\mathbb{R}^3 \times \mathbb{R} & \xrightarrow{\ \ F\ \ } & \mathbb{R}^2 \times \mathbb{R} \\
((\tilde{F}_1(x, y, z, u), y, z, u) & & (\tilde{F}_1(x, y, z, u), \tilde{F}_2(x, y, z, u), u)
\end{array}
$$

where $F_2(x, y, z, u) = \tilde{F}_2(\pi_1 \circ \tilde{F}_1^{-1}(x, y, z, u), y, z, u)$ and π_1 is the projection to the first coordinate. Observe that $\pi_1 \circ \tilde{F} \circ H^{-1} = \pi_1$, therefore $\tilde{F} \circ H^{-1} = (x, \tilde{F}_2(x,y,z,u), u)$ and hence F and $\tilde{F}$ are equivalents unfoldings (see [16]). Therefore we can set

$$F(x, y, z, u) = (x, g_1(x, y, z) + u g_2(x, y, z, u), u).$$

The germ $(x, g_1(x, y, z))$ has a lips/beaks singularity at the origin, consequently, changes of coordinates in the source and target reduce F_0 to the form $(x, y^3 \pm x^2 y \pm z^2)$. In this new system of coordinates

$$F(x, y, z, u) = (x, y^3 \pm x^2 y \pm z^2 + u \bar{g}_2(x, y, z, u), u).$$

Since a lips/beaks singularity is 3-determined, we can work in the $J^3(x, y, z, u)$, and write

$$j^3 F(x, y, z, u) = (x, y^3 \pm x^2 y \pm z^2 + \alpha(x, z, u)u + \beta(z, u)uy + a_1 u y^2 + a_2 u x y, u).$$

Now successive explicit changes of coordinates in the source and target reduce this 3-jet to the form

$$j^3 F(x, y, z, u) = (x, y^3 \pm x^2 y \pm z^2 + \psi(u)y, u).$$

Then F is a versal unfolding of F_0 if and only if $\psi'(0) \neq 0$ (see [16]). We have

$$\Sigma = \{(x, y, z, u) : z = 0 \text{ and } \sigma(x, y, z, u) = 3y^2 \pm x^2 + \psi(u) + \cdots = 0\}.$$

Then $\frac{\partial \sigma}{\partial u}(0) = \psi'(0)$, and hence F is versal if and only if $\frac{\partial \sigma}{\partial u}(0) \neq 0$, if and only if σ is a versal unfolding of the Morse singularity of Σ. $\qquad\square$

Let $\Sigma_F^{1,1}$ denotes the critical set of the restriction of F to Σ. Then proceeding as above, we show the following.

THEOREM 4.2 *Let $F(x,y,z,u)$ be an unfolding of a swallowtail singularity. Then F is $\mathcal{A}_e$-versal if and only if $\Sigma_F^{1,1}$ is a smooth curve.*

REMARK 4.1 The results in Sections 3 and 4 can be extended to A_k-singularities of map-germs $f : \mathbb{R}^n, 0 \to \mathbb{R}^p, 0, n \geq p$, with $\mathcal{A}_e$-codimension 1. This can be done by following the same procedure as above and using the normal forms for $\mathcal{A}_e$-codimension 1 singularities given by Goryunov in [9]. We are currently looking for similar characterizations for wider classes of singularities.

5 A STRATIFICATION OF THE SPACE OF SINGULAR PLANES

Recall from the introduction that there is a natural 4-parameter family of projections

$$\Pi: \quad M \times G(2,4) \quad \to \quad \mathbb{R}^2$$
$$(p,u) \quad \mapsto \quad (<p,a>,<p,b>)$$

where a and b form an orthonormal basis of the plane of projection $u \in G(2,4)$ (Π is a 4-parameter family because the $\mathcal{A}$-equivalence class of Π_u does not depend on the choice of orthonormal basis a,b). We start with an elementary lemma that provides a simple way of writing the family Π.

LEMMA 5.1 *The map Π can be written locally in the form*

$$\Pi: \mathbb{R}^3 \times \mathbb{R}^4, 0 \quad \to \quad \mathbb{R}^2, 0$$
$$(x,y,z,\beta_1,\gamma_1,\beta_2,\gamma_2) \quad \mapsto \quad (x+\beta_1 y+\gamma_1 z, \beta_2 y + \gamma_2 z + f(x,y,z)).$$

Proof: Given an embedding $F : M \to \mathbb{R}^4$, we can put F in Monge form in a neighbourhood of any point p, that is we can consider $F : \mathbb{R}^3, 0 \to \mathbb{R}^4, 0$ with $F(x,y,z) = (x,y,z,f(x,y,z))$. We assume, without loss of generality, that the initial plane of projection u_0 is generated by the vectors $a = (1,0,0,0)$ and $b = (0,0,0,1)$. Then vectors $(1,\bar{\alpha}_1,\bar{\beta}_1,\bar{\gamma}_1)$ and $(\bar{\alpha}_2,\bar{\beta}_2,\bar{\gamma}_2,1)$ generate all the planes near u_0. We claim that there are always linearly independent vectors of the form $(1,\beta_1,\gamma_1,0)$ and $(0,\beta_2,\gamma_2,1)$ that generate these planes. Indeed, by setting $\xi_1 = \xi_4 = 1/(1-\bar{\alpha}_2\bar{\gamma}_1)$, $\xi_2 = -\bar{\gamma}_1/(1-\bar{\alpha}_2\bar{\gamma}_1)$, and $\xi_3 = -\bar{\alpha}_2/(1-\bar{\alpha}_2\bar{\gamma}_1)$, we have

$$\xi_1(1,\bar{\alpha}_1,\bar{\beta}_1,\bar{\gamma}_1)+\xi_2(\bar{\alpha}_2,\bar{\beta}_2,\bar{\gamma}_2,1) = (1,(\bar{\alpha}_1-\bar{\gamma}_1\bar{\beta}_2)/(1-\bar{\alpha}_2\bar{\gamma}_1),(\bar{\beta}_1-\bar{\gamma}_1\bar{\gamma}_2)/(1-\bar{\alpha}_2\bar{\gamma}_1),0),$$

$$\xi_3(1,\bar{\alpha}_1,\bar{\beta}_1,\bar{\gamma}_1)+\xi_4(\bar{\alpha}_2,\bar{\beta}_2,\bar{\gamma}_2,1) = (0,(\bar{\beta}_2-\bar{\alpha}_2\bar{\alpha}_1)/(1-\bar{\alpha}_2\bar{\gamma}_1),(\bar{\gamma}_2-\bar{\beta}_1\bar{\alpha}_2)/(1-\bar{\alpha}_2\bar{\gamma}_1),1).$$

Since Π_u depends only on the plane u, we can construct an orthonormal basis of u from the vectors $a = (1,\beta_1,\gamma_1,0)$ and $b = (0,\beta_2,\gamma_2,1)$. Then, by changing coordinates in the parameter space $(\alpha_1, \beta_1, \alpha_2, \beta_2)$ and by scaling, we can write Π in the form given above. $\square$

Given a generic embedded hypersurface M in Monge form $(x, y, z, f(x, y, z))$ at some point p, we can find the algebraic conditions on the coefficients of the Taylor expansion of f for Π_u to have one of the singularities in Theorems 2.3 and 2.4 and for Π to be a versal unfolding of these singularities (see [22]). These conditions, in turn, give infinitesimal information concerning the embedding of M at p. However, they are in general too complicated and difficult to interpret geometrically. But for the singularities of codimension 1, we have the following corollary of Theorems 4.1 and 4.2. We observe that using Lemma 5.1 we can show that in the lips/beaks case Π always induces a K-versal unfolding of the Morse singularity of Π_u and for the swallowtail singularity $\Sigma_\Pi^{1,1}$ is always smooth.

COROLLARY 5.2 *The family Π is always a versal unfolding of the lips/beaks and swallowtail singularities of Π_u.*

If a generic embedded hypersurface $M \subset \mathbb{R}^4$ is given in Monge form $w = f(x, y, z)$, then Π_u is singular if and only if the normal direction to M at the origin, $(0, 0, 0, 1)$, belongs to the plane u. So we can take $a = (0, 0, 0, 1)$ and $b = (\alpha_1, \alpha_2, \alpha_3, 0)$, with $\|b\| = 1$, as the generators of the plane u. Therefore, the singular planes of projections can be identified with the sphere $S^2 = \{b = (\alpha_1, \alpha_2, \alpha_3, 0), \|b\| = 1\}$. We shall stratify this sphere according to the singularity type of the projection Π_u.

We rotate the coordinates axis so that $j^2 f = a_1 x^2 + a_2 y^2 + a_3 z^2$, where $a_i = \frac{\kappa_i}{2}$, $i = 1, 2, 3$, with κ_i the principal curvatures of M at the origin.

In Proposition 5.3, we obtain a geometric characterization of the singularities of the projections at elliptic and hyperbolic points of the embedding.

PROPOSITION 5.3 *Let M be a generic embedded hypersurface in $\mathbb{R}^4$, p a point on M, and u a singular plane of the projection determined by a direction $b = (\alpha_1, \alpha_2, \alpha_3, 0) \in S^2$.*

(i) If p is an elliptic point, then the projection Π_u has a singularity of type fold for any direction on S^2.

(ii) If p is a hyperbolic point, then Π_u has a fold singularity for any direction on S^2 except on two circles where the singularity is in general of cusp type. On these circles, and at most hyperbolic points on M, there are up to $2k$ directions, $k = 0, \cdots 6$, where the singularities are of swallowtail type. At certain hyperbolic points on M these directions could yield singularities of higher codimensions (generically of type 6,7, 8 of Theorem 2.3, respectively on a surface in M, on a curve on this surface and at isolated points on this curve).

Proof: Let $b = (\alpha_1, \alpha_2, \alpha_3, 0) \in S^2$. Then, as before, the projection to the plane u generated by b and $(0, 0, 0, 1)$ is given by

$$\Pi_u = (\alpha_1 x + \alpha_2 y + \alpha_3 z, f(x, y, z)).$$

We can assume without loss of generality that $\alpha_1 \neq 0$. We can then change coordinates in the source and target and write $\Pi_u = (x, \alpha_1^2 f((x - \alpha_2 y - \alpha_3 z)/\alpha_1, y, z))$, so that

$$j^2 \Pi_u = (x, a_1 x^2 + (a_2 \alpha_1^2 + a_1 \alpha_2^2) y^2 + (a_1 \alpha_3^2 + a_3 \alpha_1^2) z^2 + 2a_1 \alpha_2 \alpha_3 yz - 2a_1 \alpha_3 xz - 2a_1 \alpha_2 xy).$$

Let $q(y,z) = (a_2\alpha_1^2 + a_1\alpha_2^2)y^2 + 2a_1\alpha_2\alpha_3 yz + (a_1\alpha_3^2 + a_3\alpha_1^2)z^2$ be the quadratic form obtained by setting $x = 0$ in the second component of $j^2\Pi_u$. This form is degenerate if and only if

$$a_2 a_3 \alpha_1^2 + a_1 a_3 \alpha_2^2 + a_1 a_2 \alpha_3^2 = 0. \tag{5.1}$$

At an elliptic point (case (i)), the coefficients a_i have the same sign, so equation (5.1) does not hold. Therefore $\Pi_u \sim (x, y^2 \pm z^2)$ which is a fold singularity.

At a hyperbolic point (case (ii)) equation (5.1) determines a cone. This cone intersects the sphere S^2 in two circles (we only need to consider one of them as b and $-b$ define the same projection). Away from these circles, the singularity of Π_u is clearly of type fold. On the circles the quadratic q is degenerate. As the point in consideration is not parabolic, we can suppose without loss of generality that the coefficient of z^2 in q is different from zero. Then the change coordinates $Z = z + \frac{a_1\alpha_2\alpha_3}{a_1\alpha_3^2 + a_3\alpha_1^2}y$ and further changes of coordinates in the source and target yield

$$j^2\Pi_u \sim (x, -\frac{2a_1 a_3 \alpha_1^2 \alpha_2}{a_1\alpha_3^2 + a_3\alpha_1^2}xy + (a_1\alpha_3^2 + a_3\alpha_1^2)z^2).$$

As the point is hyperbolic, the coefficient of xy is different from zero (note that we assumed $\alpha_1 \neq 0$, and if $\alpha_2 = 0$ then the coefficient of z^2 in q must be zero, which is not the case from our assumption above.)

We now need to analyze the coefficient of y^3 in the second component of Π_u. To simplify the calculations, we set $\alpha_1 = 1$, so equation (5.1) determines a circle (a curve of degree 2). Then the coefficient of y^3 is a polynomial of degree 6 in α_2 and α_3. Therefore this polynomial can vanish at $0, 2, 4, 6, 8, 10$, or 12 directions on the circle. Away from these directions the singularity of Π_u is a cusp. At one of these directions we need to analyze the higher jets of the second component of Π_u which depend only on the Taylor expansion of the function f giving the surface M in Monge form. (Note that the values of α_i, $i = 1, 2, 3$ are now fixed). Without imposing any condition on the surface M, the coefficient of y^4 in Π_u does not vanish and the singularity of Π_u is of type swallowtail. When the coefficient of y^4 vanishes (this determines a 2-dimensional surface on M) the singularity of Π_u is generically of type 6 (Theorem 2.3). There might be a curve on this surface where the singularity becomes of type 7 and a point on this curve where it is of type 8.   □

With the same hypothesis of the previous Proposition, we can proceed as above to obtain the geometric characterization at parabolic and partial umbilic points of the embedding.

PROPOSITION 5.4 *(i) If p is a parabolic point (say $\kappa_2 = 0$), then Π_u is a fold except on the circle $\alpha_2 = 0$ where it has generically a singularity of type 4_2. For 0 or 2 directions on this circle the singularity is generically of type 4_3. These directions could yield singularities of type 4_4 and 4_5 at certain points (respectively on a curve on the parabolic set and at isolated points on this curve).*

Also on the circle $\alpha_2 = 0$, and for some curve on the parabolic set (labelled the 11_5-curve), the singularity is generically of type 11_5. For 0, 4 or 8 directions on the circle the singularity is of type 11_7; for 0 or 4 directions the singularity is of type 12; and for 0 or 2 directions it is of type 16. For these directions and at possible isolated points on the 11_5-curve the singularity of the projection becomes respectively of type 11_9, 13 or 17.

Furthermore, if $\kappa_1\kappa_3 < 0$, there are two more circles (which intersect transversely the circle $\alpha_2 = 0$) where the singularity is generically of type N_1. At possible directions on these circles and on some curve on the parabolic set (resp. isolated points on this curve) the singularity becomes of type $N_2, \cdots, N_5$ (resp. $N_6, \cdots, N_{11}$).

(ii) If p is a partial umbilic point (say $\kappa_1 = \kappa_2 = 0$ and $\kappa_3 \neq 0$), then the projection has generically a singularity of type 4_2. There are 0 or 2 (resp. 0, 1 or 3) circles where the singularity is of type 4_3 (resp. 11_5). At isolated points on these circles the singularity becomes of type 4_4 (resp. 11_7, 12 or 16). There is an unique direction (not on the above circles) where the singularity is of type N_{12}.

Acknowledgements: This work is part of my Ph.D. thesis. I would like to thank FAPESP for financial support. I am very grateful to Neil Kirk for letting me use his software "Transversal" and especially to Farid Tari for his extremely stimulating supervision. I would also like to thank Maria A. S. Ruas and the referee for useful comments.

REFERENCES

[1] T. Banchoff, T. Gaffney, C. McCrory. Cusps of Gauss Mappings. Pitman, New York, 1982.

[2] J.W. Bruce, Generic geometry and duality. In: Singularities, ed. J.P. Brasselet, Lille 1991, London Math. Soc. Lecture Note Series 201, Cambridge University Press. pp. 29-60, 1994.

[3] J.W. Bruce, N. Kirk, A.A. du Plessis. Complete Transversals and the classifications of singularities. Nonlinearity. 10 1:253-275, 1997.

[4] J.W. Bruce, N.P. Kirk, J.M. West. Classification of maps-germs from surfaces to four-space. Preprint. University of Liverpool, 1995.

[5] J.W. Bruce, A.C. Nogueira. Surfaces in $\mathbb{R}^4$ and duality, Quart. J. Math. Oxford Ser. 49 2:196:433-443, 1998.

[6] J.W. Bruce, A.C. Nogueira. Surfaces in $\mathbb{R}^4$ and duality II. Preprint 2000.

[7] J.W. Bruce, A.A. du Plessis, C.T.C. Wall. Determinacy and unipotency. Invent. Math. 88:521-554, 1987.

[8] J.W. Bruce, F. Tari. Families of surfaces in $\mathbb{R}^4$. To appear.

[9] V.V. Goryunov. Singularities of projections of full intersections. Journal of Soviet Mathematics. 27:2785-2811, 1984.

[10] W. Hawes. Multi-dimensional motions of the plane and space. Ph.D. Thesis, University of Liverpool, 1994.

[11] C.A. Hobbs, N. P. Kirk. On the classification and bifurcation of multigerms of maps from surfaces to 3-space. To appear in Mathematica Scandinavica.

[12] K. Houston, N.P. Kirk. On the classification and geometry of map-germs from 3-space to 4-space. In: Proccedings of the European Singularities Conference in honour of C.T.C. Wall on the occasion of his 60th Birthday, eds. J.W. Bruce, D. Mond, Singularity Theory, London Math. Soc. Lecture Note Series 263, Cambridge University Press. pp.325-351, 1999.

[13] N.P. Kirk. Computational aspects of singularity theory. Ph.D. Thesis, University of Liverpool, 1993.

[14] N.P. Kirk. Transversal, A maple package for singularity theory, Version 3.1. University of Liverpool, 1998.

[15] Yung-Chen Lu. Singularity theory and an introduction to catastrophe theory. Springer-Verlag, 1976.

[16] J. Martinet. Singularities of Smooth Functions and Maps. London Math. Soc. Lecture Note Series 58, Cambridge University Press, 1982.

[17] D.K.H. Mochida. Geometria genérica de subvariedades em codimensão maior que um em $\mathbb{R}^n$. Ph.D. Thesis, University of São Paulo, ICMC - São Carlos, 1993.

[18] D.K.H. Mochida, M. C. Romero Fuster, M. A. S. Ruas. The geometry of surfaces in 4-space from a contact viewpoint. Geom. Dedicata 54 3:323-332, 1995.

[19] D. Mond. On the classification of germs of maps from $\mathbb{R}^2$ to $\mathbb{R}^3$. Proc. London Math. Soc. 50 3:333-369, 1985.

[20] J.A. Montaldi. On contact between submanifolds. Michigan Math. J. 33:195-199, 1986.

[21] J.A. Montaldi. On generic composite of maps. Bull. Lond. Math. Soc. 23:81-85, 1991.

[22] A.C. Nabarro. Sobre a geometria local de hipersuperfícies em $\mathbb{R}^4$. Ph.D. Thesis, University of São Paulo, ICMC - São Carlos, 2000.

[23] A.C. Nabarro. Duality and contact of hypersurfaces in $\mathbb{R}^4$ with hyperplanes and lines. Preprint, 2002.

[24] A.C. Nogueira. Superfícies em $\mathbb{R}^4$ e dualidade. Ph.D. Thesis, University of São Paulo, ICMC - São Carlos, 1998.

[25] J.H. Rieger. Families of maps from the plane to the plane. J. London Math. Soc. 36 2:351-369, 1987.

[26] J.H. Rieger, M.A.S. Ruas. Classification of $\mathcal{A}$-simple germs from k^n to k^2. Compositio Math. 79 1:99-108, 1991.

[27] J.E. Rycroft. A geometrical investigation into the projections of surfaces and space curves. Ph.D. Thesis, University of Liverpool, 1992.

[28] F. Tari. Some applications of singularity theory to the geometry of curves and surfaces. Ph.D. Thesis, University of Liverpool, 1990.

[29] J.M. West. The differential geometry of the crosscap. Ph.D. thesis, University of Liverpool, 1995.

Frobenius Manifolds and Hypersurface Singularities

CLAUS HERTLING Mathematisches Institut der Universität Bonn Beringstraße 1, 53115 Bonn, Germany
E-mail: hertling@math.uni-bonn.de

This is a survey article on Frobenius manifolds and hypersurface singularities. It covers the four lectures of a minicourse at the 6th Workshop on Real and Complex Singularities (17 - 21 July 2000, ICMC-USP, Sao Carlos, Brazil).

1st lecture: Multiplication on the tangent bundle
2nd lecture: Frobenius manifolds and their structure connections
3rd lecture: Construction in singularity theory
4th lecture: Global moduli spaces for hypersurface singularities

A Frobenius manifold is a manifold with a metric and a multiplication on the tangent bundle which harmonize in the most natural way. This notion was defined by Dubrovin 1991, motivated by quantum cohomology, topological field theory and mirror symmetry. But the first big class of Frobenius manifolds had been constructed implicitly already in 1983 by K. Saito and M. Saito in singularity theory: the base space of a semiuniversal unfolding of an isolated hypersurface singularity can be equipped with the structure of a Frobenius manifold.

For a long time this construction had not been taken up. Even now results which put it in a precise way into the picture of mirror symmetry have yet to be found. But recently I found an application within singularity theory: a construction of global moduli spaces for isolated hypersurface singularities.

The four lectures will focus on the general theory of Frobenius manifolds (1st and 2nd), the construction in singularity theory (3rd) and this application (4th).

More detailed comments about the contents of the lectures can be found at their beginnings. The 2nd lecture presupposes some familiarity with meromorphic connections, the 3rd and 4th also with hypersurface singularities.

The material in this survey article is taken from [14][15] and aims at the role of Frobenius manifolds in singularity theory. For many different aspects of Frobenius manifolds and their relations to meromorphic connections, topological field theory, quantum cohomology, and mirror symmetry we recommend the books [4][20][21]. We hope that this survey article will stimulate the search for other applications and for stronger links between singularity theory on the one hand and topological field theory and mirror symmetry on the other hand.

1 MULTIPLICATION ON THE TANGENT BUNDLE

The subject of the first lecture is the local structure of a complex manifold with a multiplication on the holomorphic tangent bundle. An extensive study of this has been made in [14]. In this lecture central results and examples from [14] are put together.

First, the pointwise structure is discussed: the structure of the tangent spaces as finite dimensional $\mathbb{C}$-algebras. Then we introduce the analytic spectrum of a manifold with multiplication, a subvariety of the cotangent bundle which determines the multiplication.

If the multiplication satisfies a certain integrability condition then the manifold is called an F-manifold. This notion is most central and was defined first in [13] (cf. [20] I§5). It has a good local decomposition property. If the multiplication is generically semisimple then the analytic spectrum is a Lagrange variety if and only if the manifold is an F-manifold. This makes it possible to interpret many results in [8] as results on F-manifolds.

Important classes of F-manifolds are the base spaces of semiuniversal unfoldings of hypersurface and boundary singularities (cf. the 3rd lecture) and the complex orbit spaces of finite Coxeter groups. One can distinguish the last ones in a way which extends the correspondence between certain Coxeter groups and the simple hypersurface or boundary singularities.

A *manifold with multiplication* $(M, \circ, e)$ will always mean following: M is a complex manifold, $\dim M = m \geq 1$, TM its holomorphic tangent bundle, $\mathcal{T}_M$ the sheaf of germs of holomorphic sections of TM,

$$\circ : \mathcal{T}_M \times \mathcal{T}_M \to \mathcal{T}_M \ , \ (X, Y) \mapsto X \circ Y$$

an $\mathcal{O}_M$-bilinear commutative and associative multiplication. Then $\circ : \mathcal{T}_M \otimes \mathcal{T}_M \to \mathcal{T}_M$ is a $(2,1)$-tensor. The multiplication is supposed to have a unit everywhere, and $e \in \Gamma(M, \mathcal{T}_M)$ is the global unit field.

Fix $t \in M$. Then $(T_t M, \circ, e|_t)$ is an algebra of dimension m. It has a unique decomposition (cf. e.g. [14] Lemma 1.1) into local algebras,

$$T_t M = \bigoplus_{k=1}^{l(t)} (T_t M)_k \quad \text{with}$$

$$e = \sum_{k=1}^{l(t)} e_k \, ,$$

$(T_t M, \circ, e_k)$ a local algebra with unit e_k,

$(T_t M)_j \circ (T_t M)_k = 0$ for $j \neq k$

This is the simultaneous eigenspace decomposition for the commuting endomorphisms $X \circ : T_t M \to T_t M$ for $X \in T_t M$.

EXAMPLE 1.1 Consider a polynomial $f \in \mathbb{C}[x_0, ..., x_n]$ such that $f : \mathbb{C}^{n+1} \to \mathbb{C}$ has only isolated singularities.

$$J_f := \left(\frac{\partial f}{\partial x_0}, ..., \frac{\partial f}{\partial x_n} \right) \subset \mathcal{O}_{\mathbb{C}^{n+1}} \, ,$$

is an ideal sheaf, and $(J_f)_{pol} \subset \mathbb{C}[x]$ denotes the corresponding ideal in $\mathbb{C}[x]$. Then

$$\frac{\mathbb{C}[x]}{(J_f)_{pol}} = \bigoplus_{x \in Sing(f)} \frac{\mathcal{O}_{\mathbb{C}^{n+1},x}}{(J_f)_x}$$

is the decomposition into local algebras.

Let $(M, \circ, e)$ be as above. For any $k = 1, ..., l(t)$, the map $\lambda_k : T_t M \to \mathbb{C}$ defined by

$$
\begin{aligned}
e_k &\mapsto 1 \, , \\
\text{maximal ideal in } (T_t M)_k &\to 0 \, , \\
(T_t M)_j &\to 0 \quad \text{for } j \neq k
\end{aligned}
$$

is an algebra homomorphism. For $X \in T_t M$ one obviously has

$$\lambda_k(X) = \text{ eigenvalue of } X \circ : (T_t M)_k \to (T_t M)_k.$$

Then (cf. [14] Lemma 1.1)

$$\{\lambda_1, ..., \lambda_{l(t)}\} = \text{Hom}_{alg}(T_t M, \mathbb{C}).$$

So its points correspond one-to-one to the local subalgebras in the tangent spaces of M as algebras.

The *analytic spectrum* of $(M, \circ, e)$ is as a set

$$L := \text{Specan}(T_M, \circ) = \bigcup_{t \in M} \text{Hom}_{alg}(T_t M, \mathbb{C}) \subset T^* M \, .$$

It has a canonical complex structure $(L, \mathcal{O}_L)$, which is locally defined as follows. Let $e = \delta_1, ..., \delta_m$ be a local basis of vector fields, and $y_1, ..., y_m$ the induced fiberwise linear functions on T^*M. Setting $\delta_i \circ \delta_j = \sum_k a_{ij}^k \cdot \delta_k$ with $a_{ij}^k \in \mathcal{O}_M$, the analytic spectrum is defined by the ideal

$$I = (y_1 - 1, y_i y_j - \sum_k a_{ij}^k \cdot y_k)$$

with $\mathcal{O}_L := O_{T^*M}/I|_L$.

The analytic spectrum L has good properties ([14] Theorem 2.1): The projection $\pi : L \to M$ is finite and flat of degree $m = \dim M$. The map

$$\begin{aligned} \mathbf{a} : \mathcal{T}_M &\to \pi_* \mathcal{O}_L \\ X &\mapsto \text{(by } X \text{ induced function on } T^*M)|_L \\ \circ &\quad mult. \end{aligned}$$

is an isomorphism of free $\mathcal{O}_M$-modules of rank m and of $\mathcal{O}_M$-algebras. Therefore the multiplication $\circ$ on TM is determined by L. The function $\mathbf{a}(X)$ can be called the "eigenvalue function" of $X\circ$ because of $\mathbf{a}(X)(\lambda_k) = \lambda_k(X)$.

The *caustic*

$$\mathcal{K} := \{t \in M \mid l(t) < \text{ generic value}\} \subset M$$

is a hypersurface or empty (the proof in [14] (Prop. 2.4) for the case $m = $ *generic value of $l(t)$* works for any generic value of $l(t)$).

The multiplication is called *semisimple* if $l(t) = m$ $\forall t \in M$. Then there is locally a basis $e_1, ..., e_m$ of vector fields with $e_i \circ e_j = \delta_{ij} e_i$.

$(M, \circ, e)$ is called *massive* if $l(t) = m$ for $t \in M - \mathcal{K}$.

DEFINITION 1.2 *([13], [20] I§5, [14] Def. 3.2)*
a) $(M, \circ, e)$ is an F-manifold if the integrability condition holds:

$$\forall\, X, Y \in \mathcal{T}_M \quad \text{Lie}_{X \circ Y}(\circ) = X \circ \text{Lie}_Y(\circ) + Y \circ \text{Lie}_X(\circ) . \tag{1.1}$$

b) Let $(M, \circ, e)$ be an F-manifold. Then $E \in \mathcal{T}_M$ is an Euler field of weight $c \in \mathbb{C}$ if

$$\text{Lie}_E(\circ) = c \cdot \circ \tag{1.2}$$

(Euler field of weight 1 =: Euler field).

(1.1) for $X = Y = e$ gives $\text{Lie}_e(\circ) = 2 \cdot \text{Lie}_e(\circ)$, so $\text{Lie}_e(\circ) = 0$, and e is an Euler field of weight 0.

(1.2) can be written with vector fields X, Y and Lie brackets as

$$[E, X \circ Y] - [E, X] \circ Y - X \circ [E, Y] = c \cdot X \circ Y .$$

For $X = Y = e$ it yields $[e, E] = c \cdot e$. One can write (1.1) analogously with four vector fields and nine Lie brackets.

PROPOSITION 1.3 *([14] Prop. 4.1) If M_1, M_2 are F-manifolds, then $M_1 \times M_2$ is an F-manifold. If E_1, E_2 are Euler fields of weight c on M_1, M_2, then $E_1 + E_2$ is an Euler field of weight c on $M_1 \times M_2$.*

THEOREM 1.4 *([14] Thm. 4.2) Let $(M, \circ, e)$ be an F-manifold. Fix $t \in M$. The decomposition*

$$T_t M = \prod_{k=1}^{l(t)} (T_t M)_k$$

extends to a decomposition

$$(M, t) = \prod_{k=1}^{l(t)} (M_k, t)$$

of (M, t) into a product of germs of F-manifolds with

$$T_t M_k \cong (T_t M)_k .$$

An Euler field of weight c decomposes accordingly.

The proof of Theorem 1.4 in [14] uses (1.1) in a way which justifies the name *integrability condition*.

THEOREM 1.5 *([14] Thm. 6.2) Let $(M, \circ, e)$ be a manifold with massive multiplication.*
Then the complex structure $(L, \mathcal{O}_L)$ is reduced everywhere.
The following conditions are equivalent:

i) $(M, \circ, e)$ is an F-manifold.

ii) The vector fields $e_1, \ldots, e_m$ defined locally in $M - \mathcal{K}$ with $e_i \circ e_j = \delta_{ij} e_i$ satisfy

$$[e_i, e_j] = 0 .$$

iii) L is a Lagrange variety.

Locally each 1-dimensional F-manifold is isomorphic to $(\mathbb{C}, \circ, e)$ with $e = \frac{\partial}{\partial u}$, u a coordinate on $\mathbb{C}$. This F-manifold or a germ of it is called A_1.

Let $(M, \circ, e)$ be a massive F-manifold. By Theorem 1.4, a germ in $M - \mathcal{K}$ is isomorphic to A_1^m. The corresponding local coordinates $u_1, ..., u_m$ satisfy

$$e_i = \frac{\partial}{\partial u_i} \quad \text{and} \quad e_i \circ e_j = \delta_{ij} \cdot e_i .$$

They are the *canonical coordinates* (Dubrovin). There each Euler field of weight 1 takes the form

$$E = \sum_i (u_i + r_i) e_i \quad \text{with } r_i \in \mathbb{C} .$$

Let $(M, \circ, e, E)$ be a massive F-manifold with Euler field E. The *discriminant* of the Euler field E is

$$\mathcal{D} := \det(E\circ)^{-1}(0) \subset M$$

$\mathcal{D}$ is a free divisor with $\mathrm{Der}_M(\log \mathcal{D}) = E \circ T_M$ ([14] Thm. 13.1). $\mathcal{D}$ is everywhere transverse to e (consider the development of $\mathcal{D}$ in $\mathbb{P}T^*M$ to make this precise). It is a rich geometric object. The pair $(\mathcal{D}, e)$ determines the multiplication in a simple explicit way ([14] Cor. 11.6), which is explained in the Examples 1.6 (2) and (3).

EXAMPLES 1.6 1) The base spaces of semiuniversal unfoldings of hypersurface singularities and of boundary singularities are massive F-manifolds (cf. the 3rd lecture and [14] Chapters 16 and 17).

2) $I_2(m)$, $m \geq 3$: $M := \mathbb{C}^2$ with coordinates $(t_1, t_2) = t$, $e := \frac{\partial}{\partial t_1}$,

$$\mathcal{D} := \{ t \in M \mid t_1^2 - \frac{4}{m^2} t_2^m = 0 \} .$$

In $M - \{ t \mid t_2 = 0 \}$ define two vector fields e_1, e_2 as in the picture: For $t \in M - \{ t \mid t_2 = 0 \}$ the e-orbit of t intersects $\mathcal{D}$ in 2 points. One shifts the tangent hyperplanes to $\mathcal{D}$ at these points with the flow of e into $T_t M$. There are unique vectors $e_1, e_2 \in T_t M$ which are tangent to them and satisfy $e_1 + e_2 = e$.

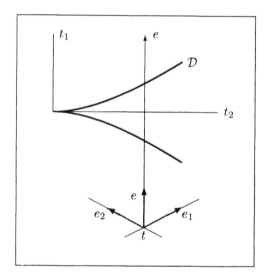

Figure 1

Define the multiplication by $e_i \circ e_j := \delta_{ij} e_i$ on $M - \{ t \mid t_2 = 0 \}$. Then one finds with some calculations ([14] Ch. 12): $e = \frac{\partial}{\partial t_1}$ is the unit field,

$$\frac{\partial}{\partial t_2} \circ \frac{\partial}{\partial t_2} = t_2^{m-2} \cdot e ,$$

and $(M, \circ, e)$ is a massive F-manifold with Euler field

$$E = t_1 \frac{\partial}{\partial t_1} + \frac{2}{m} t_2 \frac{\partial}{\partial t_2}$$

and caustic

$$\mathcal{K} = \{ t \mid t_2 = 0 \} .$$

3) Let W be a finite irreducible Coxeter group acting on $\mathbb{R}^m$ and $\mathbb{C}^m$.

$$\mathbb{C}^m \longrightarrow \mathbb{C}^m/W =: M \cong \mathbb{C}^m$$

$$\bigcup \text{ refl. hyperplanes} \longrightarrow \mathcal{D}$$

$$\mathbb{C}[x_1,...,x_m]^W \longrightarrow \mathbb{C}[t_1,...,t_m] \text{ with } \deg t_i = \tilde{d}_i \; ,$$

$$h = \tilde{d}_1 > \tilde{d}_2 \geq ... \geq \widetilde{d_{m-1}} > \tilde{d}_m = 2 \; ,$$

$$e := \frac{\partial}{\partial t_1} \; , \qquad E := \sum_i \frac{\tilde{d}_i}{h} \cdot t_i \frac{\partial}{\partial t_i} \; .$$

$(M, \mathcal{D}, e)$ determines a multiplication $\circ$ in a way which generalizes example 2).

For generic t one defines $e_1,...,e_m$ as in 2): The e-orbit of t intersects $\mathcal{D}$ in m points. One shifts the tangent hyperplanes to $\mathcal{D}$ with e into $T_t M$. There they are in general position. There are unique vectors $e_1,...,e_m \in T_t M$ which are each tangent to the intersections of $m-1$ hyperplanes and which satisfy together $e_1 + ... + e_m = e$.

The multiplication is defined by $e_i \circ e_j := \delta_{ij} e_i$ on $T_t M$. It extends to TM. $(M, \circ, e, E)$ is a massive F-manifold with Euler field E.

All these are nontrivial statements which follow implicitly from [8] as well as from [4] (Lecture 4), cf. [14] (Ch. 18).

Dubrovin proved, moreover, that the complex orbit spaces are massive Frobenius manifolds (cf. the next lecture). Givental characterized the corresponding analytic spectra in a way which can be translated into Theorem 1.7 a),b) (cf. [8] Thm. 14 and [14] Thm. 18.4).

A massive F-manifold $(M, \circ, e)$ is a called *simple* if it contains only finitely many isomorphism classes of germs of F-manifolds.

THEOREM 1.7 *a) (Givental) The F-manifold of a finite irreducible Coxeter group is simple with analytic spectrum*

$$L \cong \mathbb{C}^{m-1} \times \{(x,y) \in \mathbb{C}^2 \mid x^2 - y^n = 0\} \; .$$

$n = 1:\; A_k,\; D_k,\; E_k \; ,$
$n = 2:\; B_k,\; F_4 \; ,$
$n = 3:\; H_3,\; H_4 \; ,$
$n :\quad I_2(n+2) \; .$

b) (Givental) An irreducible germ (M,t) of a simple F-manifold with

$$(L, \lambda) \cong \prod (\text{ plane curve singularities})$$

is isomorphic to the germ $(\mathbb{C}^m/W, 0)$ for some finite irreducible Coxeter group W.

c) ([14] Thm. 18.3) An irreducible germ (M,t) of a simple F-manifold with

$$T_t M \text{ a Frobenius algebra}$$

is isomorphic to the germ $(\mathbb{C}^m/W, 0)$ for some finite irreducible Coxeter group W.

2 FROBENIUS   MANIFOLDS   AND   THEIR STRUCTURE CONNECTIONS

The notion of a Frobenius manifold was introduced by Dubrovin [3], motivated by topological field theory. It turns up now in many places, especially in quantum cohomology (cf. [20]), mirror symmetry, and in singularity theory (cf. the 3rd lecture).

The first part of this lecture presents the definition and elementary properties of Frobenius manifolds. A Frobenius manifold is an F-manifold with a metric satisfying certain conditions. The integrability condition of the F-manifold and another condition can be recast as the *potentiality*, which gives rise to a potential, the Gromov-Witten potential in the case of quantum cohomology.

One distinguished class of Frobenius manifolds, which can be described in a relatively elementary way, are the complex orbit spaces of finite irreducible Coxeter groups. This description is due to Dubrovin ([4] Lecture 4). The metric had already been defined by K. Saito [22]. A conjecture of Dubrovin concerning them is proved in [14] (Thm. 19.3) using Theorem 1.7.

Associated to a Frobenius manifold are two kinds of meromorphic connections: the *1st structure connections* have irregular singularities along a smooth hypersurface, the *2nd structure connections* have logarithmic singularities along a discriminant. In the singularity case some of them correspond to (rigidified, enriched versions of) oscillating integrals (the 1st) and the Gauß-Manin connection (the 2nd). Therefore in Lecture 3 it will be very useful to be aware of the 2nd structure connections.

The 2nd and 1st structure connections are treated in much more detail in [15] (Ch. 4).The presentation there unifies results of K. Saito [23][24], Dubrovin [4] (Lecture 3 + Appendix G), and Manin [20] (II §§1,2). Some general material about meromorphic connections is collected in [15] (Ch. 3), and much more in [21].

Let $(M, \circ, e)$ be a manifold with multiplication. Then $\circ : \mathcal{T}_M \otimes \mathcal{T}_M \to \mathcal{T}_M$ is a $(2,1)$-tensor. For any vector field $X \in \mathcal{T}_M$, the Lie derivative $\mathrm{Lie}_X(\circ)$ is also a $(2,1)$-tensor. If ∇ is a connection on TM, then $\nabla_X(\circ)$ is also a $(2,1)$-tensor, and $\nabla(\circ)$ is a $(3,1)$-tensor.

In the following, a *metric g* on M is a symmetric and nondegenerate $(2,0)$-tensor

$$g : \mathcal{T}_M \otimes \mathcal{T}_M \to \mathcal{O}_M ,$$
$$g : T_t M \times T_t M \to \mathbb{C} .$$

Its *Levi-Civita connection* ∇ is the unique connection which is torsion free, $\nabla_X Y - \nabla_Y X = [X, Y]$, and satisfies $\nabla g = 0$, i.e.

$$X \, g(Y, Z) = g(\nabla_X Y, Z) + g(Y, \nabla_X Z) .$$

THEOREM 2.1 *([14] Thm. 5.3) Let $(M, \circ, e, g)$ be a manifold with multiplication $\circ$, unit field e, and metric g, such that g is multiplication invariant, i.e.*

$$g(X \circ Y, Z) = g(X, Y \circ Z).$$

Define the (symmetric) $(3,0)$-tensor A by

$$A(X, Y, Z) = g(X \circ Y, Z),$$

and let ∇ be the Levi-Civita connection of g and $\varepsilon := g(.,e)$, a 1-form, be the coidentity (with $\varepsilon(X \circ Y) = g(X,Y)$).

Then the following conditions are equivalent:

i) $(M,\circ,e)$ is an F-manifold and ε is closed.

ii) (Potentiality) The $(4,0)$-tensor ∇A is symmetric in all 4 arguments.

iii) The $(3,1)$-tensor $\nabla(\circ)$ is symmetric in all 3 arguments.

DEFINITION 2.2 *(Dubrovin) A Frobenius manifold is a tuple $(M,\circ,e,E,g)$ with*

a) $(M,\circ,e,g)$ as in Theorem 2.1 with i)–iii).

b) g flat.

c) e flat, i.e. $\nabla e = 0$.

d) $\mathrm{Lie}_E(\circ) = 1 \cdot \circ$, $\mathrm{Lie}_E(g) = D \cdot g$ for some $D \in \mathbb{C}$.

REMARKS 2.3 i) a)+c) implies $\mathrm{Lie}_e(\circ) = 0$ and $\mathrm{Lie}_e(g) = 0$.

ii) Suppose $(M,\circ,e,g)$ is given with g multiplication-invariant and flat. Then

$\qquad$ Potentiality

$\Longleftrightarrow \quad X\, A(Y,Z,W)$ is symmetric in X,Y,Z,W for flat fields X,Y,Z,W

$\Longleftrightarrow \quad \forall\, t \in M\ \exists$ a *potential* $\Phi \in \mathcal{O}_{M,t}$ with

$\qquad XYZ\, \Phi = A(X,Y,Z)$ for flat fields X,Y,Z .

In the case of quantum cohomology, the Gromov-Witten invariants give rise to such a potential. One also has the metric. Then one constructs the tensor A and with it the multiplication. In the case of singularities (3rd talk), the potentials of the Frobenius manifolds do not play a prominent role up to now.

iii) Let $(M,\circ,e,E,g)$ be a Frobenius manifold. Define the $\mathcal{O}_M$-linear map (a $(1,1)$-tensor)

$$\mathcal{V} : \mathcal{T}_M \to \mathcal{T}_M , \quad X \mapsto \nabla_X E - \frac{D}{2} \cdot X .$$

Then $\mathcal{V}$ is skew-symmetric with respect to g, i.e.

$$g(\mathcal{V}(X),Y) + g(X,\mathcal{V}(Y)) = 0 ,$$

and $\mathcal{V}$ maps flat fields to flat fields, i.e. $\nabla\mathcal{V} = 0$.

Let $d_1, ..., d_m$ be the eigenvalues of $\mathcal{V} + \frac{D}{2} \cdot \mathrm{id} = \nabla E$. They can be numbered such that $d_i + d_{m+1-i} = D$ and $d_1 = 1$. The latter follows from $e = [e,E] = \nabla_e E$, which is a consequence of $\mathrm{Lie}_E(\circ) = \circ$ and $\nabla e = 0$.

iv) For $(M,\circ,e,E,g)$ as above, define the $\mathcal{O}_M$-linear map (a $(1,1)$-tensor)

$$\mathcal{U} : \mathcal{T}_M \to \mathcal{T}_M , \quad X \mapsto E \circ X .$$

THEOREM 2.4 *([4] Lecture 4) Let W be a finite irreducible Coxeter group acting on $(\mathbb{R}^m, \langle, \rangle)$ and $(\mathbb{C}^m, \langle, \rangle)$, with $\langle, \rangle$ the standard metric.*

$$\mathbb{C}^m \quad \longrightarrow \quad \mathbb{C}^m/W =: M \cong \mathbb{C}^m$$

$$\bigcup \text{ reflecting hyperplanes} \quad \longrightarrow \quad \mathcal{D}$$

$$\langle, \rangle \quad \longrightarrow \quad \breve{g} \text{ a flat metric on } M - \mathcal{D} \ ,$$

$$\mathbb{C}[x_1, ..., x_m]^W \quad \longrightarrow \quad \mathbb{C}[t_1, ..., t_m] \text{ with } \deg t_i = \widetilde{d}_i \ ,$$

$$h = \widetilde{d}_1 > \widetilde{d}_2 \geq ... \geq \widetilde{d_{m-1}} > \widetilde{d}_m = 2 \ ,$$

$$e := \frac{\partial}{\partial t_1} \ , \qquad E := \sum_I \frac{\widetilde{d}_i}{h} \cdot t_i \frac{\partial}{\partial t_i} \ .$$

$(M, \mathcal{D}, e)$ determines a multiplication $\circ$ on TM (cf. 1.6. 3)).
Define g on $M - \mathcal{D}$ by

$$g(X, Y) := \breve{g}(E \circ X, Y) \ .$$

Then g extends to M and is K. Saito's [22] flat metric,
$(M, \circ, e, E, g)$ is a massive Frobenius manifold with $d_i = \frac{\widetilde{d}_i}{h}$ and $D = 1 + \frac{2}{h}$.

The next theorem was been a conjecture of Dubrovin ([4] p 268). The proof uses Theorem 1.7.

THEOREM 2.5 *([14] Thm. 19.3) Let $((M, t), \circ, e, E, g)$ be a germ of a massive Frobenius manifold with*

$$E = \sum d_i \cdot \tau_i \frac{\partial}{\partial \tau_i}$$

for τ_i flat coordinates, centered at t, and all $d_i > 0$.
Then

$$(M, t) \cong \prod (\mathbb{C}^{m_i}/W_i, 0)$$

is isomorphic to a product of germs at 0 of Frobenius manifolds as in Theorem 2.4 for finite irreducible Coxeter groups W_i with the same Coxeter numbers $h_i = \frac{2}{D-1}$.

The *2nd structure connections* of a Frobenius manifold $(M, \circ, e, E, g)$ (K. Saito, Dubrovin, Manin, H.) are certain meromorphic connections which will be defined in 2.6. For that one needs the lifted tangent bundle pr^*TM over $\mathbb{P}^1 \times M$,

$$\begin{array}{ccc} pr^*TM & \longrightarrow & TM \\ \downarrow & & \downarrow \\ \mathbb{P}^1 \times M & \xrightarrow{\ pr\ } & M \ . \end{array}$$

The tensors g, $\circ$, $\mathcal{V}$, $\mathcal{U}$ lift to pr^*TM in a canonical way. ∇ lifts to a flat connection on pr^*TM with $\nabla_{\partial_z} Y = 0$ for $Y \in pr^{-1}T_M$. Here z is a coordinate on $\mathbb{C} \subset \mathbb{P}^1$, and

$$T_{\mathbb{P}^1 \times M} = pr^*T_M \oplus \mathcal{O}_{\mathbb{P}^1 \times M} \cdot \partial_z \ .$$

There is a discriminant $\check{\mathcal{D}} \subset \mathbb{P}^1 \times M$,

$$\check{\mathcal{D}} := \{(z,t) \in \mathbb{C} \times M \mid \mathcal{U} - z\,\mathrm{id} \text{ is not invertible on } T_t M\} \subset \mathbb{P}^1 \times M,$$

The projection $\check{\mathcal{D}} \to M$ is finite and flat of degree m. Define

$$\check{M} := \mathbb{C} \times M - \check{\mathcal{D}} \ .$$

Part a) of Definition 2.6 was given first (for the semisimple case) in [20] (II (1.26)+(1.27)), part b) in [4] (Lecture 3), the whole definition can be found in [15] (Ch. 4).

DEFINITION 2.6 *a) Fix $s \in \mathbb{C}$. The 2nd structure connection $\check{\nabla}^{(s)}$ on $pr^*TM|_{\check{M}}$ is defined for $X, Y \in pr^*\mathcal{T}_M$ (here consider X as a vector field on $\check{M}$ and Y as a section in the bundle) by*

$$\check{\nabla}^{(s)}_X Y = \nabla_X Y - \left(\mathcal{V} + \frac{1}{2} + s\right)(\mathcal{U} - z)^{-1}(X \circ Y) \ ,$$

$$\check{\nabla}^{(s)}_{\partial_z} Y = \nabla_{\partial_z} Y + \left(\mathcal{V} + \frac{1}{2} + s\right)(\mathcal{U} - z)^{-1}(Y) \ .$$

b) Define

$$\check{g} : pr^*\mathcal{T}_M|_{\check{M}} \times pr^*\mathcal{T}_M|_{\check{M}} \to \mathcal{O}_{\check{M}}$$
$$(X,Y) \mapsto g((\mathcal{U} - z)^{-1}X, Y) \ .$$

c) Fix $s \in \mathbb{C}$. Define

$$\Delta_s : pr^*TM|_{\check{M}} \to pr^*TM|_{\check{M}} \ ,$$
$$X \mapsto (\mathcal{V} + \frac{1}{2} + s)(\mathcal{U} - z)^{-1}X \ .$$

The following unifies and extends results in [24] (§5), [4] (Lecture 3), [20] (II §§1,2).

THEOREM 2.7 *([15] Ch. 4)*
  a) $\check{\nabla}^{(s)}$ is flat.

  b) $\check{g}$ is symmetric, nondegenerate, multiplication invariant.
  $\check{g}(X,Y)$ is constant if X is $\check{\nabla}^{(-s)}$-flat and Y is $\check{\nabla}^{(s)}$-flat.
  Hence the connections $\check{\nabla}^{(-s)}$ and $\check{\nabla}^{(s)}$ are dual.
  $\check{\nabla}^{(0)}|(\{z\} \times M - \check{\mathcal{D}})$ is the Levi-Civita connection of $\check{g}|(\{z\} \times M - \check{\mathcal{D}})$.

  c) Δ_s is a homomorphism of flat vector bundles

$$(pr^*TM|_{\check{M}}, \check{\nabla}^{(s+1)}) \to (pr^*TM|_{\check{M}}, \check{\nabla}^{(s)}) \ .$$

It is an isomorphism iff $d_i - \frac{D-1}{2} + s \neq 0$ for all $i = 1, ..., m$.

  d) The pair $(pr^*\mathcal{T}_M, \check{\nabla}^{(s)})$ has a logarithmic pole along $\{\infty\} \times M$, i.e. $pr^*\mathcal{T}_M$ is invariant by ∇_X for all $X \in \mathcal{T}_{\mathbb{P}^1 \times M}$ tangent to $\{\infty\} \times M$ (e.g. $X \in pr^*\mathcal{T}_M$ and $\frac{1}{z}\partial_{\frac{1}{z}} = -z\partial_{\partial_z}$).

This yields on $pr^*TM|_{\{\infty\}\times M}$ a flat residual connection with respect to the coordinate $\frac{1}{z}$ (see e.g. [15] Ch. 3.2 for the definition) and the residue endomorphism $-z\check{\nabla}^{(s)}_{\partial_z}$. By the isomorphism $pr^*TM_{\{\infty\}\times M} \cong TM$ they are mapped to ∇ and $\mathcal{V} + \frac{1}{2} + s$, respectively,

$$(pr^*\mathcal{T}_M/\frac{1}{z}pr^*\mathcal{T}_M)|_{\{\infty\}\times M} \xrightarrow{\cong} \mathcal{T}_{\{\infty\}\times M} \xrightarrow{\cong} \mathcal{T}_M$$

$$\text{residue endomorphism } -z\check{\nabla}^{(s)}_{\partial_z} \longrightarrow \mathcal{V} + \frac{1}{2} + s$$

$$\text{flat residual connection w.r.t. } \frac{1}{z} \longrightarrow \nabla .$$

e) Suppose $(M, \circ, e)$ is massive.

Then the pair $(pr^*\mathcal{T}_M, \check{\nabla}^{(s)})$ has a logarithmic pole along $\check{\mathcal{D}}$, i.e. ∇_X leaves $pr^*\mathcal{T}_M$ invariant for $X \in \mathrm{Der}_{\mathbb{C}\times M}(\log \check{\mathcal{D}})$.

The residue endomorphism around a point in $\check{\mathcal{D}}_{reg}$ has eigenvalues $(-(\frac{1}{2} + s), 0, ..., 0)$. It is semisimple for $s \neq -\frac{1}{2}$. For $s = -\frac{1}{2}$ it is 0 or has one 2×2 Jordan block.

One can recover most of the structure of the Frobenius manifold from the 2nd structure connection $\check{\nabla}^{(s)}$: The discriminant $\check{\mathcal{D}} \subset \mathbb{P}^1 \times M$ determines the multiplication on TM in a similar way as the discriminant $\mathcal{D} \subset M$ in Examples 1.6 2)+3) ([15] Lemma 4.14). $\check{\nabla}^{(s)}$ around $\{\infty\} \times M$ determines the flat structure ∇ and the endomophism $\mathcal{V} + \frac{1}{2} + s$ by Theorem 2.7 d). The essential features of $\check{\nabla}^{(s)}$ are put together in the picture.

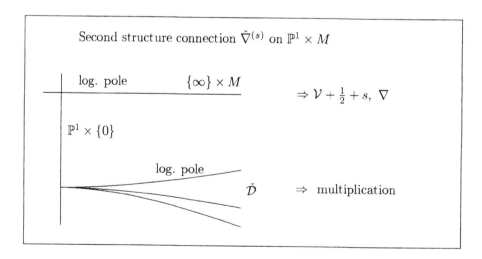

Figure 2

The first structure connections $\hat{\nabla}^{(s)}$ are meromorphic connections, where roughly the information in the discriminant $\check{\mathcal{D}}$ is replaced by a 2nd order pole of the connection along the smooth hypersurface $\{\infty\} \times M$. Many of them are Fourier-Laplace transforms of the 2nd structure connections. A huge amount of material about connections of this type is presented in [21]. Definition 2.8 and Theorem 2.9 are taken from [15] (Ch. 4), but can be found also at many other places ([4][20][21]).

DEFINITION 2.8 $\hat{M} := \mathbb{C} \times M - \{0\} \times M$. *Fix* $s \in \mathbb{C}$.

The 1st structure connection $\hat{\nabla}^{(s)}$ *on* $pr^*TM|_{\hat{M}}$ *is defined for* $X, Y \in pr^*(\mathcal{T}_M)|_{\hat{M}}$ *by*

$$\begin{aligned}\hat{\nabla}^{(s)}_X Y &= \nabla_X Y + z \cdot X \circ Y\ ,\\ \hat{\nabla}^{(s)}_{\partial_z} Y &= \nabla_{\partial_z} Y + \frac{1}{z}(\mathcal{V} + \frac{1}{2} + s)(Y) + E \circ Y\ .\end{aligned}$$

THEOREM 2.9 *a)* $\hat{\nabla}^{(s)}$ *is flat,*

the restrictions to $\{z\} \times M$ *coincide for all* $s \in \mathbb{C}$.

b) The pair $(pr^*\mathcal{T}_M, \hat{\nabla}^{(s)})$ *has a logarithmic pole along* $\{0\} \times M$,

$$\begin{array}{rcl}(pr^*\mathcal{T}_M / z\ pr^*\mathcal{T}_M)_{\{0\}\times M} \stackrel{\cong}{\longrightarrow} \mathcal{T}_{\{0\}\times M} & \stackrel{\cong}{\longrightarrow} & \mathcal{T}_M\\ \textit{residue endomorphism}\ z\check{\nabla}^{(s)}_{\partial_z} & \longrightarrow & \mathcal{V} + \frac{1}{2} + s\\ \textit{flat residual connection w.r.t. } z & \longrightarrow & \nabla\ .\end{array}$$

c) The pair $(pr^*\mathcal{T}_M, \hat{\nabla}^{(s)})$ *has a pole of order 2 along* $\{\infty\} \times M$.

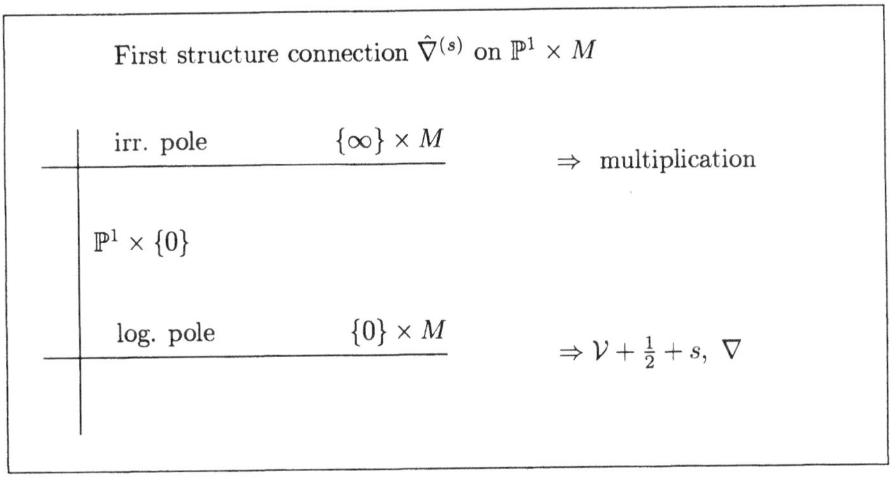
First structure connection $\hat{\nabla}^{(s)}$ on $\mathbb{P}^1 \times M$
irr. pole
$\{\infty\} \times M$
$\Rightarrow$ multiplication
$\mathbb{P}^1 \times \{0\}$
log. pole
$\{0\} \times M$
$\Rightarrow \mathcal{V} + \frac{1}{2} + s,\ \nabla$

Figure 3

3 CONSTRUCTION IN SINGULARITY THEORY

The first big class of Frobenius manifolds was constructed implicitly in singularity theory 1983, long before Dubrovin's definition. K. Saito [23][24] and M. Saito [25] showed that the base space of a semiuniversal unfolding of an isolated hypersurface singularity can be equipped with the structure of a Frobenius manifold.

The structure of an F-manifold with Euler field is canonical and not so difficult to obtain. It will be explained below.

The construction of a metric is much more difficult and not unique. It depends on the choice of a primitive form of K. Saito [23][24]. The construction of such a form depends on the choice of an opposite filtration to a Hodge filtration ([25], see Theorem 3.2 below). The original proof applies results of Malgrange [18][19] on deformations and normal forms of microdifferential systems to the Gauß-Manin system of an unfolding.

A simplified version is given in great detail in [15] (Ch. 6). It avoids Malgrange's results, stays largely within the framework of meromorphic connections, and is sufficiently explicit for applications. Still it uses polarized mixed Hodge structures and K. Saito's higher residue pairings. This lecture intends to give an outline of it. The details can be found in [15] (Ch. 6).

Let $f : (\mathbb{C}^{n+1}, 0) \to (\mathbb{C}, 0)$ be holomorphic with an isolated singularity at 0.

$$\mathcal{O}_{\mathbb{C}^{n+1}, 0} / \left(\frac{\partial f}{\partial x_0}, ..., \frac{\partial f}{\partial x_n} \right) =: \mathcal{O}/J_f$$

is the Jacobi algebra, the Milnor number is $\mu = \dim \mathcal{O}/J_f$.

A semiuniversal unfolding is a holomorphic function germ

$$F : (\mathbb{C}^{n+1} \times \mathbb{C}^\mu, 0) \quad \to \quad (\mathbb{C}, 0)$$
$$(x_0, ..., x_n, t_1, ..., t_\mu) = (x, t) \quad \to \quad F(x, t)$$

with $F|(\mathbb{C}^{n+1} \times \{0\}, 0) = f$ and such that $\frac{\partial F}{\partial t_i}|(\mathbb{C}^{n+1} \times \{0\}, 0)$, $i = 1, ..., \mu$, are representatives of a basis of $\mathcal{O}/J_f$.

A good representative of F can be chosen as follows: Choose $\varepsilon > \delta > \theta > 0$ and $\Delta := B_\delta^1 \subset \mathbb{C}$, $M := B_\theta^\mu \subset \mathbb{C}^\mu$,

$$\mathcal{X} := F^{-1}(\Delta) \cap (B_\varepsilon^{n+1} \times M) \subset \mathbb{C}^{n+1} \times \mathbb{C}^\mu,$$

Then $F : \mathcal{X} \to \Delta$ is a good representative. It can be seen as a family of functions

$$F_t : \mathcal{X} \cap (B_\varepsilon^{n+1} \times \{t\}) \to \Delta, \quad t \in M,$$

parametrized by M. The function F_t has finitely many singularities, the sum of their Milnor numbers is μ. The map

$$\varphi : \mathcal{X} \quad \to \quad \Delta \times M$$
$$(x, t) \quad \to \quad (F(x, t), t)$$

generalizes a Milnor fibration. It is a locally trivial C^∞-fiber bundle outside of a discriminant $\hat{\mathcal{D}} \subset \Delta \times M$.

The critical space $C \subset \mathcal{X}$ of φ is the zero set of $(\frac{\partial F}{\partial x_0}, ..., \frac{\partial F}{\partial x_n})$. It is the union of the singularities of all F_t, $t \in M$. The projection $pr_{C,M} : C \to M$ is finite and flat of degree μ.

The *Kodaira-Spencer map* $\mathbf{a}_C$ is an isomorphism of free $\mathcal{O}_M$-modules of rank μ,

$$
\begin{array}{ccc}
\mathbf{a}_C : \mathcal{T}_M & \longrightarrow & (pr_{C,M})_* \mathcal{O}_C \\
\dfrac{\partial}{\partial t_i} & \longmapsto & \dfrac{\partial F}{\partial t_i}|_C \\
\text{mult. } \circ & & \text{mult.} \\
e & & [1] \\
E & & [F] \, .
\end{array}
$$

It induces a vector field $E := \mathbf{a}_C^{-1}([F])$ and a multiplication $\circ$ on $\mathcal{T}_M$ by $X \circ Y := \mathbf{a}_C^{-1}(\mathbf{a}_C(X) \cdot \mathbf{a}_C(Y))$.

THEOREM 3.1 *([14] Thm. 16.3)* $(M, \circ, e, E)$ *is a massive F-manifold with Euler field E and caustic* $\mathcal{K} = \{t \mid |Sing(F_t)| < \mu\}$ *, the eigenvalues of $E \circ |T_t M$ are the critical values of F_t,*

$$
(T_t M, \circ, E|_t) \cong \left(\bigoplus_{x \in Sing(F_t)} Jacobi \ algebra \ of \ (F_t, x), \ mult. \ , [F_t] \right) \, .
$$

$\mathcal{O}/J_f$ carries a multiplication and extends to the sheaf $(pr_{C,M})_* \mathcal{O}_C$. The μ-dimensional space $\Omega_{\mathbb{C}^{n+1},0}^{n+1}/df \wedge \Omega_{\mathbb{C}^{n+1},0}^n$ carries a "metric" and extends to the sheaf

$$
\Omega_F := (pr_{\mathcal{X},M})_* \Omega_{\mathcal{X}/\Delta \times M}^{n+1} \, .
$$

Its support is the critical space C. It is a free $(pr_{C,M})_* \mathcal{O}_C$-module of rank 1 and a free $\mathcal{O}_M$-module of rank μ.

The *Grothendieck residue pairing* is [11][23][24]

$$
J_F : \Omega_F \times \Omega_F \to \mathcal{O}_M \, ,
$$

$$
([g_1 dx], [g_2 dx]) \longmapsto \mathrm{Res}_{\mathcal{X}/M} \left[\frac{g_1 g_2 dx}{\frac{\partial F}{\partial x_0} \cdot ... \cdot \frac{\partial F}{\partial x_n}} \right] (t) = \frac{1}{(2\pi i)^{n+1}} \int_\Gamma \frac{g_1 g_2 dx}{\frac{\partial F}{\partial x_0} \cdot ... \cdot \frac{\partial F}{\partial x_n}}
$$

with $\Gamma := \{(x, t) \mid |\frac{\partial F}{\partial x_i}| = \gamma\} \subset B_\varepsilon^{n+1} \times \{t\}$ for some small $\gamma > 0$.

It is symmetric and nondegenerate (Grothendieck). It is independent of the coordinates $x_0, ..., x_n$ (K. Saito).

The choice of a *volume form* $\omega = $ unit $(x, t) \cdot dx$ induces an isomorphism

$$
\begin{array}{ccccc}
\mathcal{T}_M & \xrightarrow{\mathbf{a}_C} & (pr_{C,M})_* \mathcal{O}_C & \xrightarrow{\cong} & \Omega_F \, , \\
X & \longmapsto & \mathbf{a}_C(X) & \longmapsto & [\mathbf{a}_C(X) \cdot \omega] \, , \\
\text{metric } g & & & & J_F \, .
\end{array}
$$

With it one lifts J_F to a metric g on M.

For which ω is $(M, \circ, e, E, g)$ a Frobenius manifold? For ω a primitive form of K. Saito. We will not give here K. Saito's definition [23][24]. But we will outline the simplified version in [15] (Ch. 6) of the construction in [25] of such a form.

Let $\mathcal{U} := E\circ$. By Theorem 3.1, the critical values of F_t are the eigenvalues of $\mathcal{U}|_t$. Therefore the discriminant $\check{D}$ of φ is

$$\check{D} = \{(z,t) \in \Delta \times M \mid \mathcal{U} - z\,\mathrm{id} \text{ is not invertible on } T_t M\}.$$

The cohomology bundle of φ is

$$H^n := \bigcup_{(z,t)\in\Delta\times M - \check{D}} H^n(\varphi^{-1}(z,t), \mathbb{C})$$

a vector bundle of rank μ with flat connection ∇ coming from the topology. $H^n|\Delta^* \times \{0\}$ is the cohomology bundle of a Milnor fibration of f.

We describe now the data which are needed to formulate the choice $(U_\bullet, \gamma_1)$ in Theorem 3.2. Afterwards we return to the construction of the primitive form, using this choice.

The map $\mathbf{e} : \mathbb{C} \to \mathbb{C}^*$, $\zeta \mapsto e^{2\pi i \zeta}$, restricts to a universal covering of Δ^*. The μ-dimensional space of *global, flat, multivalued* sections of $H^n|\Delta^* \times \{0\}$ is

$$H^\infty := \{A \ : \ \mathbf{e}^{-1}(\Delta^*) \to H^n|\Delta^* \times \{0\} \mid$$
$$A \text{ is holomorphic and factors locally into } \mathbf{e}$$
$$\text{and a flat section of } H^n|\Delta^* \times \{0\}\}.$$

It is obviously equipped with a lattice $H_{\mathbb{Z}}^\infty$ and a monodromy $h : H_{\mathbb{Z}}^\infty \to H_{\mathbb{Z}}^\infty$,

$$h = h_s \cdot h_u = h_u \cdot h_s \text{ with } h_u \text{ unipotent, } h_s \text{ semisimple,}$$
$$N := \log h_u,$$
$$H_\lambda^\infty := \ker(h_s - \lambda), \quad H_{\neq 1}^\infty := \bigoplus_{\lambda\neq 1} H_\lambda^\infty, \quad H^\infty = H_1^\infty \oplus H_{\neq 1}^\infty.$$

It is also equipped with a polarizing form $S : H_{\mathbb{Z}}^\infty \times \mathbb{H}_{\mathbb{Z}}^\infty \to \mathbb{Q}$ (see [12] (Ch. 3) or [15] (5.78) for its definition). S is monodromy invariant and nondegenerate. It is $(-1)^n$-symmetric on $H_{\neq 1}^\infty$, where it comes from the intersection form. It is $(-1)^{n+1}$-symmetric on H_1^∞, there it comes from the variation operator.

Steenbrink's Hodge filtration $F^\bullet$ on H^∞ satisfies (cf. [29][12][15])

$$H^\infty = F^0 \supset F^1 \supset ... \supset F^n \supset F^{n+1} = 0,$$
$$h_s(F^p) = F^p, \quad N(F^p) \subset F^{p-1},$$
$$S(F^p H_{\neq 1}^\infty, F^q H_{\neq 1}^\infty) = 0 \text{ for } p+q > n \ (n+1 \text{ for } H_1^\infty),$$

$F^\bullet$, S, and a weight filtration from N form a *polarized mixed Hodge structure* (cf. e.g. [12] or [15] (Ch 5.5) for the definition and comments).

A filtration $U_\bullet$ on H^∞ with the following properties is an *opposite filtration*:

$$0 = U_{-1} \subset U_0 \subset U_1 \subset \ldots \subset U_n = H^\infty \ ,$$

$$H^\infty = \bigoplus_p F^p \cap U_p \ ,$$

$$h_s(U_p) = U_p \ , \quad N(U_p) \subset U_{p-1} \ ,$$

$$S(U_p H^\infty_{\neq 1}, U_q H^\infty_{\neq 1}) = 0 \ \text{ for } p + q < n \ (n+1 \text{ for } H^\infty_1) \ .$$

Giving the opposite filtration $U_\bullet$ is equivalent to giving the splitting $H^\infty = \bigoplus_p F^p \cap U_p$ of the Hodge filtration.

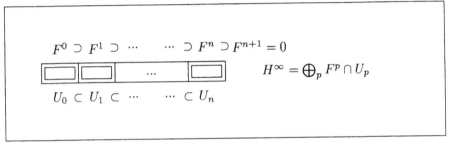

Figure 4

The *spectral numbers* $\alpha_1, \ldots, \alpha_\mu$ of f are μ rational numbers with

$$\sharp(j \mid \alpha_j = \alpha) = \dim \mathrm{Gr}_F^{[n-\alpha_j]} H^\infty_{e^{-2\pi i \alpha_j}} \ .$$

They satisfy

$$-1 < \alpha_1 \leq \ldots \leq \alpha_\mu < n \quad \text{and} \quad \alpha_i + \alpha_{\mu+1-i} = n - 1 \ .$$

By results of Varchenko and M. Saito [26] (3.11), the smallest spectral number has multiplicity 1:

$$\alpha_1 < \alpha_2 \ , \quad 1 \ = \ \dim \mathrm{Gr}_F^{[n-\alpha_1]} H^\infty_{e^{-2\pi i \alpha_1}}$$
$$= \ \dim \mathrm{Gr}_{[n-\alpha_1]}^U H^\infty_{e^{-2\pi i \alpha_1}}$$

(for $U_\bullet$ an opposite filtration).

The following is essentially due to K. Saito and M. Saito, but in this form it is taken from [15] (Thm. 6.1).

THEOREM 3.2 *Any choice $(U_\bullet, \gamma_1)$ with $U_\bullet$ an opposite filtration and γ_1 a generator of $\mathrm{Gr}_{[n-\alpha_1]}^U H^\infty_{e^{-2\pi i \alpha_1}}$ induces an isomorphism $\mathcal{T}_M \to \Omega_F$ and a metric g on M such that $(M, \circ, e, E, g)$ is a Frobenius manifold.*

∇E is semisimple with eigenvalues $d_i = 1 + \alpha_1 - \alpha_i$, $i = 1, \ldots, \mu$ and with $D = d_i + d_{\mu+1-i} = 2 - (\alpha_\mu - \alpha_1) = 2 + 2\alpha_1 - (n-1)$.

One has to construct from $(U_\bullet, \gamma_1)$ a primitive form as a very special volume form, which then induces an isomorphism $(\mathcal{T}_M, g) \to (\Omega_F, J_F)$.

A very rough idea is to extend the cohomology bundle H^n over $\Delta \times M - \check{\mathcal{D}}$ with its flat connection to a meromorphic connection over $\mathbb{P}^1 \times M$ which resembles a 2nd structure connection of a Frobenius manifold. Then one will find a distinguished global section which resembles the lifted unit field and which will be the primitive form. Details of the following outline can be found in [15] (Ch. 6).

First, the sheaf

$$\mathcal{H}^{(0)} := \varphi_* \Omega_{\mathcal{X}/M}^{n+1} / dF \wedge d\varphi_* \Omega_{\mathcal{X}/M}^{n+1}$$

is a free $\mathcal{O}_{\Delta \times M}$-module of rank μ [10] with

$$\mathcal{H}^{(0)} | (\Delta \times M - \check{\mathcal{D}}) \cong \quad \text{sheaf of hol. sections of } H^n .$$

$(\nabla, \mathcal{H}^{(0)})$ has a logarithmic pole along $\check{\mathcal{D}}$. The residue endomorphism along $\check{\mathcal{D}}_{reg}$ has eigenvalues $(\frac{n-1}{2}, 0, ..., 0)$. It is semisimple for $n \neq 1$.

The sheaf $\mathcal{H}^{(0)}$ gives a distinguished meromorphic extension of the flat connection over $\check{\mathcal{D}}$. The pair $(\mathcal{H}^{(0)}, \nabla)$ is the Gauß-Manin connection of the map φ.

Let z be a coordinate on $(\Delta \subset)\mathbb{C} \subset \mathbb{P}^1$. Varchenko [31][1] found that the limit behaviour for $z \to 0$ of the set $\{(\text{sections of } \mathcal{H}^{(0)}) | \Delta \times \{0\}\} =: H_0''$ gives rise to Steenbrink's Hodge filtration $F^\bullet$ (resp. a closely related Hodge filtration $F_{Va}^\bullet$). See [1], [29], [25], [12], [15] (Ch. 5.6) for details. This set H_0'' is called the Brieskorn lattice, because Brieskorn considered it first [2]. It can be written algebraically as $H_0'' = \Omega_{\mathbb{C}^{n+1},0}^{n+1} / df \wedge \Omega_{\mathbb{C}^{n+1},0}^{n+1}$ and is a very rich datum of the singularity f.

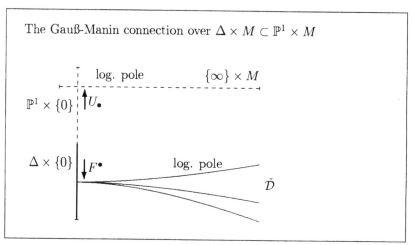

The Gauß-Manin connection over $\Delta \times M \subset \mathbb{P}^1 \times M$

Figure 5

The simplified version in [15] (Ch. 6) of M. Saito's [25] application of results of Malgrange [18][19] boils down to the following: a choice of an opposite filtration $U_\bullet$ yields an extension $\overline{\mathcal{H}^{(0)}}$ of $\mathcal{H}^{(0)}$ to a free $\mathcal{O}_{\mathbb{P}^1 \times M}$-module such that $(\nabla, \overline{\mathcal{H}^{(0)}})$ has a logarithmic pole along $\{\infty\} \times M$ and its residue endomorphism has eigenvalues $-\alpha_1, ..., -\alpha_\mu$.

To obtain this, one first extends only the sheaf $\{(\text{sections of } \mathcal{H}^{(0)}) | \Delta \times \{0\}\}$ to a free $\mathcal{O}_{\mathbb{P}^1 \times \{0\}}$-module with logarithmic pole at ∞. Then there is a unique extension to a sheaf $\overline{\mathcal{H}^{(0)}}$ with logarithmic pole along $\{\infty\} \times M$, it is automatically a free $\mathcal{O}_{\mathbb{P}^1 \times M}$-module.

Roughly, the limit behaviour for $z \to \infty$ of sections in $\overline{\mathcal{H}^{(0)}}|_{\mathbb{P}^1 \times \{0\}}$ corresponds to $U_\bullet$ in the same way as the one for $z \to 0$ corresponds to $F^\bullet$.

Let $pr : \mathbb{P}^1 \times M \to M$ be the projection. The sheaf $pr_* \overline{\mathcal{H}^{(0)}} =$ { fiberwise global sections } is a free $\mathcal{O}_M$-module of rank μ. It contains the distinguished μ-dimensional space

{ fiberwise global sections | the restrictions to $\{\infty\} \times M$

are flat w.r.t. the residual connection} .

The residue endomorphism acts on it, the eigenspace with eigenvalue $-\alpha_1$ is 1-dimensional. The choice of γ_1 yields a generator v_1 of this space,

CLAIM: v_1 is a primitive form in the sense of [23][24].
The proof uses K. Saito's higher residue pairings and their relation to the polarizing form S on H^∞.

One also obtains isomorphisms

$$\mathcal{T}_M \xrightarrow{\cong} pr_* \overline{\mathcal{H}^{(0)}} ,$$
$$X \mapsto -\nabla_X \nabla_{\partial_z}^{-1} v_1 ,$$

and

$$(\check{\nabla}^{(-\frac{n}{2})}, pr^* \mathcal{T}_M) \xrightarrow{\cong} (\nabla, \overline{\mathcal{H}^{(0)}}) .$$

($\check{\nabla}^{(-\frac{n}{2})}$ is a second structure connection of the Frobenius manifold.)

4   GLOBAL MODULI SPACES FOR HYPERSURFACE SINGULARITIES

The construction of Frobenius manifolds in singularity theory shares many properties with other constructions which are on the B-side of mirror symmetry (quantum cohomology is on the A-side). So there is good hope that it will take its place there. But precise results, which relate this construction to mirror symmetry, have still to be found.

What we can present here is a completely independent application within singularity theory: a construction of global moduli spaces for hypersurface singularities. We will state the result precisely and sketch a part of the proof, which contains the central ideas. The whole proof and a thorough discussion can be found in [15] (Ch. 8).

Another application ([15] Ch. 9) will only be mentioned at the end: with the G-function of a Frobenius manifold one can make statements on the variance of the spectral numbers of a singularity.

We need two equivalence relations for holomorphic functions germs with an isolated singularity. Consider $f, g \in \mathbf{m}^2 \subset \mathcal{O}_{\mathbb{C}^{n+1}, 0}$ with isolated singularities at 0. An analytic equivalence relation is the *right equivalence*:

$$f \sim_R g \quad :\Longleftrightarrow \quad \exists\, \varphi : (\mathbb{C}^{n+1}, 0) \to (\mathbb{C}^{n+1}, 0) \text{ biholomorphic}$$
$$\text{with } f = g \circ \varphi .$$

The group $\mathcal{R} := \{\varphi : (\mathbb{C}^{n+1}, 0) \to (\mathbb{C}^{n+1}, 0) \text{ bihol.}\}$ acts on $\mathbf{m}^2$.

A topological equivalence is the μ-*homotopy*:

$$f \sim_\mu g : \Longleftrightarrow \quad \exists\, f_t,\ t \in [0,1],\ \text{with } \mu(f_t) = \mu,\ f_0 = f,\ f_1 = g \ .$$

Here f_t, $t \in [0,1]$, is a family of function germs in $\mathbf{m}^2 \subset \mathcal{O}_{\mathbb{C}^{n+1},0}$ which depend continuously on the parameter t.

Let $\mathcal{E}$ be a μ-homotopy class with Milnor number μ. We are interested in the structure of the quotient set $\mathcal{E}/\mathcal{R}$.

The first step is to apply a result of Mather: f is $(\mu+1)$-determined, that means, its right equivalence class is determined by its k-jet $j_k f \in \mathbf{m}^2/\mathbf{m}^{k+1}$ for any $k \geq \mu+1$. This implies that the projection

$$\mathcal{E}/\mathcal{R} \xrightarrow{\ 1:1\ } j_k\mathcal{E}/j_k\mathcal{R} = \quad \text{alg. variety/alg. group,}$$

is one-to-one. So $\mathcal{E}/\mathcal{R}$ can be considered as a quotient of an algebraic variety by an algebraic group.

One even knows that all orbits have the same dimension, because each $j_k\mathcal{R}$-orbit $j_k\mathcal{R}\, j_k f$ has the codimension $\mu(f) - 1$ in $\mathbf{m}^2/\mathbf{m}^{k+1}$ for $k \geq \mu + 1$. But, of course, this alone does not at all imply that the quotient is a variety.

For example, the quotient of $\mathbb{C}^2 - \{0\}$ with the action of $\mathbb{C}^* \hookrightarrow GL(2,\mathbb{C})$, $x \mapsto \left(\begin{smallmatrix} x & 0 \\ 0 & x^{-1} \end{smallmatrix}\right)$, on it is not Hausdorff, the punctured coordinate planes cannot be separated.

The whole construction in the 3rd lecture is needed to prove the following result.

THEOREM 4.1 *([15] Thm. 8.15) The set $j_k\mathcal{E}/j_k\mathcal{R}$ is an analytic geometric quotient: the quotient topology is Hausdorff and $(\pi_*\mathcal{O}_{j_k\mathcal{E}})^{j_k\mathcal{R}}$ gives a complex structure, here π is the projection $j_k\mathcal{E} \to j_k\mathcal{E}/j_k\mathcal{R}$.*

Why is the quotient topology Hausdorff? Here we will sketch the proof of this. The details and everything else can be found in [15] (Ch. 8).

Consider the following situation: a singularity $f \in \mathcal{E}$ and a representative F of a semiuniversal unfolding with parameter space M (as in lecture 3). The μ-constant stratum in M is

$$S_\mu(f) := \{t \in M \mid Sing(F_t) = \{x\},\ F_t(x) = 0\} \ .$$

Let $(t_i)_{i \in \mathbb{N}}$ be a sequence with $t_i \in S_\mu(f)$ and $t_i \to 0$ for $i \to \infty$.

Fix also another singularity $\widetilde{f}$ with semiuniversal unfolding $\widetilde{F}$ with parameter space $\widetilde{M}$ and μ-constant stratum $S_\mu(\widetilde{f})$. Suppose that there is a sequence $(\widetilde{t}_i)_{i \in \mathbb{N}}$ with $\widetilde{t}_i \in S_\mu(\widetilde{f})$ and $\widetilde{t}_i \to 0$ for $i \to \infty$ and that there are coordinate changes φ_i with

$$F_{t_i} = \widetilde{F}_{\widetilde{t}_i} \circ \varphi_i \ .$$

In order to show that $j_k\mathcal{E}/j_k\mathcal{R}$ is Hausdorff it is sufficient to show that in this situation f and $\widetilde{f}$ are right equivalent. This reduction follows with results of Gabrielov [7] and Teissier [30] (§6). They yield the isomorphism of germs

$$(S_\mu(f), 0) \cong (j_k\mathcal{E}, j_k f) \cap (\text{ a transversal disc }) \ .$$

The following is a picture of the situation.

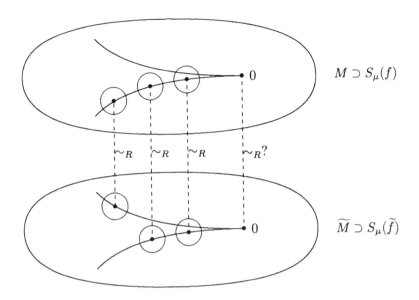

Figure 6

The parameter spaces M and $\widetilde{M}$ are F-manifolds with Euler fields.

1st CLAIM: φ_i induces an isomorphism

$$\varphi_{i,M} : (M, t_i) \to (\widetilde{M}, \widetilde{t}_i) ,$$

of germs of F-manifolds with Euler fields.

2nd CLAIM: $(\varphi_{i,M})_{i \in \mathbb{N}}$ has a limit isomophism

$$\varphi_{\infty,M} : (M, 0) \to (\widetilde{M}, 0)$$

of germs of F-manifolds with Euler fields.

Remember (Theorem 3.1)

$$(T_0 M, \circ, E|_0) \cong (\text{ Jacobi algebra of } f, \text{ mult. } , [f]) .$$

A result of Scherk [28] says that this datum determines f up to right equivalence.

There is a closely related result ([14] Thm. 16.6): the germ $((M,0), \circ, e, E)$ determines F up to right equivalence of unfoldings. But the proof is different, it is an immediate consequence of a result of Arnold and Hörmander on generating families for Lagrange maps.

The 2nd claim and Scherk's or Arnold's/Hörmander's result imply

$$f \sim_R \widetilde{f} .$$

It remains to prove the two claims. φ_i can be extended (not uniquely) to a local isomorphism of the unfoldings, and this induces on the base spaces an isomorphism of

F-manifolds. This is not surprising. But that the map on the base spaces is a unique map $\varphi_{i,M}$ (contrary to the map between the total spaces), that follows essentially from the fact that the group $\mathrm{Aut}((M,t_i),\circ,e,E)$ is finite (cf. [14] Thm 14.2, Thm. 16.4, [15] Thm. 8.2).

The proof of the 2nd claim is much deeper. It uses the construction of flat structures on M and $\widetilde{M}$. The aim is to show that for suitable flat structures the local isomorphisms $\varphi_{i,M}$ respect them and differ at most by translations. Then the existence of a limit isomorphism is not hard to see.

Some choice of $(U_\bullet,\gamma_1)$ as in Theorem 3.2 determines a flat structure on M. In order to find a good such choice for $\widetilde{f}$ and $\widetilde{M}$ one has to control the cohomologies and the isomorphisms which the φ_i induce there.

We use the notations of lecture 3: for any singularity g, the space $H^\infty(g)$ is defined as there and is equipped with a lattice, a monodromy, a polarizing form, and a Hodge filtration $F^\bullet(g)$.

There are canonical isomorphisms

$$H^\infty(F_t) \xrightarrow{\cong} H^\infty(f) \quad \text{and} \quad H^\infty(\widetilde{F}_{\widetilde{t}}) \xrightarrow{\cong} H^\infty(\widetilde{f})$$

for any $t \in S_\mu(f)$ and $\widetilde{t} \in S_\mu(\widetilde{f})$. They respect the lattices, the monodromies, and the polarizing forms, but in general not the Hodge filtrations (those vary holomorphically within the μ-constant strata). Using them, we identify $H^\infty(F_t)$ with $H^\infty(f)$ and $H^\infty(\widetilde{F}_{\widetilde{t}})$ with $H^\infty(\widetilde{f})$.

Choose an isomorphism

$$\psi : H^\infty(\widetilde{f}) \to H^\infty(f)$$

which respects the topological data.

There is a classifying space D_{PMHS} for polarized mixed Hodge structures on $H^\infty(f)$ [12], and there are holomorphic period maps

$$\begin{aligned} \Phi : S_\mu(f) &\to D_{PMHS}, \ t \mapsto F^\bullet(F_t) \in H^\infty(f), \\ \widetilde{\Phi} : S_\mu(\widetilde{f}) &\to D_{PMHS}, \ \widetilde{t} \mapsto F^\bullet(\widetilde{F}_{\widetilde{t}}) \in H^\infty(\widetilde{f}). \end{aligned}$$

The discrete group

$$G_{\mathbb{Z}} := \mathrm{Aut}(H^\infty(f),\, H^\infty_{\mathbb{Z}}(f),\, \text{mon.}\,,\, \text{pol. form}).$$

acts properly discontinuously on D_{PMHS} [12].

φ_i induces isomorphisms

$$\varphi_{i,c} : H^\infty(F_{t_i}) \to H^\infty(\widetilde{F}_{\widetilde{t}_i}),$$

which respect all the additional structure, even the Hodge filtrations. Therefore $\psi \circ \varphi_{i,c} \in G_{\mathbb{Z}}$ satisfies

$$\psi \circ \varphi_{i,c}(\Phi(t_i)) = \widetilde{\Phi}(\widetilde{t}_i).$$

As $G_{\mathbb{Z}}$ acts properly discontinuously on D_{PMHS}, there exists an isomorphism

$$\varphi_{\infty,c} : (H^\infty(f), F^\bullet(f)) \to (H^\infty(\widetilde{f}), F^\bullet(\widetilde{f}))$$

and an infinite subsequence $(\varphi_{i,c})_{i \in I}$ with $\varphi_{i,c} = \varphi_{\infty,c}$.

Choose $(U_\bullet, \gamma_1)$ with $U^\bullet$ an opposite filtration to $F^\bullet(f)$ and γ_1 a generator of $\mathrm{Gr}^U_{[n-\alpha_1]} H^\infty_{e^{-2\pi i \alpha_1}}$, as in Theorem 3.2. $(U_\bullet, \gamma_1)$ induces a flat metric g on M. The image $(\widetilde{U}_\bullet, \widetilde{\gamma}_1) := \varphi_{\infty,c}(U_\bullet, \gamma_1)$ induces a flat metric $\widetilde{g}$ on $\widetilde{M}$.

Going through the whole construction of the Frobenius manifolds one finds that the maps

$$\varphi_{i,M} : (M, t_i) \to (\widetilde{M}, \widetilde{t}_i)$$

for $i \in I$ are isomorphisms of the Frobenius manifolds and differ at most by translations.

They extend to $(M, 0)$ with $\varphi_{i,M}(0) \to 0$ for $i \to \infty$. Thus (by Scherk or Arnold/Hörmander and the fact that right equivalent singularities do not accumulate in a semiuniversal unfolding)

$$\varphi_{i,M}(0) = 0 \quad \text{for} \quad i \in I \text{ with } i \gg 0 \,.$$

This shows the second claim.   □

One can describe the local structure of the global moduli space precisely.

THEOREM 4.2 *([15] Thm 8.15) Let $f \in \mathcal{E}$ and M be as above.* *Then* $\mathrm{Aut}((M,0), \circ, e, E)$ *is finite. The germ of $\mathcal{E}/\mathcal{R}$ at the class of f is*

$$(j_k \mathcal{E}/j_k \mathcal{R}, [f]) \cong (S_\mu(f), 0)/\mathrm{Aut}((M,0), \circ, e, E) \,.$$

Here the μ-constant stratum $S_\mu(f)$ was considered as a reduced complex space (germ). But using the Frobenius manifold structure, one can equip it even with a canonical complex structure (which I expect to be nonreduced in general). It is given by the ideal in the next theorem.

Consider f, F, M, a choice $(U_\bullet, \gamma_1)$, and the Frobenius manifold $(M, \circ, e, E, g)$ as above. Choose flat coordinates $\tau_1, ..., \tau_\mu$ with

$$E = \sum_{i=1}^{\mu} (1 + \alpha_1 - \alpha_i)(\tau_i + r_i) \frac{\partial}{\partial \tau_i} \quad \text{for some } r_i \in \mathbb{C} \,.$$

Define $\varepsilon_{ij} \in \mathcal{O}_{M,0}$ for $i = 1, ..., \mu$ by

$$\frac{\partial}{\partial \tau_i} \circ E = \sum_j \varepsilon_{ij} \cdot \frac{\partial}{\partial \tau_j} \,.$$

THEOREM 4.3 *([15] Thm 7.2, Thm. 7.4)*
$S_\mu(f)$ is the zero set of the ideal $(\varepsilon_{ij} \mid \alpha_j - 1 - \alpha_i < 0)$.
The ideal is independent of the choices of $\tau_1, ..., \tau_\mu$ and of $(U_\bullet, \gamma_1)$.

Very roughly, another application [15] (Ch. 9) of the construction of Frobenius manifolds in singularity theory will be sketched. It uses the *G-function* of a semisimple Frobenius manifold, which was defined by Dubrovin and Zhang [6] and independently by Givental [9].

This function is defined as follows. Let M be a massive Frobenius manifold and $u_1, ..., u_m$ canonical coordinates locally on $M - \mathcal{K}$,

$$e_i = \frac{\partial}{\partial u_i} \ , \quad \eta_i := g(e_i, e_i) \ .$$

FACT: The 1-form

$$d \log \tau_I := \frac{1}{8} \sum_{i \neq j} (u_i - u_j) \frac{(e_i \eta_j)^2}{\eta_i \eta_j} du_i$$

is closed and comes from a function $\log \tau_I$. Then the G-function is

$$G := \log \tau_I - \frac{1}{48} \log \prod_{i=1}^{\mu} \eta_i \ .$$

It is unique up to addition of a scalar.

THEOREM 4.4 *([15] Thm 9.6, following a suggestion of Givental) In the case of a Frobenius manifold of a singularity f, the function G extends holomorphically over the caustic $\mathcal{K} \subset M$.*

THEOREM 4.5 *([6] Thm. 3) The derivative of G by E is a constant $\gamma \in \mathbb{C}$. In the case of a Frobenius manifold of a singularity f it is*

$$E\, G = \gamma = -\frac{1}{4} \sum_{i=1}^{\mu} (\alpha_i - \frac{n-1}{2})^2 + \frac{\mu(\alpha_\mu - \alpha_1)}{48} \ .$$

The spectral numbers of a singularity f satisfy

$$\alpha_i + \alpha_{\mu+1-i} = n - 1 \ ,$$

so $\frac{n-1}{2}$ is the *expected value* of the spectrum.

THEOREM 4.6 *Let f be a quasihomogeneous singularity. Then $\gamma = 0$ and the variance of the spectrum is*

$$\frac{1}{\mu} \sum_i (\alpha_i - \frac{n-1}{2})^2 = \frac{\alpha_\mu - \alpha_1}{12} \ .$$

Proof : $f \in J_f$, $E|_0 = 0$, $E\, G = 0 = \gamma$. $\qquad\qquad\qquad\qquad\qquad$ □

In the case of a quasihomogeneous singularity of degree 1 with respect to weights $w_0, ..., w_n \in (0, \frac{1}{2}] \cap \mathbb{Q}$, one has the following formula for the spectral numbers:

$$\sum_{i=1}^{\mu} t^{\alpha_i+1} = \prod_{j=0}^{n} \frac{t - t^{w_j}}{t^{w_j} - 1} \ .$$

Using this, A. Dimca found an elementary proof of Theorem 4.6 (after hearing a talk of mine on the proof above).

CONJECTURE 4.7 *Let f be any singularity. Then $\gamma \geq 0$ and*

$$\frac{1}{\mu} \sum_i (\alpha_i - \frac{n-1}{2})^2 \leq \frac{\alpha_\mu - \alpha_1}{12} \ .$$

Recently, M. Saito proved the conjecture in the case of irreducible plane curve singularities [27].

REFERENCES

[1] V.I. Arnold, S.M. Gusein-Zade, A.N. Varchenko. Singularities of differentiable maps. volume II. Boston: Birkhäuser, 1988.

[2] E. Brieskorn. Die Monodromie der isolierten Singularitäten von Hyperflächen. Manuscripta Math. 2:103–161, 1970.

[3] B. Dubrovin. Integrable systems in topological field theory. Nucl. Phys. B 379:627–689, 1992.

[4] B. Dubrovin. Geometry of 2D topological field theories. Integrable sytems and quantum groups, ed. M Francoviglia, S. Greco. Montecatini, Terme. 1993. Lecture Notes in Math. 1620, Springer-Verlag 1996, pp. 120–348.

[5] B. Dubrovin. Painlevé equations in 2D topological field theories. Painlevé property, one century later. Cargèse, math.AG/9803107, 1996.

[6] B. Dubrovin, Y. Zhang. Bihamiltonian hierarchies in 2D topological field theory at one-loop approximation. Commun. Math. Phys. 198:311–361, 1998.

[7] A.M. Gabrielov. Bifurcations, Dynkin diagrams and modality of isolated singularities. Funct. Anal. 8:94–98, 1974.

[8] A.B. Givental. Singular Lagrangian manifolds and their Lagrangian maps. J. Soviet Math. 52 4:3246–3278, 1988.

[9] A.B. Givental. Elliptic Gromov-Witten invariants and the generalized mirror conjecture. Integrable systems and algebraic geometry. Proceedings of the Taniguchi Symposium 1997, ed. M.-H. Saito, Y. Shimizu, K. Ueno. World Scientific, River Edge NJ 1998, pp. 107–155.

[10] G.-M.Greuel. Der Gauß-Manin-Zusammenhang isolierter Singularitäten von vollständigen Durchschnitten. Math. Ann. 214: 235–266, 1975.

[11] P. Griffiths, J. Harris. Principles of algebraic geometry. New York: John Wiley and sons, 1978.

[12] C. Hertling. Classifying spaces and moduli spaces for polarized mixed Hodge structures and for Brieskorn lattices. Compositio Math. 116:1–37, 1999.

[13] C. Hertling, Yu. Manin. Weak Frobenius manifolds. Int. Math. Res. Notices 6:277–286, 1999.

[14] C. Hertling. Multiplication on the tangent bundle. First part of the habilitation. 88 pages, also math.AG/9910116.

[15] C. Hertling. Frobenius manifolds and moduli spaces for hypersurface singularities. Second part of the habilitation. 141 pages. June 2000.

[16] N.J. Hitchin. Frobenius manifolds (notes by D. Calderbank). Gauge Theory and symplectic geometry, ed. J. Hurtubise and F. Lalonde. Montreal, 1995. Kluwer Academic Publishers, Netherlands 1997, 69–112.

[17] Va.S. Kulikov. Mixed Hodge structures and singularities. Cambridge tracts in mathematics 132, Cambridge University Press, 1998.

[18] B. Malgrange. Déformations de systèmes différentiels et microdifférentiels. Séminaire de'l ENS, Mathématique et Physique, 1979–1982, Progress in Mathematics vol. 37, Birkhäuser, Boston 1983, pp. 353–379.

[19] B. Malgrange. Deformations of differential systems, II. J. Ramanujan Math. Soc. 1:3–15, 1986.

[20] Yu Manin. Frobenius manifolds, quantum cohomology,and moduli spaces. American Math. Society, Colloquium Publ. 47, 1999.

[21] C. Sabbah. Déformations isomonodromiques et variétés de Frobenius, une introduction. Centre de Mathematiques, Ecole Polytechnique, U.M.R. 7640 du C.N.R.S., 2000:05.

[22] K. Saito. On a linear structure of the quotient variety by a finite reflexion group. Preprint RIMS–288, 1979, Publ. RIMS, Kyoto Univ. 29:535–579, 1993.

[23] K. Saito. Primitive forms for a universal unfolding of a function with an isolated critical point. J. Fac. Sci. Univ. Tokyo, Sect. IA Math. 28:775–792, 1982.

[24] K. Saito. Period mapping associated to a primitive form. Publ. RIMS, Kyoto Univ. 19:1231–1264, 1983.

[25] M. Saito. On the structure of Brieskorn lattices. Ann. Inst. Fourier Grenoble. 39:27–72, 1989.

[26] M. Saito. Period mapping via Brieskorn modules. Bull. Soc. math. France 119:141–171, 1991.

[27] M. Saito. Exponents of an irreducible plane curve singularity. Preprint, 14 pages, math.AG/0009133.

[28] J. Scherk. A propos d'un théorème de Mather et Yau. C. R. Acad. Sci. Paris, Série I. 296:513–515, 1983.

[29] J. Scherk, J.H.M. Steenbrink. On the mixed Hodge structure on the cohomology of the Milnor fibre. Math. Ann. 271:641-665, 1985.

[30] B. Teissier. Déformations à type topologique constant. Séminaire de géometrie analytique 1971–1972, Astérisque 16:215–249.

[31] A.N. Varchenko. The asymptotics of holomorphic forms determine a mixed Hodge structure. Sov. Math. Dokl. 22:772–775, 1980.